Food Microorganism Contribution to Fermented Foods

Food Microorganism Contribution to Fermented Foods

Guest Editors

Xinping Lin
Jing Lv

MDPI

Basel • Beijing • Wuhan • Barcelona • Belgrade • Novi Sad • Cluj • Manchester

Guest Editors

Xinping Lin
National Marine Food Engineering Technology Research Center
Dalian Polytechnic University
Dalian
China

Jing Lv
College of Food and Bioengineering
Zhengzhou University of Light Industry
Zhengzhou
China

Editorial Office
MDPI AG
Grosspeteranlage 5
4052 Basel, Switzerland

This is a reprint of the Special Issue, published open access by the journal *Foods* (ISSN 2304-8158), freely accessible at: https://www.mdpi.com/journal/foods/special_issues/IR3036LUW1.

For citation purposes, cite each article independently as indicated on the article page online and as indicated below:

Lastname, A.A.; Lastname, B.B. Article Title. *Journal Name* **Year**, *Volume Number*, Page Range.

ISBN 978-3-7258-7753-9 (Hbk)
ISBN 978-3-7258-7754-6 (PDF)
https://doi.org/10.3390/books978-3-7258-7754-6

Contents

About the Editors

Xinping Lin

Xinping Lin, Professor, Doctoral Supervisor, and Deputy Director of the Department of Food Nutrition and Health, was a visiting scholar at the University of Turin, Italy. Her research focuses on fermented foods, food microbiology, and fermentation engineering. Prof. Lin has been selected for the National High-level Young Talent Program, Liaoning's "Xingliao" Young Top-notch Talents, Dalian Outstanding Young Talents, and the "Ten Thousand Talents" tier of Liaoning's Talents Project, and was named a "Science and Technology Star of Dalian". She has won Science and Technology Progress Awards including First Prizes from Anhui Province and Dalian City, and a Second Prize from the Ministry of Education. Academically, she is a Council Member of the CIFST Traditional Fermented Foods Branch and a member of the Liaoning Food Safety Expert Committee. She serves on the editorial boards of the *Journal of Future Foods* and *Foods*, and the youth editorial boards of *Food Production, Processing and Nutrition; Science and Technology of Food Industry;* and *Transactions of the Chinese Society of Agricultural Engineering*. She has led numerous projects, including 1 project and 2 sub-tasks of the National Key R&D Program, 3 NSFC programs (2 General, 1 Youth), and 9 provincial/municipal projects. She has published over 100 papers (40+ SCI-indexed as first/corresponding author), filed over 30 invention patents as the first inventor (9 authorized), and co-drafted 1 industry, 4 group, and 4 enterprise standards.

Jing Lv

Jing Lv is a Lecturer at the College of Food and Bioengineering, Zhengzhou University of Light Industry. She holds a PhD in Food Science and Engineering from Dalian Polytechnic University. Her research focuses on fermented meat products and food microbiology. She has published more than 20 academic papers, been granted four national invention patents, and contributed two chapters to a book. She has contributed to two National Natural Science Foundation of China projects and leads several grants, including two projects in Henan Province focusing on the influence of starter cultures on the quality of fermented meat products. She also serves as a peer reviewer for journals such as *BMC Microbiology* and *Applied Food Research.*

Preface

Fermentation has accompanied the development of human food systems for centuries and remains one of the most important biological processes in food production and preservation. Beyond its traditional role in extending shelf life, fermentation is now widely recognized as a highly dynamic process through which microorganisms contribute to the transformation of raw materials, the improvement of nutritional and functional properties, the development of desirable sensory characteristics, and the enhancement of food safety. In recent years, growing scientific interest in the complex roles of food microorganisms has further expanded the significance of fermented foods in the context of health, sustainability, and technological innovation.

This Reprint, "Food Microorganism Contribution to Fermented Foods", has been assembled to provide a coherent and up to date collection of studies addressing the diverse ways in which microorganisms influence fermented food systems. Its scope encompasses research on the activity and succession of microbial communities during fermentation, the functions of lactic acid bacteria, yeasts, and other relevant microorganisms, the development and application of starter cultures, and the effects of fermentation on food quality, safety, nutritional value, and bioactive potential. The Reprint also reflects current interest in both traditional fermented foods and emerging substrates, including unconventional food materials and underutilized resources that may support more sustainable food production.

The purpose of this Reprint is to bring together representative contributions that illustrate the scientific depth and practical relevance of this field. As the demand for fermented foods continues to grow, there is an increasing need for a clearer understanding of the microbial mechanisms that govern fermentation outcomes and of the technological strategies that can be used to improve product consistency, safety, functionality, and consumer acceptance. By compiling recent research and critical perspectives in a single volume, this Reprint seeks to support ongoing progress in food microbiology and fermentation science, while also encouraging further exploration of innovative applications in food processing and product development.

This Reprint is addressed to researchers, food microbiologists, fermentation scientists, food technologists, and professionals in related sectors of the food industry. It is equally intended for readers interested in the scientific foundations of fermented foods and in the broader role of microorganisms in shaping future food systems. It is hoped that the studies brought together here will serve not only as a valuable record of current advances, but also as a useful reference for future research, technological innovation, and interdisciplinary exchange in this important area of food science.

Xinping Lin and Jing Lv
Guest Editors

foods

MDPI

Article

Development of a Novel Chocolate Utilizing Mushroom Fermentation and Associated Changes in Beneficial Components

Shiori Fukuda [1,2], Momoka Nakata [2], Yuka Sameshima [3], Naomi Takemoto [4] and Tokumitsu Matsui [1,2,*]

1 Department of Food Sciences and Nutrition, Graduate School of Food Sciences and Nutrition, Mukogawa Women's University, Nishinomiya 663-8558, Hyogo, Japan; sfuku@mukogawa-u.ac.jp
2 Department of Food Science and Nutrition, School of Food Science and Nutrition, Mukogawa Women's University, Nishinomiya 663-8558, Hyogo, Japan; 2111276@mwu.jp
3 Department of Food and Nutrition, Faculty of Human Life Studies, Hagoromo University of International Studies, Sakai 592-8344, Osaka, Japan; ysameshima@hagoromo.ac.jp
4 Department of Health and Nutrition, Faculty of Food Culture, Baika Women's University, Ibaraki 567-8578, Osaka, Japan; n-takemoto@baika.ac.jp
* Correspondence: tokamura@mukogawa-u.ac.jp

Abstract

This study investigated the secondary fermentation of cocoa beans using mushrooms to further improve the quality of beans. Cocoa beans were fermented using 42 species of basidiomycetes and ascomycetes. Mycelial growth was observed in 29 strains. When 75% cocoa chocolate was prepared using the cocoa beans in which mycelial growth was observed, theobromine concentration was higher in 17 strains compared with the control. Furthermore, caffeine concentration was similar to or lower than the control in all strains. Chocolate produced using cocoa beans fermented with particularly *Polyporus arcularius*, *Peziza vesiculosa*, and *Urnula craterium* exhibited significantly higher theobromine concentrations. Compared to the control theobromine concentration of 7.53 mg/g, *P. arcularius* showed 9.25 mg/g, 9.13 mg/g for *P. vesiculosa*, and 9.05 mg/g for *U. craterium*. Furthermore, the reducing sugar concentration and total polyphenol concentration increased, and the antioxidant activity was similar to or higher than that of the control. These results suggest that secondary fermentation using mushrooms could be used to develop chocolate characterized by high theobromine, low caffeine, and rich polyphenol content.

Keywords: mushroom; fermented foods; cacao; chocolate; theobromine; caffeine; polyphenol; antioxidant activity

1. Introduction

Chocolate is a popular confection. It is primarily made from cocoa beans and cocoa butter, to which sugar, milk powder, and emulsifiers are often added in commercial products. Cacao beans are harvested as cacao pods in producing countries. After harvesting, the pods are split open, and the seeds, still attached to the cacao pulp, are collected. The seeds are fermented for about a week, then dried and shipped.

The fermentation methods primarily consist of the Heap method, where cocoa beans are piled outdoors and covered with banana leaves or sheets, and the Box method, where beans are placed in fermentation boxes for controlled processing. In both methods, the fermentation of the cocoa beans is affected by the microorganisms adhering to the surface of the cacao pods, the tools and facilities used, and the surrounding soil [1]. Regarding microorganisms, yeast, lactic acid bacteria, acetic acid bacteria, and spore-forming bacteria are commonly used [1,2]. In the early fermentation stage, yeast and lactic acid bacteria convert sugars in the cacao pulp into ethanol and lactic acid through alcohol fermentation and lactic acid fermentation, respectively. In the middle stage, acetic acid bacteria convert ethanol into acetic acid through acetic acid fermentation. In the late stage, many microorganisms die due to the accumulation of their own metabolites and rising temperatures, and spore-forming bacteria become dominant. These fermentation processes not only alter the components of the cacao bean but also induce biochemical changes within the bean due to the lowering of pH by fermentation products and the generation of fermentation heat. These changes in pH and temperature activate endoproteases and carboxypeptidases in the cacao beans, producing free amino acids and reducing sugars in the bean. These undergo the amino-carbonyl reaction and Strecker degradation during the roasting process, producing aromatic compounds that form the characteristic flavor of chocolate [1–3].

During fermentation, polyphenols, caffeine, theobromine, and other alkaloids in cacao beans are oxidized, polymerized, and dissolved, resulting in a decrease in their content [1]. Although fermentation of cacao beans plays an important role in making chocolate, the fermentation process is often performed by farmers who use naturally occurring microorganisms that are difficult to control.

Therefore, in recent years, methods have been reported to improve cocoa bean quality by introducing carefully selected lactic acid bacteria into harvested cocoa beans and fermenting them [4], or by controlling fermentation temperature [5]. Among these, research has been conducted on improving cocoa bean quality through re-fermentation of dried, unfermented cocoa beans [6] and secondary fermentation [7], it is considered possible that even cocoa beans that have undergone fermentation processing can have their chocolate quality further enhanced and develop unique flavors through additional fermentation.

Mushrooms are the name given to the fruiting bodies formed by fungi during sexual reproduction, which are sufficiently large to be clearly visible to the naked eye. Such fruiting bodies are formed by some Ascomycetes and many Basidiomycetes; these fungi are referred to as mushrooms [8,9]. The distinction between Ascomycetes and Basidiomycetes is primarily based on the characteristics of their sexual spores. Ascomycetes form sexual spores (ascospores) inside asci, whereas Basidiomycetes produce sexual spores (basidiospores) externally on basidia [10]. However, both possess various enzymes, including carbohydrate-degrading enzymes [11]. Consequently, applications utilizing mushroom enzymes in fermented foods have been reported, irrespective of whether the fungi belong to the Ascomycetes or Basidiomycetes [12].

For example, mushrooms possess various enzymes, including amylase, protease, alcohol dehydrogenase, and lactate dehydrogenase. These enzymes have been utilized for the development of miso and alcoholic beverages [13–17]. Fermentation using mushrooms increases antioxidant activity and produces new beneficial components. Previous studies have reported that post-fermented tea prepared using the wood-decaying fungus *Schizophyllum commune* showed a 36.5% increase in antioxidant activity compared to the control, and approximately double the concentration of total polyphenols and total catechins [18].

These findings suggest that fermented foods with superior antioxidant properties and beneficial components compared with conventional fermented foods can be developed by utilizing the fermentation capabilities of mushrooms. However, to date, no research has been conducted on fermenting cacao beans with mushrooms, and no chocolate made from mushroom-fermented cacao beans is currently being sold.

Cocoa beans are rich in polyphenols, and their usefulness is being researched, but the amount of polyphenols varies depending on the origin, type of bean, and fermentation period [19].

https://doi.org/10.3390/foods15061045

However, if it is possible to increase the amount of polyphenols by fermenting cacao beans with mushrooms, it will be possible to develop chocolate rich in polyphenols regardless of the origin of the cacao beans or the fermentation period, and the quality of chocolate can be expected to improve.

Furthermore, it is unclear how components such as theobromine and caffeine contained in cacao beans change due to fermentation with mushrooms. Clarifying this could help discover the value of chocolate produced by fermentation with mushrooms.

This study aimed to develop a new chocolate by fermenting cocoa beans using mushrooms. We conducted secondary fermentation of cocoa beans that had undergone fermentation processing at the producer farm using mushrooms and analyzed changes in the beneficial components within the cocoa beans.

2. Materials and Methods

2.1. Pre-Culture of Test Strains

The test strains (mushroom mycelium) consisted of 42 species of basidiomycetes and ascomycetes, including strains distributed by the National Institute of Technology and Evaluation's Biological Resource Center (NBRC) and strains isolated by the authors. Plate media consisted of potato dextrose agar (Shimadzu Diagnostics Corporation, Kyoto, Japan), 2% malt extract agar (Nacalai Tesque, Inc., Kyoto, Japan), or Matsutake agar prepared to contain 0.5% glucose (Fujifilm Wako Pure Chemical Corporation, Osaka, Japan) and 2% yeast extract (Nacalai Tesque, Inc., Kyoto, Japan). For pre-cultivation, each mushroom mycelium was inoculated onto the appropriate plate medium recommended by NBRC and cultured at 25 °C.

2.2. Preparation of Fermented Cocoa Beans

Cocoa beans fermented in the country of origin (Ichinomiya Bussan Corporation, Osaka, Japan, Ghana origin) were autoclaved (121 °C, 20 min). After cooling under aseptic conditions, they were aseptically immersed in an autoclaved solution of 1% glucose and 0.5% peptone (Nihon Pharmaceutical Co., Ltd., Osaka, Japan) for at least 17 h.

The pre-cultured mushroom mycelium was inoculated onto fresh appropriate agar plates (5 mm squares × 10 plates). After soaking, 20 pieces of moisture-removed cacao beans were placed onto the agar plates inoculated with mushroom mycelium and cultured statically at 25 °C.

The end of cultivation was defined as the stage where mushroom mycelium covered the cocoa beans and grew over 80% of the entire Petri dish. For cultures where mushroom mycelium growth covering at least 80% of the Petri dish was not confirmed even after 100 days or more of incubation, the culture was terminated, and the preparation of a fermented chocolate was not performed. The control consisted of cacao beans autoclaved and then soaked in a 1% glucose and 0.5% peptone solution. The control and mushroom-fermented cocoa beans were each cultured in three separate Petri dishes.

2.3. Preparation of Fermented Chocolate

The preparation of chocolate from fermented cacao beans was based on the method reported by Kasai et al. [20], modified to suit this study. Fermented cocoa beans after static incubation were roasted in a household oven (110 °C, 40 min) while still covered with mushroom mycelium, then cooled overnight. After cooling, the mycelium and thin skins adhering to the cocoa beans were removed. The beans were then ground to a smooth consistency using a universal mill (Ichinomiya Bussan Corporation, Osaka, Japan). White superior soft sugar was added to achieve a 75% cocoa content, and chocolate was prepared. No other auxiliary ingredients were added.

https://doi.org/10.3390/foods15061045

2.4. Measurement of Theobromine and Caffeine Concentrations

Theobromine and caffeine concentrations were measured using high-performance liquid chromatography (HPLC). The HPLC system employed a DP-8020 pump (Tosoh Corporation, Tokyo, Japan) and a UV-8020 detector (Tosoh Corporation, Tokyo, Japan), with detection at UV 273 nm. The column used was TSKgel ODS-100V (Tosoh Corporation, Tokyo, Japan), and the mobile phase was a 30% methanol solution, with a flow rate of 1 mL/min. A sample preparation for theobromine and caffeine concentration analysis was performed according to the method reported by Saotome et al. [21]. Specifically, 0.1 g of fermented chocolate was mixed with 500 μL of 100 mmol/L carbonate buffer (pH 11) and 300 μL of hexane. The mixture was vortexed for 30 s, then heated at 50 °C for 10 min. After heating, 450 μL of 100 mmol/L carbonate buffer (pH 11) was added. The mixture was vortexed (30 s × 3 times), centrifuged (15,000 rpm, 20 °C, 10 min), and the hexane layer was removed. After hexane removal, 650 μL of 1 mol/L ammonium bicarbonate was added, the mixture was vortexed, and the supernatant was used as the test sample. The concentrations of theobromine and caffeine in each sample were determined using calibration curves prepared from their respective standards.

2.5. Measurement of Reducing Sugar Concentration

Reducing sugar concentration was measured using the Somogyi–Nelson method [22]. Specifically, an equal volume of distilled water was added to the fermented chocolate, the mixture was vortexed for 30 s, centrifuged (15,000 rpm, 4 °C, 10 min), and the resulting supernatant was used as the aqueous extraction sample. Somogyi's copper reagent (Merck Ltd., Darmstadt, Germany) was added to 120 μL of the aqueous extract sample and heated at 100 °C for 10 min. After cooling on ice for 5 min, 120 μL of Nelson color reagent (Merck Ltd., Darmstadt, Germany) and 1.0 mL of distilled water were added. After standing for 15 min, measurements were taken at a detection wavelength of 500 nm using a microplate reader (Tecan Japan Co., Ltd., Kanagawa, Japan). The reducing sugar concentration of each sample was determined using a calibration curve prepared from glucose standards.

2.6. Measurement of Total Polyphenol Concentration

The total polyphenol concentration was measured using the Folin–Ciocalteu method and expressed as epicatechin equivalents [23]. Specifically, for defatting 0.2 g of fermented chocolate was mixed with 1.0 mL of hexane, heated at 50 °C for 10 min, centrifuged (15,000 rpm, 20 °C, 5 min). The hexane layer was removed, and the extraction was repeated under the same conditions. Residual hexane was then completely removed using a centrifugal evaporator (Tokyo Rikakikai Co., Ltd., Tokyo, Japan) to make the defatted sample. A sample extracted with 50% methanol from the defatted sample was used as the test sample. Epicatechin (Fujifilm Wako Pure Chemical Corporation, Osaka, Japan) was used as the standard substance, and absorbance measurements were performed at a detection wavelength of 765 nm using a microplate reader.

2.7. Measurement of Antioxidant Activity

Antioxidant activity was measured by chemiluminescence using a microplate reader according to the method of Tabata et al. [24]. Specifically, we measured the ability to scavenge superoxide (O2-) among oxygen radicals as antioxidant capacity. Each well was loaded with 10 μL of the water-extracted fermented chocolate sample, 85 μL of 0.1 M KH_2PO_4 buffer (containing 0.05 M EDTA·2Na, pH 7.5; hereafter referred to as KH_2PO_4 buffer), 30 μL of 100 μg/mL xanthine oxidase solution, and 5 μL of 45 mM MPEC (2-Methyl-6-p-methoxyphenylethynylimidazopyrazinone: ATTO CORPORATION, Tokyo, Japan) solution.

https://doi.org/10.3390/foods15061045

The microplate was loaded into a microplate reader, and 25 µL of 0.72 mM hypoxanthine (containing 0.05 mM EDTA) was dispensed using the built-in pump. The accumulated luminescence was then measured for 30 s. As a positive control, 10 µL of 0.1 M KH_2PO_4 buffer was placed in the well instead of the fermented chocolate sample, and the accumulated luminescence value was measured in the same manner. Antioxidant activity was calculated as: oxidation inhibition rate (%) = [(luminescence accumulation value of positive control − luminescence accumulation value of sample)/luminescence accumulation value of positive control] × 100.

2.8. Statistical Analysis

All results were obtained from three independent experiments. Statistical analyses were performed using IBM SPSS Statistics (version 26; IBM Japan Ltd., Tokyo, Japan) with one-way analysis of variance. When significant differences were detected, Dunnett's multiple comparison test was applied for comparisons with the control group. Results are presented as mean ± standard deviation, and significance was defined as $p < 0.05$.

3. Results and Discussion

3.1. The State of Fermented Cacao Beans Using Mushrooms

Figure 1 presents the appearance of cacao beans fermented with *Polyporus arcularius* as a representative example. Cultivation was terminated when mycelium growth covered over 80% of the entire Petri dish, as observed with *P. arcularius*.

Figure 1. Appearance of cocoa beans during cultivation.

Cocoa beans were fermented using 42 mushroom mycelium. Hyphal growth was observed in 29 mushroom mycelium and no hyphal growth in 13 even after a culture period exceeding 100 days (Table 1). The incubation period during which mycelial growth on cocoa beans was confirmed was 40 to 50 days for most mushrooms, with some strains showing growth even during shorter incubation periods of 20 to 30 days. No correlation was observed between the incubation period for mycelial growth and mushroom order, family, or genus.

Table 1. Mycelium used in this study and cocoa bean fermentation period (days).

Scientific Name		Phylum	Class	Order	Family	Identification Number	Growth Medium	Culture Period (Days)
Genus	Specific Name							
Agrocybe	*cylindrica*	*Basidiomycota*	*Agaricomycetes*	*Agaricales*	*Bolbitaceae*	Ya-1 *	Matsutake Medium	33
Sarcomyxa	*edulis*	*Basidiomycota*	*Agaricomycetes*	*Agaricales*	*Sarcomyxaceae*	30526	PDA Medium	43
Hypsizigus	*marmoreus*	*Basidiomycota*	*Agaricomycetes*	*Agaricales*	*lyophyllaceae*	Bn-1 *	PDA Medium	71
Schizophyllum	*commune*	*Basidiomycota*	*Agaricomycetes*	*Agaricales*	*Schizophyllaceae*	4929	PDA Medium	47
Armillaria	*mellea*	*Basidiomycota*	*Agaricomycetes*	*Agaricales*	*Physalacriaceae*	7037	PDA Medium	55
Flammulina	*velutipes*	*Basidiomycota*	*Agaricomycetes*	*Agaricales*	*Physalacriaceae*	SA-1 *	PDA Medium	47
Hymenopellis	*radicata*	*Basidiomycota*	*Agaricomycetes*	*Agaricales*	*Physalacriaceae*	9785	PDA Medium	-

https://doi.org/10.3390/foods15061045

Table 1. *Cont.*

Scientific Name		Phylum	Class	Order	Family	Identification Number	Growth Medium	Culture Period (Days)
Genus	Specific Name							
Lentinula	*edodes*	*Basidiomycota*	*Agaricomycetes*	*Agaricales*	*Omphalotaceae*	Sh-1 *	PDA Medium	-
Coprinellus	*disseminatus*	*Basidiomycota*	*Agaricomycetes*	*Agaricales*	*Psathyrellaceae*	7550	PDA Medium	54
Agaricus	*bisporus*	*Basidiomycota*	*Agaricomycetes*	*Agaricales*	*Agaricaceae*	7214	PDA Medium	-
Calvatia	*craniiformis*	*Basidiomycota*	*Agaricomycetes*	*Agaricales*	*Agaricaceae*	30530	PDA Medium	-
Cyathus	*stercoreus*	*Basidiomycota*	*Agaricomycetes*	*Agaricales*	*Agaricaceae*	9076	Matsutake Medium	-
Pleurotus	*ostreatus*	*Basidiomycota*	*Agaricomycetes*	*Agaricales*	*Pleurotaceae*	Hr-24 *	PDA Medium	49
Pleurotus	*citrinopileatus*	*Basidiomycota*	*Agaricomycetes*	*Agaricales*	*Pleurotaceae*	Ta-1 *	PDA Medium	47
Pleurotus	*eryngii*	*Basidiomycota*	*Agaricomycetes*	*Agaricales*	*Pleurotaceae*	Er-3 *	PDA Medium	54
Pleurotus	*abalonus*	*Basidiomycota*	*Agaricomycetes*	*Agaricales*	*Pleurotaceae*	Ka-1 *	PDA Medium	-
Agrocybe	*praecox*	*Basidiomycota*	*Agaricomycetes*	*Agaricales*	*Strophariaceae*	30258	PDA Medium	-
Hypholoma	*lateritium*	*Basidiomycota*	*Agaricomycetes*	*Agaricales*	*Strophariaceae*	W533 *	2% Malt Agar Medium	21
Pholiota	*microspora*	*Basidiomycota*	*Agaricomycetes*	*Agaricales*	*Strophariaceae*	Na-4 *	PDA Medium	40
Pholiota	*adiposa*	*Basidiomycota*	*Agaricomycetes*	*Agaricales*	*Strophariaceae*	Nm-1 *	PDA Medium	29
Stropharia	*rugosoannulata*	*Basidiomycota*	*Agaricomycetes*	*Agaricales*	*Strophariaceae*	30225	Matsutake Medium	-
Auricularia	*polytricha*	*Basidiomycota*	*Agaricomycetes*	*Auriculariales*	*Auricularaceae*	W9 *	2% Malt Agar Medium	-
Cyclomyces	*fuscus*	*Basidiomycota*	*Agaricomycetes*	*Hymenochaetales*	*Hymenochaetaceae*	9789	Matsutake Medium	21
Climacodon	*septentrionalis*	*Basidiomycota*	*Agaricomycetes*	*Polyporales*	*Irpicaceae*	W4 *	2% Malt Agar Medium	-
Polyporus	*arcularius*	*Basidiomycota*	*Agaricomycetes*	*Polyporales*	*Polyporaceae*	W122 *	2% Malt Agar Medium	41
Polyporus	*badius*	*Basidiomycota*	*Agaricomycetes*	*Polyporales*	*Polyporaceae*	30355	Matsutake Medium	35
Trametes	*versicolor*	*Basidiomycota*	*Agaricomycetes*	*Polyporales*	*Polyporaceae*	6516	PDA Medium	57
Lenzites	*betulinus*	*Basidiomycota*	*Agaricomycetes*	*Polyporales*	*Polyporaceae*	4963	PDA Medium	76
Ganoderma	*sichuanense*	*Basidiomycota*	*Agaricomycetes*	*Polyporales*	*Polyporaceae*	31863	PDA Medium	47
Ganoderma	*applanatum*	*Basidiomycota*	*Agaricomycetes*	*Polyporales*	*Polyporaceae*	Napa *	2% Malt Agar Medium	21
Grifola	*frondosa*	*Basidiomycota*	*Agaricomycetes*	*Polyporales*	*Polyporaceae*	Mi-h *	2% Malt Agar Medium	45
Laetiporus	*cremeiporus*	*Basidiomycota*	*Agaricomycetes*	*Polyporales*	*Laetiporaceae*	W8 *	2% Malt Agar Medium	28
Meripilus	*giganteus*	*Basidiomycota*	*Agaricomycetes*	*Polyporales*	*meripilaceae*	Tb-1 *	PDA Medium	41
Sparassis	*latifolia*	*Basidiomycota*	*Agaricomycetes*	*Polyporales*	*Sparassidaceae*	Hb-1 *	PDA Medium	-
Hericium	*erinaceus*	*Basidiomycota*	*Agaricomycetes*	*Russulales*	*Hericiaceae*	Ym-2 *	2% Malt Agar Medium	21
Hericium	*coralloides*	*Basidiomycota*	*Agaricomycetes*	*Russulales*	*Hericiaceae*	Sn-1 *	PDA Medium	33
Tremella	*fuciformis*	*Basidiomycota*	*Tremellomycetes*	*Tremellales*	*Tremellaceae*	8990	PDA Medium	43
Morchella	*esculenta*	*Ascomycota*	*Pezizomycetes*	*Pezizales*	*Morchellaceae*	33132	PDA Medium	-
Peziza	*vesiculosa*	*Ascomycota*	*Pezizomycetes*	*Pezizales*	*Pezizaceae*	30324	PDA Medium	34
Urnula	*craterium*	*Ascomycota*	*Pezizomycetes*	*Pezizales*	*Sarcosomataceae*	30137	PDA Medium	42
Wynnea	*gigantea*	*Ascomycota*	*Pezizomycetes*	*Pezizales*	*Sarcosomataceae*	8826	PDA Medium	48
Cordyceps	*militaris*	*Ascomycota*	*Sordariomycetes*	*Hypocreales*	*Cordycipitaceae*	5928	PDA Medium	-

- No mycelial growth was observed even after a cultivation period of 100 days or more. The numbers show NBRC No., * show Lab No. (A strain isolated by pure culture of wild-type bacteria in the laboratory).

This shows the appearance of cocoa beans during fermentation, comparing a control group not fermented with mushrooms and a group fermented using *P. arcularius*.

3.2. Theobromine and Caffeine Concentration

Fermented chocolate was prepared using the 29 strains that exhibited mycelial growth. We found that the theobromine concentration in fermented chocolate exceeded the control in 17 strains, suggesting the potential for new theobromine production during fermentation (Figure 2). In the control group without mushroom fermentation, the theobromine concentration was 7.53 mg/g. However, in fermented chocolate using *P. arcularius*, it was 9.25 mg/g, *Peziza vesiculosa*-fermented chocolate showed 9.13 mg/g, and *Urnula craterium*-fermented chocolate showed 9.05 mg/g, indicating significantly higher theobromine concentrations.

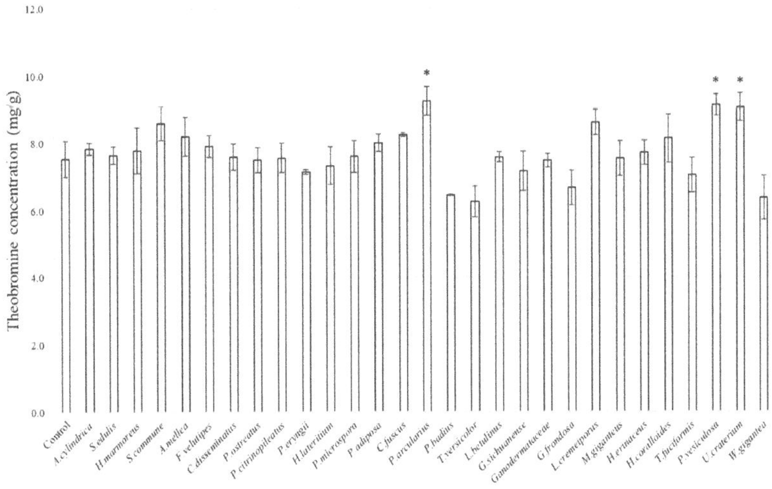

Figure 2. Theobromine concentration in fermented chocolate (mg/g). The error bars in the figure represent the standard deviation ($n = 3$). Asterisks (*) indicate significant differences compared with the control ($p < 0.05$).

P. arcularius is a basidiomycete belonging to the order *Polyporales*, family *Polyporaceae*, and genus *Polyporus badius*. However, chocolate fermented using *Polyporus badius*, a species within the same genus, exhibited lower theobromine concentrations compared with the control. Similarly to the cultivation period for cocoa beans, no correlation was observed based on the mushroom family.

Caffeine concentration in fermented chocolate was lower than the control for all mushrooms, except chocolate fermented with *P. arcularius* (Figure 3). Chocolate fermented with *P. arcularius* showed a caffeine concentration similar to that of the control. This suggests that fermentation with certain mushroom strains breaks down the caffeine in cocoa beans. In this study, compared to the control sample that underwent no fermentation, fermenting with mushrooms resulted in a decrease in caffeine content and a tendency toward increased theobromine content. This suggests that mushroom fermentation may have broken down the caffeine in the cacao beans and generated newly theobromine.

Theobromine and caffeine are types of alkaloids called methylxanthines, found in cocoa, tea leaves, coffee, and other plant sources. Caffeine and theobromine are closely linked in their respective biosynthetic pathways. In fungi and bacteria, the caffeine degradation pathway converts caffeine into theobromine, which is subsequently metabolized to 7-methylxanthine and finally to xanthine [25,26]. These steps involve sequential demethylation reactions catalyzed by specific demethylating enzymes. This suggests that mushrooms possess demethylating enzymes. We hypothesize that the activity of demethylating enzymes enables the conversion of caffeine into theobromine and 7-methylxanthine.

This finding supports a previous study that reported that when *Pleurotus ostreatus* fruiting bodies were cultivated on sawdust medium containing coffee grounds, and the mycelium was grown on potato dextrose agar supplemented with coffee grounds, the caffeine degradation products theobromine, 3-methylxanthine, and 7-methylxanthine were detected in both the fruiting bodies and the mycelium [27]. However, there are still no reports of *P. arcularius*, *P. vesiculosa*, or *U. craterium*—which are also mushrooms—degrading caffeine to produce theobromine, and the mechanism by which *P. arcularius*, *P. vesiculosa*, and *U. craterium* generate theobromine remains unknown.

https://doi.org/10.3390/foods15061045

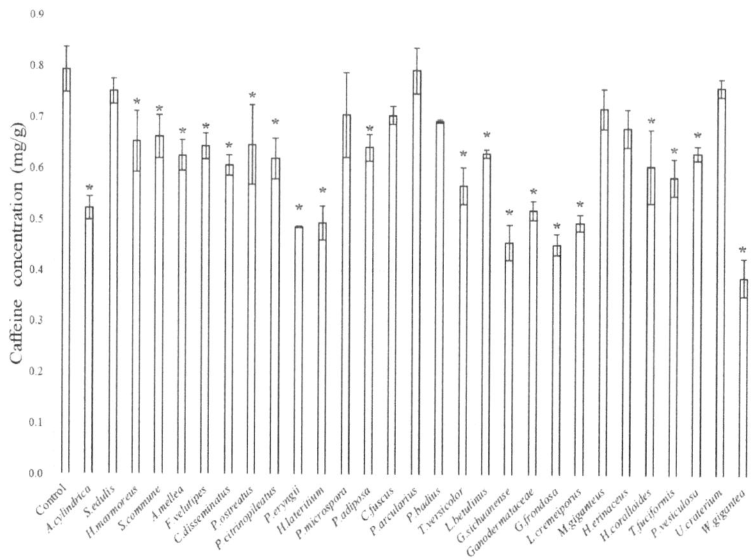

Figure 3. Caffeine concentration in fermented chocolate (mg/g). The error bars in the figure represent the standard deviation (n = 3). Asterisks (*) indicate significant differences compared with the control ($p < 0.05$).

Nevertheless, based on reports of caffeine degradation pathways and theobromine production in the same mushroom species, *P.ostreatus*, the reduction in caffeine concentration and the increase in theobromine concentration observed in chocolate fermented with mushrooms in this study were likely attributable to the presence of demethylating enzymes in the mushrooms *P. arcularius*, *P. vesiculosa*, and *U. craterium* possess demethylating enzymes. It is thought that during the cultivation period, these enzymes decomposed the caffeine contained in the cacao beans and generated theobromine.

In this study, the caffeine concentration of cocoa beans fermented with *P. arcularius* remained unchanged; however, the theobromine concentration increased. This suggests that theobromine was synthesized via a different pathway rather than being derived from the breakdown of caffeine. In the caffeine degradation pathway, demethylating enzymes convert caffeine into theobromine, 7-methylxanthine, and xanthine. The reverse reaction pathway occurs in the caffeine biosynthesis pathway: xanthine is converted to 7-methylxanthosine and then to 7-methylxanthine, ultimately forming theobromine (3,7-dimethylxanthine), which is then converted to caffeine (1,3,7-trimethylxanthine). In this process, methylation reactions occur in which methyl groups are added to the purine ring in the order N-7, N-3, and N-1 by the action of methyl transferase (N-methyltransferase), leading to the synthesis of theobromine and caffeine from xanthine [28].

If methylating enzyme (N-methyltransferases) are present in the mushroom along this pathway, it is conceivable that during the fermentation period of cacao beans, methyl groups could be added to the N-7 and N-3 positions of xanthine contained within the beans, enabling the production of theobromine. To date, there have been no reports of *P. arcularius* or other mushrooms possessing methyl transferase enzymes (N-methyltransferases). However, since the caffeine concentration in the cacao beans remained unchanged while the theobromine concentration increased, it was considered possible that *P. arcularius* possesses methyl transferase enzymes and newly produced theobromine from xanthine during the fermentation period of the cacao beans.

https://doi.org/10.3390/foods15061045

Therefore, another possibility is that the mushrooms possess both demethylase enzymes that break down caffeine and methylase enzymes that produce theobromine and caffeine from xanthine, both of which were active during the cultivation period.

There have been no reports of caffeine breakdown or theobromine production in *P. arcularius*, *P. vesiculosa*, or *U. craterium*, but the results of this study suggest the presence of demethylating and methylating enzymes, and further investigation of these enzymes is necessary to follow up on the results obtained in this study. However, by investigating and clarifying the mechanism of action of the enzymes, it may be possible to develop fermented chocolate with high theobromine and low caffeine concentrations through mushroom fermentation.

We focused on the species *P. arcularius*, *P. vesiculosa*, and *U. craterium*, which showed high theobromine concentrations, and conducted further analysis of their other components.

3.3. Production of Reducing Sugars Through Fermentation

Mushrooms possess various enzymes including amylase and cellulase. These enzymes are also active in fermented foods produced using mushroom mycelium where they increase the concentration of reducing sugars in the fermentation products [29]. We hypothesized that fermentation with mushrooms reduces the reducing sugar concentration in cocoa beans due to the amylase and cellulase present in the mushrooms.

Measurement of the reducing sugar concentration showed that the concentrations in beans fermented by *P. arcularius*, *P. vesiculosa*, and *U. craterium* were higher than those of the control (Figure 4). In particular, *U. craterium* had significantly higher reducing sugar concentration (36.9 mg/g) than the control (9.29 mg/g).

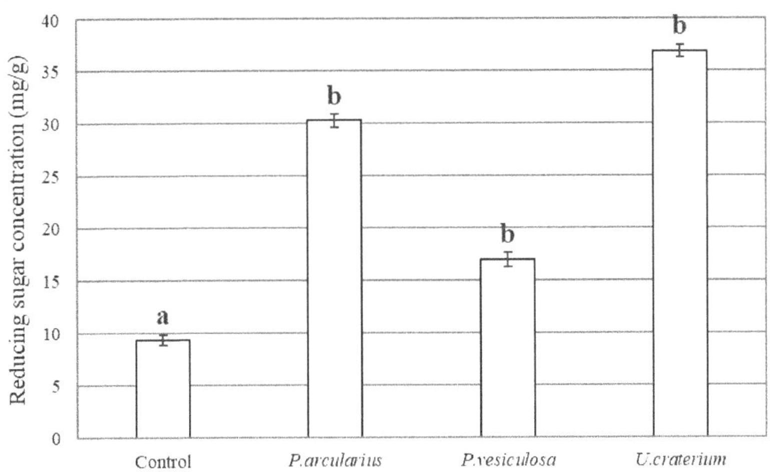

Figure 4. Reducing sugar concentration in fermented chocolate (mg/g). The error bars in the figure represent the standard deviation (n = 3). Different letters indicate $p < 0.05$ between groups.

Cocoa beans contain carbohydrates, primarily dietary fiber, but also including lignin, hemicellulose, and cellulose, in that order [30]. On the other hand, many basidiomycetes are wood-decay fungi and possess hemicellulose and cellulase that degrade hemicellulose and cellulose [11]. *P. arcularius* belongs to the Basidiomycota phylum. It is classified as a white rot fungus and can decompose hemicellulose and cellulose [31]. On the other hand, *P. vesiculosa* and *U. craterium* belong to the Ascomycota phylum, are not classified as wood-decay fungi, and it has not been reported that they possess hemicellulose or cellulase. However, some ascomycetes behave as soft-rot fungi and decompose wood with high moisture content and have decompositional activity against cellulose [32]. This suggests that *P. vesiculosa* and *U. craterium* may also possess cellulolytic activity, despite the fact that Ascomycota are typically classified as wood-decay fungi that lack hemicellulase and cellulase.

In this study, fermentation using *P. arcularius*, *P. vesiculosa*, and *U. craterium* significantly increased the reducing sugar concentration compared with the control. This suggests that cellulose and hemicellulose in the cocoa beans were decomposed into reducing sugars, such as glucose, by cellulase and hemicellulose enzymes present in the mushrooms. Although fermentation with these species increases the concentration of theobromine, which contributes to bitterness, as the concentration of reducing sugars, which contribute to sweetness, also increased; therefore, the fermented chocolate may possess both bitterness and sweetness.

3.4. Production of Total Polyphenols by Fermentation

Chocolate fermented using mushrooms had higher total polyphenol concentrations than the control, with *P. arcularius* and *P. vesiculosa* in particular showing polyphenol concentrations more than twice that of the control (Figure 5). Cocoa beans are rich in lignin, a polymer of polyphenols, and polyphenols are released when lignin is decomposed.

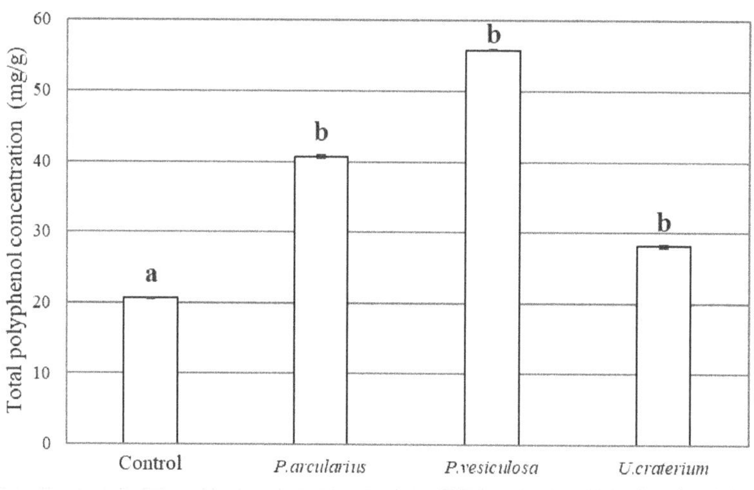

Figure 5. Total polyphenol concentration in fermented chocolate (mg/g). Total polyphenol concentration was quantified as epicatechin equivalents. The error bars in the figure represent the standard deviation (n = 3). Different letters indicate $p < 0.05$ between groups.

P. arcularius is a white rot fungus capable of decomposing cellulose, hemicellulose, and lignin [11,31]. *P. vesiculosa* and *U. craterium* belong to the Ascomycota phylum. Some ascomycetes have been reported to decompose lignin [32]. These findings support this study as fermentation with *P. arcularius*, *P. vesiculosa*, and *U. craterium* decomposed the lignin in the cocoa beans, leading to an increase in the total polyphenol concentration.

3.5. Increased Antioxidant Activity Through Fermentation

The antioxidant activity measurement results showed higher values for *P. arcularius* and *P. vesiculosa*, which had high total polyphenol content, compared with the control

https://doi.org/10.3390/foods15061045

(Figure 6). On the other hand, *U. craterium* showed antioxidant activity similar to that of the control.

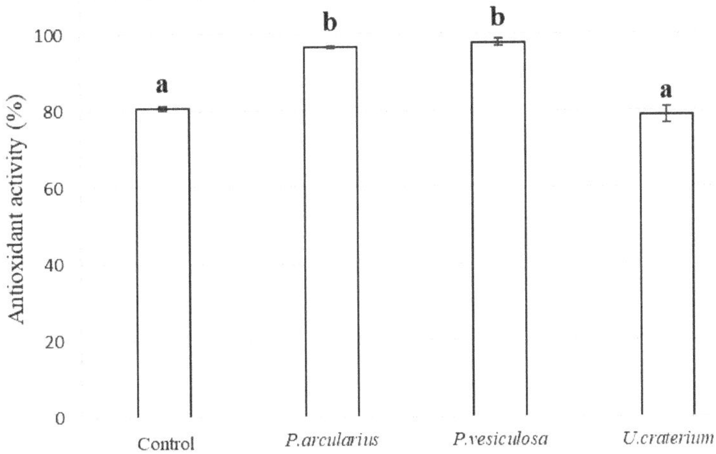

Figure 6. Antioxidant activity (%) in fermented chocolate. The error bars in the figure represent the standard deviation (n = 3). Different letters indicate $p < 0.05$ between groups.

The high antioxidant activity observed in chocolate fermented with *P. arcularius* and *P. vesiculosa* may be attributable to the release of antioxidant-active polyphenols associated with the increased total polyphenol concentration. On the other hand, antioxidant activity did not increase in chocolate fermented with *U. craterium*, despite its high total polyphenol content. This discrepancy may be explained by the presence of free polyphenols with inherently weak antioxidant activity or by oxidative degradation during the cultivation period, which may have diminished their antioxidant potential.

Polyphenols are compounds containing two or more phenol groups in a benzene ring. Polyphenols found in cacao beans include epicatechin and catechin, as well as procyanidin, which is a combination of epicatechin and catechin.

It has been reported that polyphenols contained in cacao beans exhibit higher antioxidant activity than black tea and green tea [33]. However, the antioxidant activity of each polyphenol, such as epicatechin and catechin, differs [34]. Therefore, even when the total polyphenol concentration is high, it may not directly lead to an increase in antioxidant activity.

The total polyphenol concentration measured in this study was not obtained by measuring the total polyphenol concentration in fermented chocolate using the Folin–Ciocalteu method; however, it was obtained by fractionating the obtained polyphenols and measuring each polyphenol individually. Furthermore, since the antioxidant activity was measured by determining the oxidation inhibition rate in fermented chocolate, the antioxidant effects of each polyphenol were not measured. Therefore, although lignin was decomposed in chocolate fermented with *U. craterium*, and the total polyphenol concentration increased, this may not lead to the further production of polyphenols that contribute to improved antioxidant activity.

Chocolate fermented with *U. craterium* did not exhibit increased antioxidant activity; however, its antioxidant activity was comparable to that of the non-fermented control, suggesting that mushroom fermentation did not diminish antioxidant capacity. Chocolate fermented with *P. arcularius*, *P. vesiculosa*, and *U. craterium* showed increases in theobromine

https://doi.org/10.3390/foods15061045

concentration, reducing sugar content, and total polyphenol content; moreover, the antioxidant activity was similar to or higher than that of the control.

Collectively, these findings suggest that fermenting cacao beans with mushrooms enables the development of chocolate enriched in theobromine, reducing sugars, and polyphenols, with antioxidant activity that is maintained or enhanced compared with conventional chocolate.

4. Conclusions

This study investigated the use of mushrooms for the secondary fermentation of cocoa beans to further improve quality. When cocoa beans were fermented using 42 species of basidiomycetes and ascomycetes, mycelial growth was observed in 29 mushroom strains.

As a result of preparing 75% cocoa chocolate, the chocolate made with 17 strains showed higher theobromine concentrations compared to the control group, with chocolate fermented with *P. arcularius*, *P. vesiculosa*, and *U. craterium* showing significantly higher levels. Caffeine concentration was similar to or lower than the control in all mushroom strains. The increase in theobromine concentration and decrease in caffeine concentration may indicate theobromine synthesis from xanthine via methylation enzymes, or the generation of theobromine through the breakdown of caffeine by demethylation.

Furthermore, fermentation by *P. arcularius*, *P. vesiculosa*, and *U. craterium* increased the reducing sugar concentration and total polyphenol concentration. Fermentation by *P. arcularius* and *P. vesiculosa* showed excellent antioxidant activity.

Overall, our findings suggest that fermenting cacao beans with mushrooms produces chocolate that is richer in theobromine, reducing sugars, and polyphenols than conventional chocolate and has excellent antioxidant properties. However, for the efficient production of chocolate with high theobromine and low caffeine content through secondary fermentation, further research is required to characterize the methylation and demethylation enzyme in mushrooms.

Author Contributions: Conceptualization, Data curation, Formal analysis, Funding acquisition, Investigation, Resources, Supervision, Validation, Visualization, Writing—original draft, Writing—review and editing, S.F.; Conceptualization, Data curation, Investigation, Validation, M.N.; Supervision, Y.S.; Supervision, N.T.; Conceptualization, Funding acquisition, Methodology, Project administration, Supervision, T.M. All authors have read and agreed to the published version of the manuscript.

Funding: This work was supported by MEXT's Subsidy Program for Human Resource Development for Science and Technology, Initiative for Realizing Diversity in the Research Environment (Type of Women's Leadership Development), Program for Developing the Next Generation of Aspiring Women Leaders at a Women's University.

Institutional Review Board Statement: Not applicable.

Informed Consent Statement: Not applicable.

Data Availability Statement: The original contributions presented in this study are included in the article. Further inquiries can be directed to the corresponding author.

Conflicts of Interest: The authors declare no conflicts of interest.

References

1. Ohnishi, A. Chocolate and fermentation. *Seibutsu-Kogaku Kaishi* **2025**, *103*, 256–260. (In Japanese) [CrossRef]
2. Schwan, R.F.; Wheals, A.E. The microbiology of cocoa fermentation and its role in chocolate quality. *Crit. Rev. Food Sci. Nutr.* **2004**, *44*, 205–221. [CrossRef]
3. Ohnishi, A. Microorganisms and fermentation of cocoa beans. *Trop. Agric. Res.* **2022**, *15*, 114–115. (In Japanese) [CrossRef]

https://doi.org/10.3390/foods15061045

4. Korcari, D.; Fanton, A.; Ricci, G.; Rabitti, N.S.; Laureati, M.; Hogenboom, J.; Pellegrino, L.; Emide, D.; Barbiroli, A.; Fortina, M.G. Fine Cocoa Fermentation with Selected Lactic Acid Bacteria: Fermentation Performance and Impact on Chocolate Composition and Sensory Properties. *Foods* **2023**, *12*, 340. [CrossRef]
5. Calvo, A.M.; Montenegro, A.C.; Monroy, D.M.; Rodriguez-Silva, L.G.; Carreño-Olejua, A.R.; Camargo, I.D. The Impact of Controlled Fermentation Temperature on Chemical Composition and Sensory Properties of Cacao. *Foods* **2025**, *14*, 1441. [CrossRef] [PubMed]
6. Sukmawati, D.; Al Husna, S.N.; Mahfuzh, S.F.; Nursari, P.I.; Izzati, A.N.; Sari, A.B.T.; El-Enshasy, H.A. Effect of yeast consortium on chemical properties of fermented dried cocoa beans (*Theobroma cacao* L.). In *AIP Conference Proceedings*; AIP Publishing LLC: Melville, NY, USA, 2024; Volume 2982, p. 050020. [CrossRef]
7. Kamarisima; Fathima, A.M.; Aditiawati, P.; Yusantri, S.D.; Kumalasari, D.P.; Mutsla, D.; Rachmawati, N.; Dewi, M.; Pratama, F. Enhancing fermented cocoa bean quality and flavor through short secondary fermentation using *Candida tropicalis* as starter and *Citrus limon* peel. *J. Sci. Food Agric.* **2025**, *105*, 8616–8624. [CrossRef] [PubMed]
8. Mizuno, T.; Kawai, M. (Eds.) *Chemistry and Biochemistry of Mushrooms*; Gakkai Shuppan Center: Tokyo, Japan, 1992; p. III. (In Japanese)
9. Nakamura, K. (Ed.) *Encyclopedia of Mushrooms*; Asakura Publishing Co., Ltd.: Tokyo, Japan, 1982; p. 4. (In Japanese)
10. Yaguchi, T. Classification and identification of fungi. *Jpn. J. Food Microbiol.* **2010**, *27*, 47–55. (In Japanese) [CrossRef]
11. Konno, N. Carbohydrate-active enzymes produced by mushrooms. *Mokuzai Hozon.* **2013**, *39*, 60–70. (In Japanese) [CrossRef]
12. Homma, H. Production of mold-ripened cheese using edible mushrooms and Cordycipitaceae mycelium. *Mushroom Sci. Biotechnol.* **2024**, *32*, 22–30. (In Japanese) [CrossRef]
13. Matsui, O.T.; Ohsugi, M. Production of Alcoholic Beverage by Mushroom Fermentation. *J. Brew. Soc. Jpn.* **2002**, *97*, 766–773. (In Japanese) [CrossRef]
14. Tabata, M.; Hayashi, T.; Matsui, T. Coagulometer assay for determination of milk clotting activity in mushroom enzymes. *Mushroom Sci. Biotechnol.* **2009**, *17*, 151–154. (In Japanese) [CrossRef]
15. Nakamura, K. Quinoa Miso Production by Edible Mushroom Grown on the Quinoa Seeds Cultivated in Japan. *J. Brew. Soc. Jpn.* **2015**, *110*, 388–393. (In Japanese) [CrossRef]
16. Homma, H.; Nakamura, K.; Kawamura, T.; Tokuda, H.; Nakanishi, T. Miso production using mushrooms and analysis of chemical composition. *Mushroom Sci. Biotechnol.* **2013**, *21*, 23–29. (In Japanese) [CrossRef]
17. Matsui, T. Development of functional foods using by mushrooms fermentation. *Mushroom Sci. Biotechnol.* **2016**, *24*, 169–175. (In Japanese) [CrossRef]
18. Sameshima, Y.; Suzuki, F.; Tabata, M.; Matsui, T. Properties of post-fermented tea prepared by *Schizophyllum commune* fermentation. *Mushroom Sci. Biotechnol.* **2017**, *25*, 66–69. (In Japanese) [CrossRef]
19. Natsume, M. Comprehensive study of polyphenols. *Kagaku Seibutsu* **2018**, *56*, 490–495. (In Japanese) [CrossRef]
20. Kasai, M.; Ishikawa, Y.; Sakamaki, A.; Okuyama, T.; Ashitani, H.; Kamiwaki, T.; Iida, F. Relationship between origin of cocoa beans and palatability of chocolates. *Nippon Shokuhin Kagaku Kogaku Kaishi* **2007**, *54*, 332–338. (In Japanese) [CrossRef]
21. Saotome, W.; Ikeda, H.; Ogawa, H.; Umagoshi, S.; Matsumoto, K. Quantitative analysis of theobromine and caffeine in cocoa containing prepared foods (the second report). *Cent. Cust. Lab. Rep. Inf.* **2021**, *61*, 55–60. (In Japanese)
22. Somogyi, M. Micromethods for the estimation of diastase. *J. Biol. Chem.* **1938**, *125*, 399–414. [CrossRef]
23. Goto, T. Determination of total polyphenols. In *Manual for Evaluation of Food Functionality III*; Committee for Examination of Technical Extension Materials, Ed.; Japan Society for Food Engineering: Tokyo, Japan, 2009; pp. 1–7.
24. Tabata, M.; Fukuda, S.; Osugi, M.; Sato, Y.; Yamakawa, T.; Hatano, K.; Noike, T.; Matsui, T. Changes in components and function of soybean milk by fermented by *Schizophyllum commune*. *Mushroom Sci. Biotechnol.* **2008**, *16*, 159–163. (In Japanese) [CrossRef]
25. Dash, S.S.; Gummadi, S.N. Catabolic pathways and biotechnological applications of microbial caffeine degradation. *Biotech. Lett.* **2006**, *28*, 1993–2002. [CrossRef] [PubMed]
26. Ashengroph, M. *Salinivibrio costicola* GL6, a Novel Isolated Strain for Biotransformation of Caffeine to Theobromine Under Hypersaline Conditions. *Curr. Microbiol.* **2017**, *74*, 34–41. [CrossRef] [PubMed]
27. Carrasco-Cabrera, C.P.; Bell, T.L.; Kertesz, M.A. Caffeine metabolism during cultivation of oyster mushroom (*Pleurotus ostreatus*) with spent coffee grounds. *Appl. Microb. Cell Physiol.* **2019**, *103*, 5831–5841. [CrossRef]
28. Kato, M. Caffeine biosynthesis in plants. *Radioisotopes* **2001**, *50*, 377–379. (In Japanese) [CrossRef]
29. Homma, H. Production of Shio-koji-like seasoning using mushrooms that produce salt-tolerant amylase and protease. *Mushroom Sci. Biotechnol.* **2024**, *32*, 89–94. (In Japanese) [CrossRef]
30. Tsuji, K.; Mori, B.; Hisakazu, I.; Kiyoshi, E.; Kazuhiro, O.; Tatsuya, K.; Masamichi, K.; Seiichi, S.; Kiwamu, S.; Takashi, T.; et al. *Science of Dietary Fiber (Food Ingredients Series)*; Asakura Publishing Co., Ltd.: Tokyo, Japan, 1997; pp. 31–32. (In Japanese)
31. Imazeki, R.; Otani, Y.; Hongo, T. *Mushrooms of Japan*; Yama-kei Publishers: Tokyo, Japan, 1988; p. 451. (In Japanese)
32. Fukasawa, Y. The influences of wood decay fungi and their decay types on communities of diverse organisms developed on coarse woody debris. *Jpn. J. Ecol.* **2013**, *63*, 311–320. (In Japanese) [CrossRef]

https://doi.org/10.3390/foods15061045

33. Lee, K.W.; Kim, Y.J.; Lee, H.J.; Lee, C.Y. Cocoa has more phenolic phytochemicals and a higher antioxidant capacity than teas and red wine. *J. Agric. Food Chem.* **2003**, *51*, 7292–7295. [CrossRef]
34. Hatano, T.; Miyatake, H.; Natsume, M.; Osakabe, N.; Takizawa, T.; Ito, H.; Yoshida, T. Proanthocyanidin glycosides and related polyphenols from cacao liquor and their antioxidant effects. *Phytochemistry* **2002**, *59*, 749–758. [CrossRef]

https://doi.org/10.3390/foods15061045

foods

MDPI

Article

Genetic Modulation of *ATF1* in *Saccharomyces cerevisiae* for Enhanced Acetate Ester Production and Flavor Profile in a Sour Meat Model System

Ning Zhao [1], Ying Yue [1], Shufeng Yin [1], Hao Liu [1], Xiaohan Jia [1], Ning Wang [1], Chaofan Ji [1], Yiwei Dai [1], Liguo Yin [2], Huipeng Liang [1] and Xinping Lin [1,*]

[1] State Key Laboratory of Marine Food Processing & Safety Control, National Engineering Research Center of Seafood, School of Food Science and Technology, Dalian Polytechnic University, Dalian 116034, China; ningzhao0221@163.com (N.Z.); nicoleyue1996@163.com (Y.Y.); yinshufeng1220@163.com (S.Y.); 18763666325@163.com (H.L.); jiaxiaohan1129@163.com (X.J.); ningwang0221@163.com (N.W.); jichaofan@outlook.com (C.J.); daiyiwei@dlpu.edu.cn (Y.D.); lhpdxx@126.com (H.L.)

[2] Solid-State Fermentation Resource Utilization Key Laboratory of Sichuan Province, Yibin University, No. 5, Section 3 of Daxue Road, Sanjiang New Area, Yibin 644000, China; lgying22@126.com

* Correspondence: yingchaer@163.com; Tel.: +86-158-9816-0352

Abstract

Acetate esters, synthesized by alcohol acyltransferase (AATases) encoded primarily by the *ATF1* gene, are pivotal for the desirable fruity aroma in fermented foods. However, the role and regulatory impact of *ATF1* in solid-state fermented meat remain largely unexplored. This study engineered *Saccharomyces cerevisiae* by knocking out and overexpressing *ATF1* to investigate its influence on flavor formation in a sour meat model system. Compared to the wild-type strain, *ATF1* overexpression (SCpA group) increased ethyl acetate content by 70.15% and uniquely produced significant levels of isoamyl acetate. Conversely, *ATF1* deletion (SCdA group) led to a 61.23% reduction in ethyl acetate. Transcriptomic analysis revealed that *ATF1* overexpression triggered a systemic metabolic shift, not only activating the final esterification step but also upregulating key genes in central carbon metabolism (*SUC2*, *ICL1*), amino acid biosynthesis, and precursor supply pathways (*ACS2*, *ADH1*). This synergistic regulation redirected metabolic flux towards the accumulation of both alcohol and acyl-CoA precursors, thereby amplifying acetate ester synthesis. Our findings demonstrate that *ATF1* is a critical engineering target for flavor enhancement in fermented meats and uncover a broader metabolic network it influences, providing a robust strategy for the targeted modulation of food flavor profiles.

Keywords: flavor formation; *Saccharomyces cerevisiae*; *ATF1*; acetate esters; fermented meat; transcriptomics

1. Introduction

Fermented meat products are processed through microbial fermentation and biochemical actions. This not only imparts a unique flavor but also effectively extends the shelf life while improving the nutritional profile throughout the fermenting procedure. [1]. Sour meat, a traditional ethnic fermented meat product in China, is produced through a unique fermentation process. It is primarily made from pork, rice flour, and salt, and undergoes fermentation under anaerobic conditions. The final product is characterized by a firm texture and a distinct sour flavor with ester-like aroma notes. However, the characteristics of sour meat are influenced by a multitude of factors spanning ambient conditions, manufacturing procedures, duration, and material composition, which can lead to significant instability in quality [2]. The inoculation of starter cultures is an effective approach to enhance the stability and flavor profile of sour meat. For instance, Shang [3] inoculated *Lactiplantibacillus plantarum* and *Pediococcus pentosaceus* into fermented meat, significantly reducing the content of biogenic amines, enhancing its safety, and simultaneously stimulating the generation of volatile organic compounds. Guo [4] inoculated *Yarrowia lipolytica* into fermented meat, stimulating the synthesis of esters, aldehydes, and alcohols—with a notable increase in ethyl esters—thereby imparting a richer flavor profile to the product. Zhang [5] employed co-fermentation with *Latilactobacillus curvatus* and *Pediococcus pentosaceus* in fermented meat, which improved the color, total lactobacilli count, and free amino acid content, while the concentrations of nitrite, biogenic amines, total volatile base nitrogen (TVB-N), and malondialdehyde were effectively lowered. This led to a significant elevation in the safety and quality attributes of the fermented meat.

The ester content in fermented meat plays a dominant role in shaping its flavor profile. Crucial contributors include acetate and ethyl esters, with examples being ethyl acetate, ethyl octanoate, and ethyl hexanoate. These ester compounds typically contribute fruity and creamy flavors to fermented meat. Research has indicated that inoculating *S. cerevisiae* LXPSC1 in sour meat significantly improved the flavor quality, identifying it as an ideal fermentation agent for the enhancement of sour meat quality. *S. cerevisiae* shows a direct association with elevated concentrations of esters, alcohols, and total VOCs, with ester content in particular increasing by 4.36-fold. [6]. Research shows that S. cerevisiae is instrumental in ester compound synthesis due to the presence of AATases. AATases mediate the conversion of ethanol and acetyl-CoA into acetate ester. In *S. cerevisiae*, these AATases are primarily encoded by the *ATF1* and *ATF2* genes [7]. The synthesis of AATases is promoted by the overexpression of both the *ATF1* and *ATF2* genes [8,9]. This enhancement efficiently enables the condensation of acetyl-CoA with ethanol, yielding ethyl acetate as well as isoamyl acetate. Marullo assessed the roles of the *ATF1* and *ATF2* genes in regulating the synthesis of ethyl acetate and isoamyl acetate. The results indicated that *ATF1* plays a more critical role and is primarily responsible for producing both esters. In contrast, the absence of *ATF2* did not significantly affect the yield of either compound [10].

Traditional studies on microorganisms frequently employ intentional inoculation with specific microbial strains to investigate their impact on the attributes of fermented foods. However, the approach of adding microbial variants with gene knockouts or overexpressed genes is more conducive to thoroughly studying and clarifying the role of key genes in shaping the microenvironment. For instance, Shi [11] enhanced the production of the key enzymes acetyl-CoA synthase (*ACS1/2*) and aldehyde dehydrogenase (*ALD6*) in the Ac-CoA biosynthesis module by employing various terminators. Additionally, they achieved co-expression of the *ATF1* gene within the *ACS1/2* and *ALD6* cassettes, which led to a further increase in the ethyl acetate content of the liquor. Zhang [12] achieved a nine-fold increase in ethyl acetate production by overexpressing the *ATF1* gene through the transformation of a multicopy plasmid in industrial brewer's yeast. The study by Dong [13] revealed that overexpression of the *ATF1* gene significantly boosts the strain's capacity to produce ethyl acetate, thereby enriching the aroma of the liquid fermentation of corn hydrolysate and further improving its taste and flavor. The regulation of acetate esters by *ATF1* has been validated in liquid-state fermented foods, such as wine and beverages, particularly in alcoholic drinks. However, research on solid-state fermented foods, such as fermented meats, is exceedingly scarce.

Therefore, based on our preliminary findings regarding the impact of added *S. cerevisiae* on ester production in sour meat, we further investigated the influence of *ATF1*, a gene closely associated with ester synthesis in *S. cerevisiae*, on its flavor. We hypothesize

https://doi.org/10.3390/foods15020378

that modulation of the expression level of the *ATF1* gene in *Saccharomyces cerevisiae* can significantly alter the synthesis efficiency of acetate esters, thereby directionally modulating the flavor profile of sour meat in a solid-state sour meat model system. In this study, we performed both knockout and overexpression of the *ATF1* gene, which encodes AATases in *S. cerevisiae*. The aim was to modulate the expression of ester synthesis-related genes at the genetic level in *S. cerevisiae*, and to examine the function of the *ATF1* gene in ester production in solid-state fermented foods. Furthermore, we utilized transcriptomic analysis to assess the effects of *ATF1* gene knockout and overexpression on the expression levels of other genes in *S. cerevisiae*. This aimed to uncover potential pathways and genes involved in ester synthesis, providing precise target information for further optimization of fermentation processes.

2. Materials and Methods

2.1. Experimental Materials, Strains and Plasmids

The pork belly and lard were obtained from the Dashui Market in Dalian City. Table 1 lists all the strains and plasmids used in this study.

Table 1. Strains and plasmids.

Strains and Plasmids		Genotype or Sequence	Source
Strains	*Saccharomyces cerevisiae* SC	*MATa; MAL2-8c; SUC2; his 3△1; ura3-52; gal80△;XI-5::*P_{TEF1}*-Cas9-*T_{CYC1}	Yongjin Zhou's lab; Dalian Institute of Chemical Physics, CAS
	S. cerevisiae SCdA	*MATa; MAL2-8c; SUC2; his3△1; ura3-52; gal80△; XI-5::*P_{TEF1}*-Cas9-*T_{CYC1}*; △atf1*	This research
	S.cerevisiae SCpA	*MATa; MAL2-8c; SUC2; his3△1; ura3-52; gal80△; XI-5::*P_{TEF1}*-Cas9-*T_{CYC1}*;* X3::P_{TPI1}-ATF1-T_{CYC1}	This research
	Escherichia coli DH5a	*F-, φ80d/lacZ△M15,△(lacZYA-argF) U169, deoR, recAl, endAl, hsdR17(rk, mk*$^+$*), phoA, supE44, λ*$^-$*, thi-1, gyrA96, relA1*	Takara
Plasmid	pgRNA-X3	*URA3, 2 μm, Amp*^{R}, gRNA1-X3, gRNA2-X3	This research
	pgRNA-ATF1	*URA3, 2 μm, Amp*^{R}, gRNA1-ATF1 gRNA2-ATF1	This research

2.2. Chemicals

All chemicals used were of analytical or specified grade. YPD liquid medium, agar powder, and LB liquid medium were obtained from Qingdao Hope Bio-Technology Co., Ltd. (Qingdao, China). Acetyl-CoA, amino acid mixture standard, and the following analytical standard esters: isoamyl acetate (CAS: 123-92-2), ethyl hexanoate (CAS: 123-66-0), ethyl octanoate (CAS: 106-32-1), ethyl butyrate (CAS: 105-54-4), ethyl acetate (CAS: 141-78-6), ethyl heptanoate (CAS: 106-30-9), ethyl decanoate (CAS: 110-38-3), phenethyl acetate (CAS: 103-45-7), and ethyl propionate (CAS: 105-37-3), as well as chromatographic grade cyclohexanone and C7–C30 saturated alkanes, were obtained from Sigma-Aldrich (St. Louis, MO, USA). Anhydrous ethanol and isoamyl alcohol (both HPLC grade) were supplied by Kemio Chemical Reagent Co., Ltd. (Tianjin, China). Trichloroacetic acid and acetone (chromatographic grade) were purchased from Aladdin Reagent (Shanghai) Co., Ltd. (Shanghai, China). The malondialdehyde (MDA) assay kit was provided by Nanjing Jiancheng Bioengineering Institute (Nanjing, China).

https://doi.org/10.3390/foods15020378

2.3. Knockout and Overexpression of the ATF1 Gene

The primers used for knockout and overexpression of the *ATF1* gene in *S. cerevisiae* SC are listed in Table S1. The *ATF1* knockout plasmid was constructed by amplifying fragments using primer pairs Ag-1 & Pg-1, Ag-2 & Pg-2, and P-6005 & P-6005. Subsequently, the three amplified products were cloned into vectors using the Vazyme ClonExpress II One Step Cloning Kit (C113). Transformation of the recombinant plasmids was subsequently carried out in *Escherichia coli* DH10B. Correct transformants were selected, and the plasmids were validated to confirm the successful construction of the *ATF1* knockout plasmid. Similarly, the construction of the *ATF1* overexpression plasmid was achieved by amplifying fragments using primer pairs X3g-1 & Pg-1, X3g-2 & Pg-2, and P-6005 & P-6005 (where "X3" refers to the third neutral site on chromosome X of *Saccharomyces cerevisiae*, a site known to contribute to expanding the metabolic capacity of the yeast [14]), followed by ligation, transformation, and validation to obtain the correct plasmid.

S. cerevisiae transformation was performed according to the protocol described by Cao [15]. The exogenous DNA fragments used for knockout were obtained using fusion PCR technology. The knockout cassette was obtained by fusing the upstream and downstream gene fragments of *ATF1*, which were amplified separately. The expression cassette for the *ATF1* gene overexpression is shown in Figure S1. After transforming *S. cerevisiae*, the transformants were selected on SD (HIS) plates. The successful knockout and overexpression of *ATF1* were verified by extracting genomic DNA from the colonies. The *ATF1* knockout strain was designated as SCdA, and the *ATF1* overexpression strain was designated as SCpA.

2.4. Strains and Culture Conditions

The selected *S. cerevisiae* strains included the parental strain SC, the gene-knockout strain SCdA, and the overexpressing strain SCpA, all stored at −80 °C. Before use, strains were first activated on YPD solid medium and then inoculated into YPD liquid medium, followed by a 24 h incubation on a shaker at 28 °C. Following centrifugation of the culture (10,000 rpm, 10 min, 4 °C), the harvested cells were rinsed, reconstituted in sterile saline, and standardized to 7 log CFU/mL.

2.5. Preparation of the Sour Meat Model System

The sour meat model system was formulated according to the study by Cano-García [16].

Extraction of myofibrillar proteins [17]: Pork was trimmed of fat and connective tissue before being minced. The minced meat was combined with a five-fold volume of extraction buffer (0.1 M Tris-HCl, 20 mM EDTA, pH 7.0), followed by thorough homogenization until no visible particles remained. After centrifugation at 10,000 rpm for 30 min at 4 °C, the supernatant was collected and discarded.

Preparation of the sour meat model system [16]: Each 100 g of the model system contains 35 g of myofibrillar protein, 25 g of pork fat, 5 g of glucose, and 35 g of 0.1 M phosphate-buffered solution (pH 7.5). The mixture is blended and homogenized to prepare a sour meat model system. The inoculated group is treated by inoculating the model system with *S. cerevisiae* at a concentration of 1×10^7 CFU/g, followed by fermentation at 30 °C for 7 days. The experimental groups were designated as follows: F, the sour meat model system prior to fermentation; C, a control group without *S. cerevisiae* inoculation; SCdA, a group inoculated with the *ATF1*-knockout strain (SCdA); SC, a group inoculated with the wild-type *S. cerevisiae*; and SCpA, a group inoculated with the *ATF1*-overexpressing strain (SCpA).

https://doi.org/10.3390/foods15020378

2.6. The Determination of pH, TBARS, Protein Concentration, and TCA-Soluble Peptide Content

Following homogenization and centrifugation of a 2 g sample mixed with 20 mL of water, the supernatant was collected for pH measurement with a benchtop pH meter (FE28; Mettler Toledo, Greifensee, Switzerland). The determination of TBARS was conducted as described by Cheng [18]. Homogenization of a 2 g sample was performed in 20 mL of phosphate buffer, followed by centrifugation (12,000× g, 6 min). We collected the resulting supernatant and quantified the protein concentration via the Bradford method. As described by Wang [19], a 2 g aliquot of the sample was mixed with 8 mL of 10% TCA solution, blended for 1 min, and then subjected to centrifugation to collect the supernatant for measuring the content of TCA-soluble peptides. Using L-tyrosine as the standard, a standard curve was plotted. The content of TCA-soluble oligopeptides was expressed as μmol Tyr/g.

2.7. Determination of Fatty Acids

According to Pei's method [20], total lipids were extracted from a 5 g sample. The sample was thoroughly combined with the extraction solution, concentrated at 35 °C by rotary evaporation, and the total lipid content was subsequently determined. Following derivatization of lipid fatty acids into fatty acid methyl esters with boron trifluoride-methanol, analysis was carried out on an Agilent 7890A GC-5975C MSD system (Santa Clara, CA, USA) equipped with an HP-5-MS capillary column (30 m × 0.25 mm × 0.25 μm). The specific conditions and parameters are detailed in the reference. The identification of fatty acids is combined with a comparative analysis of GC-MS retention times against those in a fatty acid standard by searching the NIST11 library. These compounds were then quantified based on their corresponding peak areas.

2.8. The Determination of Amino Acids and Electronic Tongue

A 3.0 g sample was blended with 12 mL of deionized water for homogenization before being treated with acetone. The mixture was allowed to stand for 10 min before centrifugation to collect the supernatant. Evaporated to dryness at 60 °C using a water bath, the supernatant was subsequently reconstituted in acetone and passed through a 0.22 μm membrane filter in preparation for subsequent analysis. Amino acid analysis was performed using the fully automatic amino acid analyzer LA8080 according to the manufacturer's instructions. The standard used was an amino acid mixed standard (Sigma-Aldrich, St. Louis, MO, USA; Catalog number: AAS18). According to the formula TAV = C/T [21], the Taste Activity Value (TAV) is calculated to reflect the taste contribution of flavor amino acids in the sour meat model system. A TAV greater than 1 denotes a significant contribution to the overall taste. Here, C is the absolute concentration (mg/100g) of the compound, and T is its taste threshold (mg/100g).

The electronic tongue measurements were conducted according to the method of Liu [22]. The electronic tongue model TS-5000Z was purchased from Insent Inc. (Atsugi, Japan). Homogenization of a 30 g sample aliquot was performed with 100 mL of water. The resulting mixture was then passed through a 0.22 μm water-compatible filter membrane to prepare it for analysis.

2.9. Determination of Volatile Compounds

Analysis of volatile compounds in the samples was conducted through an SPME-GC-MS system. Full-scan (SCAN) mode was first employed to detect all flavor compounds in the sour meat, followed by selected ion monitoring (SIM) mode for precise quantification of esters. The procedure was performed as follows: We placed 2 g of the sample in a 20 mL headspace vial, added 20 μL of cyclohexanone internal standard (50 mg/L), and heated

https://doi.org/10.3390/foods15020378

it at 60 °C for 30 min. We then collected volatiles using an SPME fiber for 40 min, which was subsequently desorbed for 10 min in the injector of an Agilent 7890A-5975C GC-MS system (Agilent Technologies, Inc.; Santa Clara, CA, USA) and analyzed on an HP-5MS capillary column (30 m × 250 μm × 0.25 μm). The specific conditions were the same as those described by Liu [23]. Substance identification was performed using the NIST 11 standard library for automatic retrieval. Calculation of the Kovats retention indices (RI) for all volatile compounds was performed using the retention time data of a C7–C30 n-alkane homolog mixture. Quantification was performed via comparison of individual peak areas to that of the internal standard, with results given in micrograms per 100 g (μg/100g).

2.10. Transcriptomics Sequencing

After fermentation at 30 °C for 7 d, aliquots (5 g) of each sample from the SCdA and SCpA groups were aseptically transferred into sterile 10 mL centrifuge tubes and cryopreserved at −80 °C. The samples were transported to Biomarker Technologies Corporation (Beijing, China) using dry ice for Illumina transcriptomics sequencing. The purity and concentration of RNA were assessed using a NanoDrop 2000 spectrophotometer (Thermo Fisher Scientific Inc; Wilmington, DE, USA), while the integrity of RNA was precisely evaluated using an Agilent 2100 Bioanalyzer/LabChip GX (Agilent Technologies, Inc.; Santa Clara, CA, USA). After sample qualification, libraries were constructed. Following library quality inspection, sequencing was performed on the Illumina NovaSeq 6000 platform in PE150 mode.

After obtaining the raw data, initial filtering was performed. Subsequently, the genome of *S. cerevisiae* S288C (GCF_000146045.2_R64) was used as a reference for alignment analysis using TopHat2. Following quality control of the transcriptomic data using RSeQC (v2.6.3) to analyze sequencing saturation, gene coverage, and duplicate reads, differentially expressed genes (DEGs) were identified. Gene expression levels, calculated as FPKM, were compared using the EdgeR package (v3.10) with screening thresholds set at $|\text{Fold Change}| \geq 1.5$ and FDR < 0.05. The final DEGs were then functionally annotated and subjected to KEGG enrichment analysis.

2.11. Statistical Analysis

A completely randomized experimental design was employed in this study, and all experiments were independently repeated at least three times. Data were analyzed using SPSS version 27.0 (International Business Machines Corp., Armonk, NY, USA), and results are presented as means ± standard deviation (SD). Statistically significant differences among samples were determined by distinct alphabetic characters. Statistical significance was analyzed by one-way ANOVA followed by Duncan's test at a significance level of $p < 0.05$. Figures and charts were created using Origin 2022 (OriginLab Corp., Northampton, MA, USA). The associated heatmaps were generated with TBtools-II v2.4 software.

3. Results and Discussion

3.1. Changes in pH and TBARS in the Sour Meat Model System

Figure 1a demonstrated a marked decline in pH levels ($p < 0.05$) throughout the fermentation process. At the end of fermentation, the pH in every group had dropped to levels below 4.5. The drop in pH results from the bacterial metabolism of carbohydrates by lactic acid bacteria, which in turn drives lactic acid production [24]. A pH below 4.5 can suppress the proliferation of spoilage and pathogenic bacteria [25]. Therefore, the lower pH values of the four sample groups can delay food spoilage, facilitate preservation, and ensure food safety. Relative to the control (group C), a more pronounced reduction in pH was observed in the inoculated groups over the fermentation period ($p < 0.05$). The possible reasons are: on one hand, the inoculation of *S. cerevisiae* affects the growth of lactic acid bacteria by promoting their proliferation [24], while, on the other hand, the inoculation of *S. cerevisiae* makes carbohydrates more readily metabolized by lactic acid bacteria [26]. Therefore, by contributing to increased levels of organic acids, the inoculation of *S. cerevisiae* helps maintain a lower pH in the system.

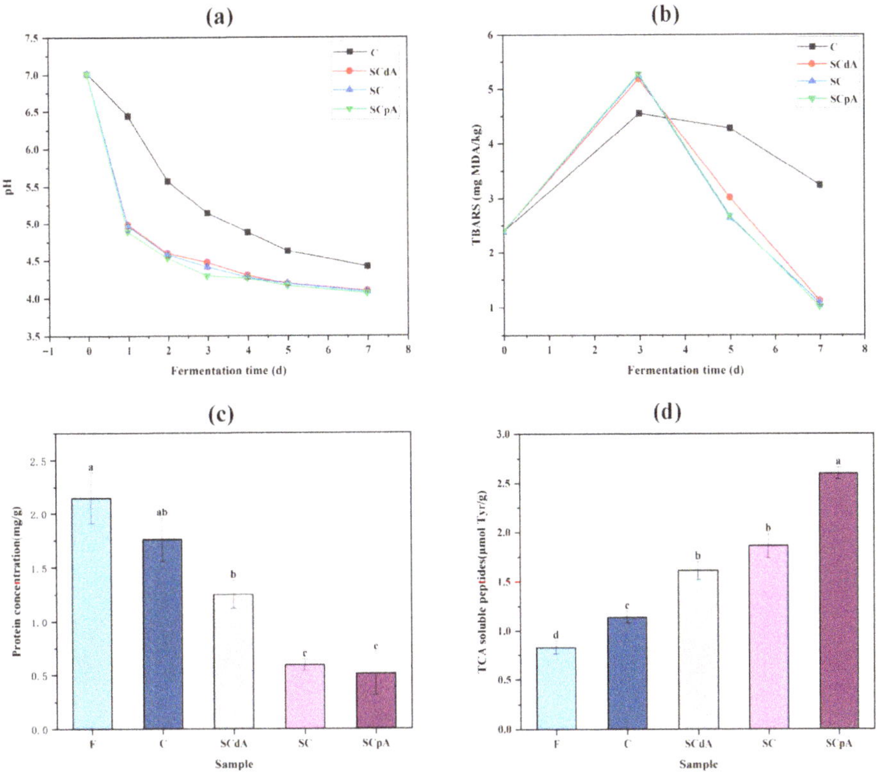

Figure 1. Changes of pH (**a**) and TBARS (**b**), content of protein (**c**) and TCA soluble peptide (**d**) in the sour meat model system. Note: F: the sour meat model system prior to fermentation; C: a control group without inoculation with SC; SCdA: a group inoculated with SCdA; SC: a group inoculated with untreated SC; SCpA: a group inoculated with SCpA. Different letters a–d indicated significant differences among different groups ($p < 0.05$).

TBARS is an indicator employed to assess the extent of lipid oxidation throughout fermentation [27]. Figure 1b shows that the TBARS level in the four fermentation groups initially increased and then decreased during fermentation. The TBARS values in the *S. cerevisiae*-inoculated samples showed a marked increase compared to the control at the mid-fermentation stage ($p < 0.05$), but subsequently declined to levels considerably below the control by the conclusion of the fermentation process ($p < 0.05$). Similar findings were also reported by Chaves-López [28]: in a myofibrillar protein sausage model inoculated with *S. cerevisiae*, protein consumption was significantly enhanced, leading to a more complex and intensified flavor profile. During the later stages of fermentation, *S. cerevisiae* was able to inhibit lipid oxidation to some extent, as evidenced by lower MDA content in *S. cerevisiae*-inoculated groups. This may be because *S. cerevisiae* promotes the production of MDA during the early stages of fermentation, which then interacts with other components such as nucleotides, nucleic acids, proteins, and other aldehydes to promote the accumulation

https://doi.org/10.3390/foods15020378

of aromatic substances [29,30]. Therefore, *S. cerevisiae* promotes fat oxidation to enhance aroma in the beginning of the sour meat model fermentation, while it inhibits fat oxidation during the later stages to prevent the development of off-flavors.

3.2. Changes in Protein Concentration and TCA Content in the Sour Meat Model System

As shown in Figure 1c, the protein concentration significantly decreased ($p < 0.05$) in all groups after fermentation. The decrease in protein concentration is attributed to the breakdown of proteins into peptides, amino acids, aldehydes, and other low-molecular-weight components [31]. The protein concentration in *S. cerevisiae*-inoculated groups was significantly lower than that in control group C ($p < 0.05$). This may be due to the introduction of *S. cerevisiae* promoting protein degradation [30]. In summary, inoculation with *S. cerevisiae* facilitates the decomposition and transformation of proteins in the sour meat model system.

TCA-soluble peptides are important components contributing to the umami flavor of acid meat [32]. Figure 1d shows that the content of TCA-soluble oligopeptides significantly increased ($p < 0.05$) in all groups after fermentation. The TCA-soluble oligopeptide content in *S. cerevisiae*-inoculated groups was significantly higher than that in control group C, indicating that more myofibrillar proteins were degraded by endogenous proteases and microbial proteases in *S. cerevisiae*-inoculated groups. Jones's research found that *S. cerevisiae* contains proteases and aminopeptidases, which can hydrolyze proteins to produce peptides and amino acids [33], aligning with our study results. The continuous accumulation of soluble peptides in *S. cerevisiae*-inoculated groups may enhance the potential for achieving desirable flavor in meat products.

3.3. Changes in Free Fatty Acid Content in the Sour Meat Model System

Lipases catalyze the hydrolysis of lipids in meat products, generating free fatty acids (FFA). [34]. The fatty acid content was found to be significantly elevated in all sour meat model systems following the fermentation process. Moreover, the *S. cerevisiae*-inoculated groups yielded a significantly greater fatty acid content than the control group C ($p < 0.05$) (Table S2). In general, within each group, palmitic acid, stearic acid, and oleic acid accounted for relatively high proportions. Additionally, the fatty acid compositional profile demonstrated a distinct pattern: polyunsaturated fatty acids (PUFA) showed the highest concentration, followed by saturated fatty acids (SFA), with monounsaturated fatty acids (MUFA) accounting for the lowest proportion. Consistent findings have been reported in other studies. For instance, in Olivares' research [35], as the fermentation of sausages progressed, the content of FFA significantly increased ($p < 0.01$). Notably, sausages inoculated with *S. cerevisiae* exhibited the highest free fatty acid (FFA) content. Additionally, an observable trend emerged across treatment groups, indicating that unsaturated fatty acid levels exceeded those of saturated fatty acids. Palmitic acid and oleic acid were identified as the predominant fatty acids in all samples. During the fermentation process, unsaturated fatty acids undergo oxidative degradation, generating volatile compounds that influence the flavor profile of meat products. Gianelli [36] demonstrated that Mallorca Black Pork Sobrassada contains a significant amount of unsaturated fatty acids. As fermentation progresses, these FFAs are oxidized, resulting in a marked rise in the levels of volatile compounds within the product. In our sour meat model system, the SCpA group exhibited the highest content of unsaturated fatty acids, indicating that this group may produce higher levels of volatile compounds, which could potentially influence the flavor profile of the sour meat.

https://doi.org/10.3390/foods15020378

3.4. Changes in Free Amino Acid Content and Electronic Tongue Analysis in the Sour Meat Model System

As shown in Figure 2a, relative to the fresh sample F, the total free amino acid content significantly increased ($p < 0.05$) in all groups after fermentation. Moreover, *S. cerevisiae*-inoculated groups demonstrated significantly elevated levels ($p < 0.05$) relative to control group C. Our results corroborate the findings of Nie [37], who attributed this phenomenon to the proteolytic or aminopeptidase activity of *S. cerevisiae* on meat proteins, leading to the breakdown of proteins and production of free amino acids. In *S. cerevisiae*-inoculated groups, the SCpA group exhibited the highest total free amino acid (TAA) content, followed by the SC group, and then the SCdA group. This phenomenon may be attributed to the overexpression of the *ATF1* gene, which leads to the upregulation of certain genes involved in the regulation of amino acid biosynthesis pathways, thereby promoting amino acid synthesis and indirectly resulting in significant accumulation of amino acids.

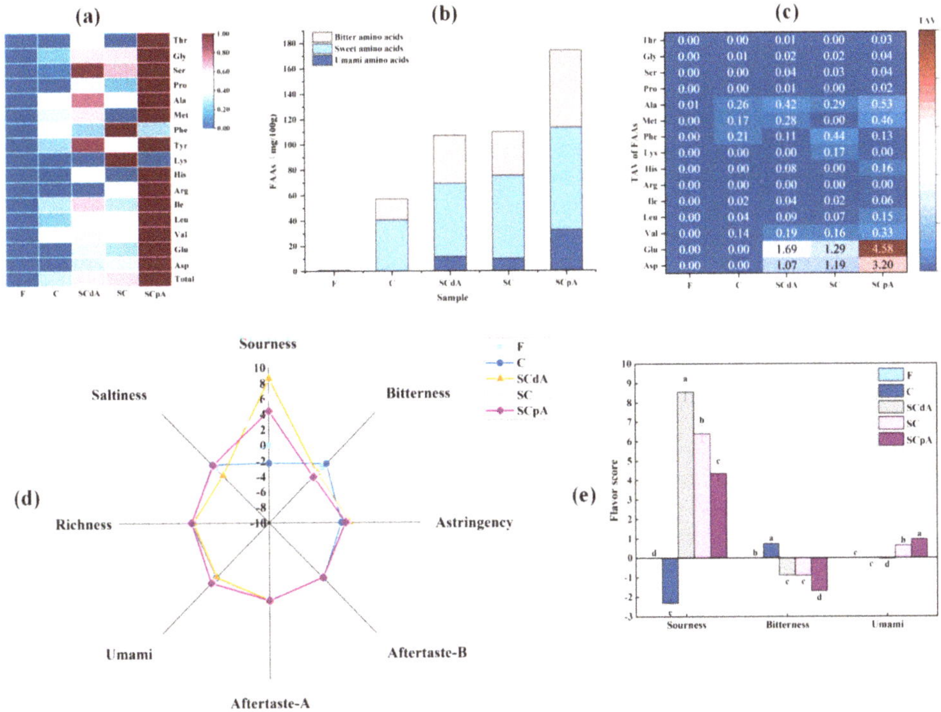

Figure 2. (**a**) Changes of FAAs content before and after fermentation in different groups; (**b**) Taste activity value (TAV) of FAAs; (**c**) Classification of taste-presenting FAAs; (**d**,**e**) Electronic tongue. Note: Total: Total free amino acid. Different letters a–d indicated significant differences among different groups ($p < 0.05$).

In the study by Chaves-López, the proteolytic activity of *S. cerevisiae* strains in the myofibrillar protein model led to the accumulation of amino acids such as Glu and Ala [28]. Similarly, as shown in Figure 2b, our findings demonstrate that the content of sweet amino acids increased post-fermentation compared to pre-fermentation levels. Notably, the concentration in the S. cerevisiae-inoculated groups was significantly higher than that in the control group ($p < 0.05$). After fermentation, compared to the control group, the *S. cerevisiae*-inoculated groups produced substantial amounts of Asp and Glu, two free amino acids that impart a savory (umami) taste. As shown in Figure 2c, both amino acids showed a taste activity value (TAV) > 1, indicating that Asp and Glu are the primary taste-active amino acids in this simulation system. Asp and Glu exhibited the highest concentrations in the SCpA group, and similarly, the content of amino acids such as Met and Val was also highest in the SCpA group ($p < 0.05$). The accumulation of these free amino acids will contribute to improving the flavor and nutritional value of the products during the sour meat fermentation process [38].

Electronic tongue analysis results for pre- and post-fermentation stages in the model system are shown in Figure 2d,e. As can be seen from the figure, after fermentation, the sourness of *S. cerevisiae*-inoculated groups was significantly higher than that of the control group C, while the bitterness was significantly lower compared to the control group. In *S. cerevisiae*-inoculated groups, the umami taste was highest in the SCpA group, while sourness and bitterness were the lowest ($p < 0.05$). Overall, the SCpA group exhibited the highest level of overall acceptability, which may be attributed to its rich composition of flavor substances.

3.5. Changes in Volatile Compounds in the Sour Meat Model System

Figure 3a shows that a total of 35 volatile compounds were identified in the model system before and after fermentation, including 9 alcohols, 5 aldehydes, 8 acids, 11 esters, and 2 other compounds. These substances, which contribute significantly to the unique flavor profile of sour meat, are produced through lipid oxidation, proteolysis, and amino acid catabolism [39]. Prior to fermentation, 8 compounds were detected in sample F. After 7 days of fermentation, 10, 21, 20, and 21 compounds were identified in groups C, SCdA, SC, and SCpA, respectively. Compared with the pre-fermentation stage, both the variety and the quantity of compounds increased. According to Figure 3b, the total content of volatile substances in *S. cerevisiae*-inoculated group samples was significantly elevated compared to the control group C and the pre-fermentation sample F, with the SCpA group showing the most significant difference ($p < 0.05$). Additionally, the ester content in the SCpA group was also significantly higher compared to the other groups. Figure 3c shows that post-fermentation flavor compounds in the control group were predominantly alcohols, whereas esters were the dominant compounds in *S. cerevisiae*-inoculated groups. Notably, the SCpA group displayed the highest relative abundance of esters among all *S. cerevisiae*-inoculated groups. This indicates that inoculating with *S. cerevisiae* can alter the production of volatile substances in the model system, particularly having a significant effect on the esters.

The semi-quantitative results obtained in SCAN mode are susceptible to matrix interference. Therefore, we further performed precise quantification of key esters in sour meat under the selected ion monitoring (SIM) mode, particularly focusing on ethyl acetate and isoamyl acetate, which are produced by the AATases encoded by the *S. cerevisiae ATF1* gene. According to Figure 3d, esters in the *S. cerevisiae*-inoculated groups predominantly comprised ethyl decanoate, ethyl octanoate, and ethyl hexanoate. Additionally, in the SCpA group, significant amounts of ethyl phenylacetate, isoamyl acetate, and ethyl acetate were detected. These esters contribute a rich aroma profile (including fruity and floral notes) to the model system [40]. After fermentation, the ethyl acetate content among the groups was ranked as SCpA > SC > SCdA. Specifically, compared to the SC group, the ethyl acetate content in the SCdA group decreased by 61.23%, while in the SCpA group, it increased by 70.15%. Moreover, isoamyl acetate was detected only in the SCpA group ($p < 0.05$). This indicates that an increase in the expression level of the *S. cerevisiae ATF1* gene can enhance the content of ethyl acetate and isoamyl acetate in the sour meat model system. These two esters impart a fruity sweetness to the product. Previous studies have reported that overexpression of the *ATF1* gene in *S. cerevisiae* can increase the acetate ester

https://doi.org/10.3390/foods15020378

content in alcoholic beverages [41]. Our experimental results also confirm a similar effect in fermented meat systems.

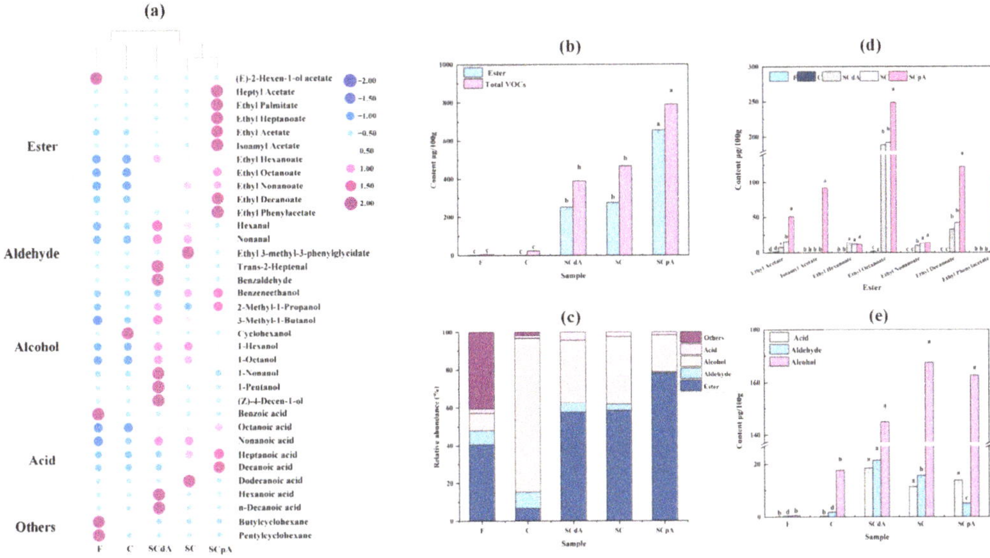

Figure 3. (**a**) Heat map of volatile compounds in the sour meat model system; (**b**) the contents of ester and total VOCs in different groups before and after fermentation; (**c**) the relative abundance of different classes of VOCs; (**d**) the contents of key esters in different groups before and after fermentation; (**e**) the contents of acid, aldehyde and alcohol in different groups before and after fermentation. Note: Data used in figures (**a**–**c**,**e**) were obtained using the SCAN mode; data in figure (**d**) were acquired using the SIM mode. Different letters a–d indicated significant differences among different groups ($p < 0.05$).

Alcohols are essential components of fermented meat flavor, primarily originating from carbohydrate metabolism, lipid oxidation, and amino acid catabolism [42]. According to Figure 3a,e, compared to the pre-fermentation stage, both the variety and the quantity of alcohols increased in each group post-fermentation. Specifically, the SCpA and SC groups exhibited the highest levels of alcohols, particularly benzeneethanol ($p < 0.05$). Benzeneethanol, as one of the alcohols that influence the flavor of fermented meat, is a volatile alcohol with a rose-like aroma [43]. However, in the SCdA group, the content of other alcohols was significantly higher than that in the SCpA and SC groups, particularly 3-methyl-1-butanol ($p < 0.05$). This may be due to the absence of the *S. cerevisiae ATF1* gene in the SCdA group, which prevents the production of isoamyl acetate in the simulated system, leading to a significant accumulation of 3-methyl-1-butanol. This is consistent with the findings of Li et al. in their study on liquor, which showed that *ATF1* deletion significantly decreases the content of ethyl esters while increasing the content of alcohols [44]. After fermentation, the content and variety of aldehydes and acids increased to different extents in each group, with significantly higher levels observed in *S. cerevisiae*-inoculated groups compared to group C ($p < 0.05$). This indicates that inoculating with *S. cerevisiae* can enhance the content of aldehydes and acids in the sour meat model system. In sour meat, the primary aldehydes are aliphatic aldehydes, which in meat products are typically the oxidation products of unsaturated fatty acids caused by endogenous enzymes within fat tissues or muscle tissues [43,45]. In the sour meat model system inoculated with *S. cerevisiae*, the aldehyde content across the groups was ranked as SCdA > SC > SCpA ($p < 0.05$). This is likely because the overexpression of *ATF1* in the SCpA group promotes the conversion

https://doi.org/10.3390/foods15020378

of more fatty acids into fatty acyl-CoA, which in turn increases the production of fatty acid esters. This inhibits the autoxidation of fatty acids, leading to a reduction in aliphatic aldehyde content.

Overall, inoculation with *S. cerevisiae* significantly altered the production profile of volatile compounds in the model system, particularly enhancing both the concentration and diversity of esters. Notably, the *SCpA* group exhibited the most pronounced effect. The upregulation of the *ATF1* gene in *S. cerevisiae* significantly elevates the concentrations of ethyl acetate and isoamyl acetate in the simulated sour meat system, thereby imparting a distinct fruity-sweet aroma profile to the final product. Additionally, inoculation of *S. cerevisiae* enhances the concentrations of alcohols, aldehydes, and organic acids in the system.

3.6. Transcriptomic and Major Metabolic Pathway Analysis of the Sour Meat Model System

Transcriptomic analysis of *S. cerevisiae* fermentations in the SCdA and SCpA groups was performed, with a Fold Change ≥ 1.5 and FDR < 0.05 set as the criteria for identifying differentially expressed genes (DEGs). A total of 3197 DEGs were detected, as shown in Figure 4a, including 1639 upregulated genes (red) and 1558 downregulated genes (green). As shown in Figure 4b, the top 20 significantly enriched pathways were characterized based on KEGG enrichment analysis. The DEGs were mainly enriched in pathways such as carbon metabolism, biosynthesis of amino acids, glycolysis/gluconeogenesis, and starch and sucrose metabolism. The secondary products of these metabolic pathways, such as fatty acids, alcohols, esters, and amino acids, all contribute to the production of aroma compounds in *S. cerevisiae* [46].

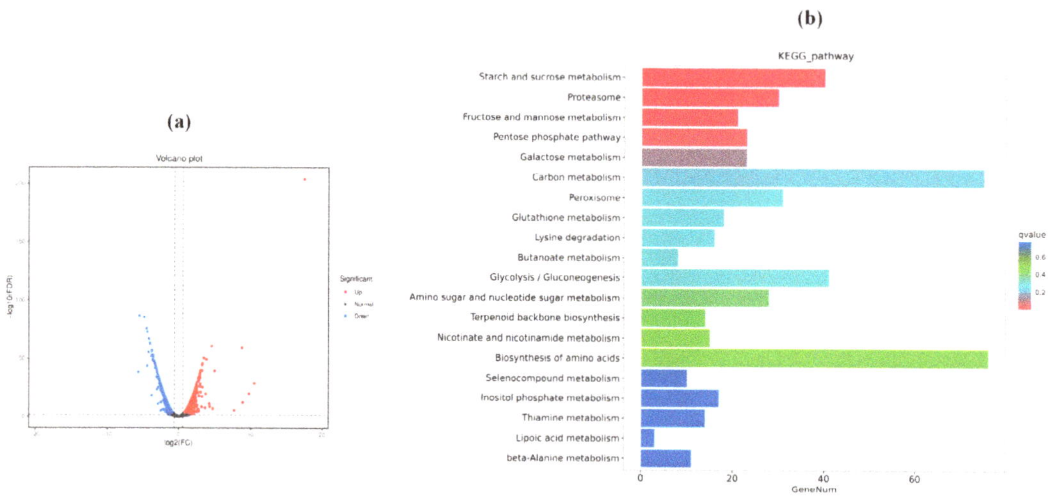

Figure 4. (**a**) Volcano plot of DEGs between the *ATF1*-overexpressing strain (SCpA) and *ATF1*-knockout strain (SCdA) in *S. cerevisiae*; (**b**) KEGG enrichment histogram of DEGs between SCdA and SCpA.

To systematically elucidate the formation mechanism of unique flavor compounds in the sour meat model system by the SCpA group, we conducted an in-depth analysis of key enriched metabolic pathways associated with DEGs, based on the detection results of volatile compounds (Figure 5). Analysis revealed that through systematic regulation of key pathways such as central carbon metabolism, the Ehrlich pathway, and ester synthesis, the SCpA group redirected the metabolic flux of flavor precursors, ultimately leading to the formation of its distinct aromatic profile.

https://doi.org/10.3390/foods15020378

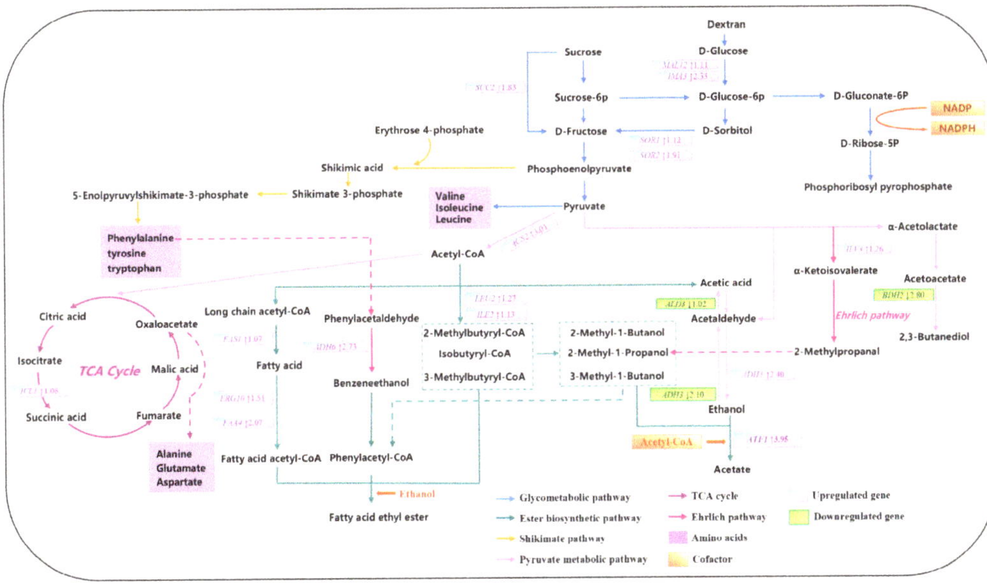

Figure 5. Fold change in DEGs in the aroma metabolism pathway between SCdA and SCpA groups. Note: The unit for differential gene expression fold change is log2 FC.

The enhancement of central carbon metabolism provided a substantial material basis for the synthesis of flavor precursors. As shown in Figure 5, in the sugar metabolic pathways, the sucrase gene *SUC2*, as well as genes involved in glucose and fructose utilization (including *MAL12*, *IMA3*, *SOR1*, and *SOR2*), were significantly upregulated in the SCpA group. The synergistic action of these genes markedly improved the strain's efficiency in carbon source utilization, thereby supplying more abundant phosphoenolpyruvate (PEP) and pyruvate for downstream metabolism. These two key metabolic nodes serve as a "crossroads" for flavor compound synthesis: on one hand, PEP provides precursors for the synthesis of aromatic amino acids such as phenylalanine via the shikimate pathway; on the other hand, pyruvate not only serves as the starting point for the synthesis of branched-chain amino acids like valine and leucine, but also acts as the entry point to TCA cycle. It is noteworthy that the upregulation of *ICL1*, a key gene in the TCA cycle, in the SCpA group promoted the accumulation of oxaloacetate, thereby ensuring adequate substrate availability for the synthesis of amino acids such as aspartate and glutamate [47]. This synergistic enhancement of upstream metabolic pathways established an abundant precursor pool for the subsequent synthesis of flavor compounds, including higher alcohols and esters, which is consistent with the significantly higher total free amino acid (FAA) content observed in the SCpA group compared to other groups.

Differential regulation of the Ehrlich and pyruvate metabolic pathways precisely shaped the composition of higher alcohols. Regarding alcohol biosynthesis, alcohol dehydrogenases encoded by *ADH* genes catalyze the conversion between aldehydes and alcohols. Increased expression of genes such as *ADH6* has been reported to be associated with elevated benzeneethanol production, while enhanced expression of genes including *ADH3* is linked to increased synthesis of higher alcohols such as 3-methyl-1-butanol and 2-methyl-1-propanol [48]. Overexpression of the *ILV3* gene facilitates the conversion of pyruvate to α-ketoisovalerate, which is subsequently decarboxylated to 2-methylpropanal and ultimately reduced to 2-methyl-1-propanol [49]. In the SCpA group, the Ehrlich pathway and pyruvate metabolic pathway exhibited a highly specific regulatory pattern.

https://doi.org/10.3390/foods15020378

On one hand, both *ILV3*, which is closely associated with 2-methyl-1-propanol synthesis, and the critical gene *ADH6* responsible for benzeneethanol production were significantly upregulated. This finding is closely aligned with the experimental results showing the highest detected levels of benzeneethanol and 2-methyl-1-propanol in the SCpA group. On the other hand, the alcohol dehydrogenase gene *ADH3*, which catalyzes the formation of various higher alcohols such as 3-methyl-1-butanol, was markedly downregulated. This differential expression pattern, characterized by both increases and decreases among members of the alcohol dehydrogenase gene family, precisely modulated the conversion efficiency of different precursors into their corresponding higher alcohols, thereby reprogramming the composition profile of higher alcohols.

The acetyl-CoA-centered ester biosynthesis network was significantly activated. Esters represent important flavor contributors in the system, and their synthesis efficiency depends on the supply of two major precursor types: alcohols and acyl-CoAs. In the SCpA group, both of these supply lines were enhanced. First, regarding the supply of acyl-CoAs, the upregulation of *ACS2* directly enhanced the synthesis capacity of acetyl-CoA [50]. Concurrently, upregulated expression of *FAS1*, *ERG10*, and *FAA4* promoted the generation of long-chain fatty acids and their activated products—long-chain acyl-CoAs. Meanwhile, increased expression of *LEU2* and *ILE2* boosted the production of precursors such as 2-methylbutyryl-CoA. The downregulation of *BDH2* reduced the conversion of acetoacetate to 2,3-butanediol, leading to the accumulation of acetoacetate [51]. We speculate that this may result in more pyruvate being converted into acetyl-CoA, thereby increasing the supply of acetyl-CoA available for ester synthesis. Furthermore, regarding alcohol supply, the upregulation of *ADH1* ensured sufficient ethanol substrate. Ultimately, in the esterification step, the expression of *ATF1*—a key gene encoding alcohol acetyltransferase—was significantly upregulated. This enzyme efficiently utilizes the aforementioned acetyl-CoA and alcohols to catalyze the extensive synthesis of acetate esters such as ethyl acetate and isoamyl acetate [52]. This result is also consistent with the previously reported inverse correlation between *ALD3* expression and ethyl acetate content [53]. In summary, through the coordinated enhancement of diversified acyl-CoA supply, generation of key alcohol substrates, and the final catalytic step of esterification, the SCpA group significantly elevated the overall synthesis capacity of ester flavor compounds via the synergistic effects of multiple genes. This finding is closely aligned with the flavor detection results.

4. Conclusions

Through a sour meat model system, this research demonstrates that modifying the *ATF1* gene in *Saccharomyces cerevisiae* substantially affects acetate ester synthesis. Overexpression of *ATF1* increased the strain's esterification ability, raising levels of ethyl acetate and isoamyl acetate, while its knockout notably reduced ethyl acetate content. Transcriptomics further showed that *ATF1* overexpression not only enhanced esterification directly but also upregulated key genes in central carbon metabolism, amino acid synthesis, and precursor supply (e.g., *SUC2*, *ICL1*, *ACS2*, *ADH1*). This indicates that changes in *ATF1* expression disturb a wider metabolic network, working together to boost alcohol and acyl-CoA precursor accumulation and ultimately increase acetate ester yield. These findings position *ATF1* as a useful genetic target for steering acetate ester flavor in the sour meat model system. However, because the model differs from actual sour meat fermentation in matrix structure, microbial interactions, and environmental factors, industrial applicability will require further testing at pilot and production scales.

https://doi.org/10.3390/foods15020378

Supplementary Materials: The following supporting information can be downloaded at: https://www.mdpi.com/article/10.3390/foods15020378/s1, Figure S1: Construction of *ATF1* gene deletion and overexpression strains. Table S1: Oligonucleotide primers. Table S2: Free fatty acid content in model system of sour meat.

Author Contributions: N.Z.: Writing—original draft, Formal analysis, Investigation, Data curation, Writing—reviewing and editing. Y.Y.: Data curation, Methdology. S.Y.: Formal analysis, Methodology. H.L. (Hao Liu): Investigation. X.J.: Data curation. N.W.: Formal analysis. C.J.: Investigation, Resources. Y.D.: Methodology, Resources. L.Y.: Resources. H.L. (Huipeng Liang): Methodology, Resources. X.L.: Writing-review and editing, Conceptualization, Supervision, Project administration. All authors have read and agreed to the published version of the manuscript.

Funding: This work was supported by the National Natural Science Foundation of China (32372363); Xing Liao Talent Program-Outstanding Young Talents (XLYC2403129); Dalian Outstanding Young Scientific and Technological Talents Program (2024RJ019); Solid-state Fermentation Resources Utilization Key Laboratory of Sichuan Province (2023GTYY01).

Institutional Review Board Statement: Not applicable.

Informed Consent Statement: Not applicable.

Data Availability Statement: The original contributions presented in the study are included in the article/Supplementary Material, further inquiries can be directed to the corresponding author.

Acknowledgments: We would like to thank Liang Huipeng from the Institute of Technology, China Resources Beer (Holdings) Company Limited (Room 306, China Resources Building, No. 8 Jianguomen North Avenue, Dongcheng District, Beijing 100005, China) for his assistance in the transcriptomic analysis.

Conflicts of Interest: The authors declare no conflicts of interest.

References

1. Meng, D.; Zhao, L.; Zhu, Y.; Sun, L.; Li, M.; Wu, H.; Liang, D.; Liu, Y.; Ma, Y.; Zhao, G.; et al. Analysis of structural composition and antioxidant activity of traditional fermented sour meat peptides. *Food Chem.* **2024**, *460*, 140697. [CrossRef] [PubMed]
2. Lv, J.; Xu, W.; Ji, C.; Liang, H.; Li, S.; Yang, Z.; Zhang, S.; Lin, X. Relationships between the bacterial diversity and metabolites of a Chinese fermented pork product, sour meat. *Int. J. Food Sci. Technol.* **2021**, *56*, 2742–2750. [CrossRef]
3. Shang, H.; Yue, Y.; Guo, B.; Ji, C.; Zhang, S.; Dong, L.; Ferrocino, I.; Cocolin, L.S.; Lin, X. The effects of *Lactiplantibacillus plantarum* 3-19 and *Pediococcus pentosaceus* 18-1 on preventing the accumulation of biogenic amines and promoting the production of volatile organic compounds during sour meat fermentation. *Int. J. Food Microbiol.* **2024**, *421*, 110806. [CrossRef] [PubMed]
4. Guo, B.; Wu, Q.; Jiang, C.; Chen, Y.; Dai, Y.; Ji, C.; Zhang, S.; Dong, L.; Liang, H.; Lin, X. Inoculation of *Yarrowia lipolytica* promotes the growth of lactic acid bacteria, *Debaryomyces udenii* and the formation of ethyl esters in sour meat. *Food Microbiol.* **2024**, *119*, 104447. [CrossRef]
5. Zhang, Y.; Hu, P.; Xie, Y.; Wang, X. Co-fermentation with *Lactobacillus curvatus* LAB26 and *Pediococcus pentosaceus* SWU73571 for improving quality and safety of sour meat. *Meat Sci.* **2020**, *170*, 108240. [CrossRef]
6. Lv, J.; Lin, X.; Liu, M.; Yan, X.; Liang, H.; Ji, C.; Li, S.; Zhang, S.; Chen, Y.; Zhu, B. Effect of *Saccharomyces cerevisiae* LXPSC1 on microorganisms and metabolites of sour meat during the fermentation. *Food Chem.* **2022**, *402*, 134213. [CrossRef]
7. Zhang, J.; Zhang, C.; Dai, L.; Dong, J.; Liu, Y.; Guo, X.; Xiao, D. Effects of overexpression of the alcohol acetyltransferase-encoding gene *ATF1* and disruption of the esterase-encoding gene *IAH1* on the flavour profiles of Chinese yellow rice wine. *Int. J. Food Sci. Technol.* **2012**, *47*, 2590–2596. [CrossRef]
8. Lilly, M.; Bauer, F.F.; Lambrechts, M.G.; Swiegers, J.H.; Cozzolino, D.; Pretorius, I.S. The effect of increased yeast alcohol acetyltransferase and esterase activity on the flavour profiles of wine and distillates. *Yeast* **2006**, *23*, 641–659. [CrossRef]
9. van Wyk, N.; Kroukamp, H.; Espinosa, M.I.; von Wallbrunn, C.; Wendland, J.; Pretorius, I.S. Blending wine yeast phenotypes with the aid of CRISPR DNA editing technologies. *Int. J. Food Microbiol.* **2020**, *324*, 108615. [CrossRef]
10. Philippe, M.; Marine, T.; Rémy, V.; Lucas, H.; Sabine, G.; Benoit, C.; Warren, A.; JeanChristophe, B. Metabolic, Organoleptic and transcriptomic impact of *Saccharomyces cerevisiae* genes involved in the biosynthesis of linear and substituted esters. *Int. J. Mol. Sci.* **2021**, *22*, 4026. [CrossRef]
11. Shi, W.; Li, J.; Chen, Y.; Liu, X.; Chen, Y.; Guo, X.; Xiao, D. Metabolic engineering of *Saccharomyces cerevisiae* for ethyl acetate biosynthesis. *ACS Synth. Biol.* **2021**, *10*, 495–504. [CrossRef] [PubMed]

12. Zhang, C.; Liu, Y.; Qi, Y.; Zhang, J.; Dai, L.; Lin, X.; Xiao, D. Increased esters and decreased higher alcohols production by engineered brewer's yeast strains. *Eur. Food Res. Technol.* **2013**, *236*, 1009–1014. [CrossRef]
13. Dong, J.; Xu, H.; Zhao, L.; Chen, Y.; Zhang, C.; Guo, X.; Hou, X.; Chen, D.; Zhang, C.; Xiao, D. Enhanced acetate ester production of Chinese liquor yeast by overexpressing *ATF1* through precise and seamless insertion of *PGK1* promoter. *J. Ind. Microbiol. Biotechnol.* **2014**, *41*, 1823–1828. [CrossRef] [PubMed]
14. Kong, S.; Yu, W.; Gao, N.; Zhai, X.; Zhou, Y.J. Expanding the neutral sites for integrated gene expression in *Saccharomyces cerevisiae*. *FEMS Microbiol. Lett.* **2022**, *369*, fnac081. [CrossRef]
15. Cao, C.; Zhang, H.; Cao, X.; Kong, S.; Zhu, B.; Lin, X.; Zhou, Y.J. Construction and optimization of nonclassical isoprenoid biosynthetic pathways in yeast peroxisomes for (+)-valencene production. *J. Agric. Food Chem.* **2023**, *71*, 11124–11130. [CrossRef]
16. Cano-García, L.; Rivera-Jiménez, S.; Belloch, C.; Flores, M. Generation of aroma compounds in a fermented sausage meat model system by *Debaryomyces hansenii* strains. *Food Chem.* **2014**, *151*, 364–373. [CrossRef]
17. Li, S.; Guo, X.; Shen, Y.; Pan, J.; Dong, X. Effects of oxygen concentrations in modified atmosphere packaging on pork quality and protein oxidation. *Meat Sci.* **2022**, *189*, 108826. [CrossRef]
18. Cheng, W.; Sørensen, K.M.; Engelsen, S.B.; Sun, D.-W.; Pu, H. Lipid oxidation degree of pork meat during frozen storage investigated by near-infrared hyperspectral imaging: Effect of ice crystal growth and distribution. *J. Food Eng.* **2019**, *263*, 311–319. [CrossRef]
19. Wang, W.; Xia, W.; Gao, P.; Xu, Y.; Jiang, Q. Proteolysis during fermentation of Suanyu as a traditional fermented fish product of China. *Int. J. Food Prop.* **2017**, *20*, S166–S176. [CrossRef]
20. Pei, Q.; Guo, B.; Yue, Y.; Zhang, X.; Zhang, S.; Ji, C.; Ferrocino, I.; Cocolin, L.; Lin, X. Exploring the synergistic effect of *Lactiplantibacillus plantarum* 1–24-LJ and lipase on improving Quality, Flavor, and safety of Suanzharou. *Food Res. Int.* **2025**, *200*, 115432. [CrossRef]
21. Zhai, Y.; Luan, A.; Yang, Z.; Rong, Z.; Liu, Y.; Wang, F.; Li, X. The impacts of cold plasma on the taste and odor formation of dried silver carp products. *Food Chem.* **2024**, *454*, 139775. [CrossRef]
22. Liu, A.; Yan, X.; Shang, H.; Ji, C.; Zhang, S.; Liang, H.; Chen, Y.; Lin, X. Screening of *Lactiplantibacillus plantarum* with high stress tolerance and high esterase activity and their effect on promoting protein metabolism and flavor formation in Suanzhayu, a Chinese fermented fish. *Foods* **2022**, *11*, 1932. [CrossRef]
23. Liu, M.; Lin, X.; Yang, J.; Yan, X.; Ji, C.; Liang, H.; Zhang, S.; Dong, L. Effects of salt and rice flour concentration on microbial diversity and the quality of sour meat, a Chinese traditional meat. *Food Sci. Hum. Wellness* **2024**, *13*, 2790–2798. [CrossRef]
24. Li, X.-A.; Kong, B.; Wen, R.; Wang, H.; Li, M.; Chen, Q. Flavour compensation role of yeast strains in reduced-salt dry sausages: Taste and odour profiles. *Foods* **2022**, *11*, 650. [CrossRef] [PubMed]
25. Paludan-Müller, C.; Madsen, M.; Sophanodora, P.; Gram, L.; Møller, P.L. Fermentation and microflora of plaa-som, a Thai fermented fish product prepared with different salt concentrations. *Int. J. Food Microbiol.* **2002**, *73*, 61–70. [CrossRef] [PubMed]
26. Campagnol, P.C.B.; dos Santos, B.A.; Wagner, R.; Terra, N.N.; Pollonio, M.A.R. The effect of yeast extract addition on quality of fermented sausages at low NaCl content. *Meat Sci.* **2011**, *87*, 290–298. [CrossRef]
27. Chen, Q.; Kong, B.; Han, Q.; Xia, X.; Xu, L. The role of bacterial fermentation in lipolysis and lipid oxidation in Harbin dry sausages and its flavour development. *LWT—Food Sci. Technol.* **2017**, *77*, 389–396. [CrossRef]
28. Chaves-López, C.; Paparella, A.; Tofalo, R.; Suzzi, G. Proteolytic activity of *Saccharomyces cerevisiae* strains associated with Italian dry-fermented sausages in a model system. *Int. J. Food Microbiol.* **2011**, *150*, 50–58. [CrossRef]
29. Gao, P.; Wang, W.; Xia, W.; Xu, Y.; Jiang, Q. Lipolysis and lipid oxidation caused by *Staphylococcus xylosus* 135 and *Saccharomyces cerevisiae* 31 isolated from Suan yu, a traditional Chinese low-salt fermented fish. *Int. J. Food Sci. Technol.* **2016**, *51*, 419–426. [CrossRef]
30. Özyurt, G.; Gökdoğan, S.; Şimşek, A.; Yuvka, I.; Ergüven, M.; Boga, E.K. Fatty acid composition and biogenic amines in acidified and fermented fish silage: A comparison study. *Arch. Anim. Nutr.* **2016**, *70*, 72–86. [CrossRef]
31. Ji, L.; Wang, S.; Zhou, Y.; Nie, Q.; Zhou, C.; Ning, J.; Ren, C.; Tang, C.; Zhang, J. Effects of *Saccharomyces cerevisiae* and *Kluyveromyces marxianus* on the physicochemical, microbial, and flavor changes of sauce meat during storage. *Foods* **2024**, *13*, 396. [CrossRef]
32. Sentandreu, M.A.; Stoeva, S.; Aristoy, M.C.; Laib, K.; Voelter, W.; Toldrá, E. Identification of small peptides generated in Spanish dry-cured ham. *J. Food Sci.* **2003**, *68*, 64–69. [CrossRef]
33. Minorsky, P.V. Frontiers of Plant Cell Biology: Signals and pathways, system-based approaches 22nd Symposium in Plant Biology (University of California—Riverside). *Plant Physiol.* **2003**, *132*, 428–435. [CrossRef] [PubMed]
34. Mi, R.; Chen, X.; Xiong, S.; Qi, B.; Li, J.; Qiao, X.; Chen, W.; Qu, C.; Wang, S. Predominant yeasts in Chinese Dong fermented pork (Nanx Wudl) and their aroma-producing properties in fermented sausage condition. *Food Sci. Hum. Wellness* **2021**, *10*, 231–240. [CrossRef]
35. Olivares, A.; Navarro, J.L.; Flores, M. Effect of fat content on aroma generation during processing of dry fermented sausages. *Meat Sci.* **2011**, *87*, 264–273. [CrossRef] [PubMed]

https://doi.org/10.3390/foods15020378

36. Olivares, A.; Flores, M.; Gianelli, M.P. Key aroma components of a dry-cured sausage with high fat content (Sobrassada). *Food Sci. Technol. Int.* **2011**, *17*, 63–71. [CrossRef]
37. Nie, X.; Jia, X.; Kang, X.; Pu, H.; Ling, Z.; Wang, X.; Yu, X.; Zhang, Y.; Liu, D.; Zhao, Z. Effects of isolated *Saccharomyces cerevisiae* on the metabolites and volatile organic compounds of Chinese-style sausage. *Food Res. Int.* **2024**, *197*, 115269. [CrossRef]
38. Zhong, A.; Chen, W.; Duan, Y.; Li, K.; Tang, X.; Tian, X.; Wu, Z.; Li, Z.; Wang, Y.; Wang, C. The potential correlation between microbial communities and flavors in traditional fermented sour meat. *LWT—Food Sci. Technol.* **2021**, *149*, 111873. [CrossRef]
39. Li, Y.; Cao, Z.; Yu, Z.; Zhu, Y.; Zhao, K. Effect of inoculating mixed starter cultures of *Lactobacillus* and *Staphylococcus* on bacterial communities and volatile flavor in fermented sausages. *Food Sci. Hum. Wellness* **2023**, *12*, 200–211. [CrossRef]
40. Sidira, M.; Kandylis, P.; Kanellaki, M.; Kourkoutas, Y. Effect of curing salts and probiotic cultures on the evolution of flavor compounds in dry-fermented sausages during ripening. *Food Chem.* **2016**, *201*, 334–338. [CrossRef]
41. van Wyk, N.; Michling, F.; Bergamo, D.; Brezina, S.; Pretorius, I.S.; von Wallbrunn, C.; Wendland, J. Effect of isomixing on grape must fermentations of *ATF1*–overexpressing wine yeast strains. *Foods* **2020**, *9*, 717. [CrossRef]
42. Hu, Y.; Zhang, L.; Zhang, H.; Wang, Y.; Chen, Q.; Kong, B. Physicochemical properties and flavour profile of fermented dry sausages with a reduction of sodium chloride. *LWT—Food Sci. Technol.* **2020**, *124*, 109061. [CrossRef]
43. Hu, Y.; Zhang, L.; Liu, Q.; Wang, Y.; Chen, Q.; Kong, B. The potential correlation between bacterial diversity and the characteristic volatile flavour of traditional dry sausages from Northeast China. *Food Microbiol.* **2020**, *91*, 103505. [CrossRef]
44. Li, P.; Ge, J.; Gao, Y.; Wang, J.; Zhang, C.; Xiao, D. A seamless gene deletion method and its application for regulation of higher alcohols and ester in Baijiu *Saccharomyces cerevisiae*. *BioMed Res. Int.* **2019**, *2019*, 6723849. [CrossRef] [PubMed]
45. Xiao, Y.; Liu, Y.; Chen, C.; Xie, T.; Li, P. Effect of *Lactobacillus plantarum* and *Staphylococcus xylosus* on flavour development and bacterial communities in Chinese dry fermented sausages. *Food Res. Int.* **2020**, *135*, 109247. [CrossRef]
46. Richter, C.L.; Hirst, M.B. Review of aroma formation through metabolic pathways of *Saccharomyces cerevisiae* in beverage fermentations. *Am. J. Enol. Vitic.* **2016**, *67*, 361–370. [CrossRef]
47. McCammon, M.T. Mutants of *Saccharomyces cerevisiae* with defects in acetate metabolism: Isolation and characterization of Acn– mutants. *Genetics* **1996**, *144*, 57–69. [CrossRef] [PubMed]
48. Pigeau, G.M.; Inglis, D.L. Response of wine yeast (*Saccharomyces cerevisiae*) aldehyde dehydrogenases to acetaldehyde stress during Icewine fermentation. *J. Appl. Microbiol.* **2007**, *103*, 1576–1586. [CrossRef]
49. Li, P.; Li, T.; Zhang, C.-Y.; Xiao, D.-g. Effect of *ILV2* deletion and *ILV3* or/and *ILV5* overexpression in *Saccharomyces uvarum* on diacetyl and higher alcohols metabolism during wine fermentation. *Eur. Food Res. Technol.* **2020**, *246*, 563–572. [CrossRef]
50. Dong, J.; Wang, P.; Fu, X.; Dong, S.; Li, X.; Xiao, D. Increase ethyl acetate production in *Saccharomyces cerevisiae* by genetic engineering of ethyl acetate metabolic pathway. *J. Ind. Microbiol. Biotechnol.* **2019**, *46*, 801–808. [CrossRef]
51. Cui, D.-Y.; Wei, Y.-N.; Lin, L.-C.; Chen, S.-J.; Feng, P.-P.; Xiao, D.-G.; Lin, X.; Zhang, C.-Y. Increasing Yield of 2,3,5,6-Tetramethylpyrazine in Baijiu through *Saccharomyces cerevisiae* metabolic engineering. *Front. Microbiol.* **2020**, *11*, 596306. [CrossRef]
52. Verstrepen, K.J.; Van Laere, S.D.M.; Vercammen, J.; Derdelinckx, G.; Dufour, J.; Pretorius, I.S.; Winderickx, J.; Thevelein, J.M.; Delvaux, F.R. The *Saccharomyces cerevisiae* alcohol acetyl transferase Atf1p is localized in lipid particles. *Yeast* **2004**, *21*, 367–377. [CrossRef]
53. Heit, C.; Martin, S.; Yang, F.; Inglis, D. Osmoadaptation of wine yeast (*Saccharomyces cerevisiae*) during Icewine fermentation leads to high levels of acetic acid. *J. Appl. Microbiol.* **2018**, *124*, 1506–1520. [CrossRef]

https://doi.org/10.3390/foods15020378

foods

MDPI

Article

Differences in Nutrition and Sensory Quality Between Cooked Soybeans, Fermented Natto, and Post-Ripening Natto

Yuguang He [1,†], Yuanyuan Jiang [2,†], Da Li [1], Xue Ou [3], Xinyu Miao [1], Mubai Sun [1], Honghong Niu [1], Mei Hua [1], Ying Su [1], Jinghui Wang [1,*] and Zhuo Liu [1,*]

1 Institute of Agro-Food Technology, Jilin Academy of Agricultural Sciences (Northeast Agricultural Research Center of China), Changchun 130033, China; 15044133729@163.com (Y.H.); jacken0000@163.com (D.L.); miaoxinyu@cjaas.com.cn (X.M.); 2006718850@163.com (M.S.); nhh29852@cjaas.com.cn (H.N.); huamei@cjaas.com (M.H.); suying1@163.com (Y.S.)
2 Institute of Agriculture Economy and Information, Jilin Academy of Agricultural Sciences (Northeast Agricultural Research Center of China), Changchun 130033, China; haixin-99@163.com
3 College of Agriculture, Yanbian University, Yanji 133002, China; ouxue1002@163.com
* Correspondence: wangjhjaas@163.com (J.W.); liuzhuo0726@163.com (Z.L.)
† These authors contributed equally to this work.

Abstract

Microbial fermentation is an important means to enhance the nutrition and functionality of food, and soybean fermentation has a long history and a wide variety of products. This study systematically compared the effects of fermentation and post-ripening processes of *Bacillus subtilis natto* JLCC513 on the nutritional components, active substances, and sensory characteristics of soybeans. The experimental results showed that, in terms of basic nutrition, fermentation led to a significant decrease in fat and reducing sugar content, followed by an initial increase and then a decrease in total protein content. In contrast, water-soluble protein continued to increase, and the total amount of free amino acids surged. The active nutritional indicators before and after soybean fermentation showed that nattokinase activity continued to increase during fermentation and post-ripening. At the same time, the number of viable bacteria decreased slightly during post-ripening. The increase in the proportion of easily absorbed aglycone-type isoflavones before and after soybean fermentation is accompanied by a sustained increase in vitamin K2 and gamma aminobutyric acid (GABA) content. In terms of sensory quality, color-difference analysis shows a decrease in brightness (L value) and an increase in redness (a value), resulting in the characteristic yellow-brown color of natto. In terms of texture characteristics, the hardness decreases, while the viscosity and elasticity are significantly enhanced. Through GC-IMS analysis of volatile aromas during soybean fermentation and post-ripening, it was found that esters (such as ethyl acetate) and pyrazines (such as 2,3-dimethylpyrazine) increased, and the product flavor shifted from grassy to fruity and nutty. In summary, natto bacteria enhance the digestibility, nutritional value, and sensory acceptance of soybeans through enzymatic hydrolysis and metabolic transformation. The post-ripening stage plays a key role in flavor maturation and further accumulation of active ingredients.

Keywords: fermented foods; *Bacillus subtilis natto*; texture characteristics; flavor

1. Introduction

Microbial utilization in food has a long history, dating back to before 7000 BC, when ancient civilizations used natural fermentation methods to preserve and improve food [1].

For instance, ancient Egyptians employed yeast to ferment bread [2], while ancient Chinese literature documented brewing alcohol using *Aspergillus* [3]. Today, microbial fermentation technology has become a core method in food processing [4], profoundly transforming food substrates through metabolic activities to enhance their nutritional value and sensory qualities. In today's global society, fermented foods such as natto, yogurt, and cheese not only significantly extend food shelf life but also endow products with unique functional properties [5]. Microorganisms catalyze the transformation of substrates by secreting abundant enzyme systems that generate small-molecule peptides, vitamins, and volatile flavor compounds, thereby achieving the "bio-enhancement" of food [6]. Modern research has confirmed that probiotics in fermented foods can enhance immune function and promote nutrient absorption [7,8]. These benefits highlight the key role of microbial fermentation in the functionalization of food.

Soybeans are an important plant protein source, with protein content of 35–40%, and they contain all essential amino acids required by the human body, making them a high-quality, easy-to-digest, easily absorbed protein. Furthermore, soybeans are rich in dietary fiber, isoflavones, B vitamins, and minerals (such as calcium, iron, zinc, etc.), components that play important roles in maintaining physiological functions and enhancing immunity [9]. However, the isoflavones in soybeans mostly exist in glycosidic forms and require conversion to exert better bioactivity [10]; meanwhile, the large molecular structure of soybean protein leads to its low digestibility, and the beany flavor and bitterness limit its wide application. Traditional processing methods, such as boiling or roasting, can easily cause the loss of heat-sensitive nutrients and offer only limited improvements in sensory qualities.

As a potential probiotic, Ma et al. [11] and Sun et al. [12–14] found that the natto bacterium *Bacillus subtilis natto* JLCC513 can modulate the gut microbiota and intestinal barrier function in obese rats through multiple pathways, thereby contributing to weight loss. During the fermentation process, this strain can produce a variety of enzymes (including protease, amylase, and cellulase), which effectively break down macromolecules in soybeans and enhance their nutritional value. During natto fermentation, the natto bacteria multiply rapidly, and nattokinase activity increases significantly, leading to profound changes in the nutritional components of the soybeans. Protein degradation is a key step: under the action of proteases secreted by natto bacteria, soybean protein is hydrolyzed into small-molecule peptides and free amino acids, making it easier for the body to absorb and turning natto into a functional food. This comprehensively enhances the edible experience and health value of soybean products.

Post-ripening of fermented food is a crucial step in the fermentation process, which refers to the process of storing the product under specific conditions (such as specific temperature and humidity) for a period of time after the main fermentation stage, allowing it to undergo a series of slow and complex biochemical reactions, ultimately achieving the desired quality.

Although studies have reported changes in the process of producing natto through the fermentation of soybeans with *Bacillus subtilis natto*, there is a lack of systematic analysis, especially an overall analysis of the post-ripening stage. This study utilized *Bacillus subtilis natto* JLCC513 to ferment soybeans and analyzed unfermented ripe soybeans, fermented soybeans after 24 h, and post-ripe soybeans after 24 h of ripening. The production process and testing indicators for natto and post-ripening in the experiment are shown in Figure 1. The aim was to systematically elucidate the effects of fermentation and post-ripening on the nutritional components, bioactive compounds, and sensory characteristics of soybeans, thereby filling a knowledge gap in this type of research.

https://doi.org/10.3390/foods15020237

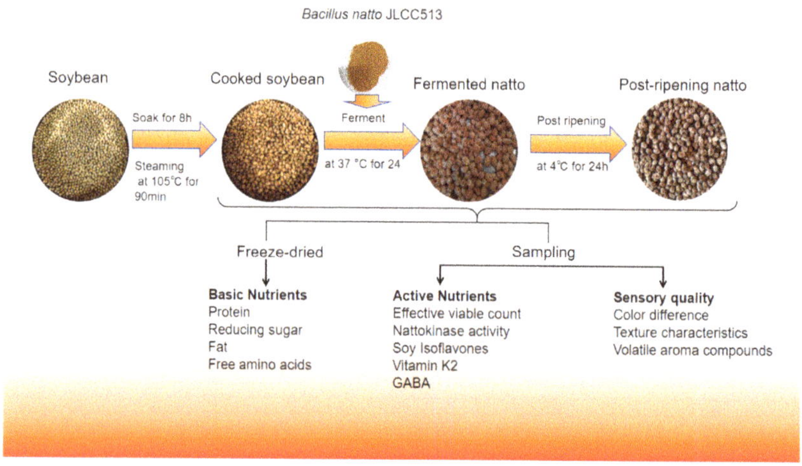

Figure 1. Natto and post-ripening production process and testing indicators.

2. Materials and Methods

2.1. Reagents and Strains

Bacillus subtilis natto JLCC513 was isolated from rice stalks in Northeast China and stored at the China General Microbial Culture and Collection Center (CGMCC, accession number: 20625). The soybeans used for fermentation are commercially available ordinary soybeans purchased from Juncheng Bean Industry (Shulan, China) in Heilongjiang Province. The natto bacterial powder is laboratory-made, with an effective live bacterial count of 100 million CUF/g. The chemical reagents were purchased from China National Pharmaceutical Group Chemical Reagent Co., Ltd. (Shanghai, China).

2.2. Natto Production

Referring to the method proposed by Miao et al. [15]. Soybeans soaked for 8 h were steamed at 105 °C for 90 min to obtain cooked soybeans as the control group, designated as the CS (Cooked Soybean) group. After cooling to room temperature, the soybeans were thoroughly mixed with the fermentation starter powder and incubated in a constant temperature incubator at 37 °C for 24 h to obtain natto, designated as the FN (Fermented Natto) group. After fermentation, the natto was placed in a 4 °C refrigerator for 24 h of post-ripening to obtain post-ripened natto, designated as the PRN (Post-Ripening Natto) group. Samples from the cooked soybean, natto, and post-ripened natto groups were collected for subsequent analyses.

2.3. Determination of Basic Nutritional Indicators During the Fermentation and Post-Ripening Process

The samples of cooked soybeans, fermented natto, and post-ripened natto were vacuum freeze-dried, pulverized, and passed through a 60-mesh sieve for subsequent analysis.

2.3.1. Fat

Use the Soxhlet extraction method to determine the sample's fat content. Dry the freeze-dried powder at 100 °C for 2 h, then place it in a dryer to cool to room temperature and weigh it. Add anhydrous ether to the Soxhlet fat extractor (Pomex, Suzhou, China) to extract the sample. After extraction, dry the sample at 100 °C for 2 h, then cool it in a dryer to a constant weight. The difference between the two masses is the fat content in the sample.

https://doi.org/10.3390/foods15020237

2.3.2. Reducing Sugars

Determination of reducing sugars by the 3,5-dinitrosalicylic acid method. Take the freeze-dried powder, mix it with distilled water and DNS reagent in a 1:1:3 ratio, and heat it in a boiling water bath for 5 min. After cooling to room temperature, measure the absorbance at 540 nm using an enzyme-linked immunosorbent assay (Molecular Devices, Sunnyvale, CA, USA) and calculate the reducing sugar content from the glucose standard curve.

2.3.3. Proteins and Water-Soluble Proteins

Measure protein content using a Kjeldahl nitrogen analyzer (Foss, Hilleroed, Denmark). Dissolve 5 g of freeze-dried powder in 50 mL of distilled water, stir thoroughly for 60 min, centrifuge at 5000 r/min for 30 min, and analyze the supernatant on a Kjeldahl nitrogen analyzer to determine water-soluble protein content.

2.3.4. Free Amino Acids

The amino acid composition was analyzed using a fully automated amino acid analyzer (Sykam, Munich, Germany). The freeze-dried sample was hydrolyzed at 110 °C for 24 h in a sealed container containing 6 M HCl. After filtration and drying, the residue was dissolved in a 0.2 M sodium citrate solution (pH 2.2) and shaken in a water bath for 1 h. After filtration, it was detected in an amino acid analyzer.

2.4. Determination of Nutrients and Active Substances During the Fermentation and Post-Ripening Process

2.4.1. Determination of Nattokinase and Effective Bacterial Count

Five grams each of cooked soybean, fermented natto, and post-ripened natto were ground and mixed with 50 mL of physiological saline. Shake and extract at 37 °C and 180 r/min for 10 min. Centrifuge at 4 °C and 10,000 r/min for 10 min. Take the supernatant to obtain the crude nattokinase enzyme solution. Take 20 μL of crude enzyme solution and spot it on a fibrin plate. Incubate at 37 °C for 16 h, measure the diameter of the transparent circle, and obtain the activity of nattokinase based on the standard curve of urokinase activity.

Determine the total bacterial count using the dilution plate method. Grind and dissolve 5 g of cooked soybeans, fermented natto, and post-cooked natto separately in 50 mL of sterile physiological saline solution. Shake and extract at 37 °C and 180 r/min for 10 min, then dilute with a 10-fold gradient. Take 100 μL of diluent and evenly spread it on the LB solid plate. Invert the plate and incubate at 37 °C for 24 h, then count the colonies.

2.4.2. Soy Isoflavones

The method for analyzing isoflavones by HPLC (Thermo Fisher, Breda, The Netherlands) is based on the methods of Murphy et al. [16], Xu, and Chang [17], with slight modifications: Take 1 g of the sample into a 50 mL stoppered conical flask, dissolve it thoroughly in 3.5 mL of distilled water, add 1 mL of 0.1 mol/L HCl, and then add 5 mL of chromatographic grade acetonitrile. Shake in a gas bath shaker (Labshark, Changde, China) at room temperature (250 r/min) and extract for 2 h. After extraction, filter and take 5 mL of the filtrate. Rotate and evaporate to dryness at a temperature below 34 °C. After rotary evaporation, the residue was dissolved in 80% chromatographic-grade methanol, filtered through a 0.22 μm membrane, and placed in a 1.5 mL liquid-phase vial for later use. The chromatographic column is a Venusil MP-C18 column (5 μm, 250 × 4.6 mm i.d.); column temperature, 34 °C; detection wavelength, 262 nm; injection volume, 10 μL. Mobile phase A is H_2O (0.3% acetic acid), mobile phase B is acetonitrile (0.3% acetic acid). The flow rate was set to 0.7 mL/min. After accurately weighing 1 g of the standard substance, dissolve it in 40 μL of dimethyl sulfoxide (DMSO) and dilute to a certain volume with 80% chromatographic-grade methanol. After being configured as a mixed standard, dilute to

https://doi.org/10.3390/foods15020237

different concentrations, analyze by HPLC, prepare a standard curve, measure the sample, record the corresponding peak area, and calculate the content of soy isoflavones in the sample based on the standard curve.

2.4.3. Vitamin K2

Isopropanol and n-hexane were mixed in a 1:2 (v/v) ratio as the extraction solution and stored. The extraction solution was mixed with the fermentation sample at a 4:1 (v/v) ratio, and the mixture was placed in an ultrasonic cleaner for 30 min to fully extract vitamin K2. The detection method of vitamin K2: HPLC was used to detect the content of vitamin K2, and the detection conditions were Shim pack Velox Biphenyl chromatographic column (2.7 μm, 4.6 mm × 100 mm); The mobile phase is methanol; Detection wavelength: 268 nm; Injection volume: 10 μL; Flow rate: 0.5 mL min^{-1}.

2.4.4. Gamma Aminobutyric Acid

Grind 5 g of cooked soybeans, fermented natto, and post-cooked natto separately and add them to 50 mL of physiological saline. Shake and extract at 37 °C and 180 r/min for 10 min, then centrifuge at 4 °C and 10,000 r/min for 10 min to obtain the extraction solution. Take 0.2 mL of crude enzyme solution and add 400.0 μL of borate buffer (pH 9.0), 200.0 μL of 6% phenol, and 1.0 mL of 10% sodium hypochlorite in sequence. After heating in a boiling water bath for 10 min, immediately cool in an ice bath to room temperature. After the solution turns blue-green, add 4.0 mL of 60% ethanol and measure the absorbance at 645 nm. Simultaneously, use the above method to determine the GABA standard curve. Draw a standard curve with GABA concentration as the horizontal axis and absorbance value as the vertical axis, and calculate the content of GABA in the fermentation broth.

2.5. Detection of Sensory Indicators During the Fermentation and Post-Ripening Process

2.5.1. Color Difference

Use a colorimeter (CR-10 Plus, KONICA MINOLTA, Japan) for detection. Select 10 particles from each group, measure their L * (brightness), a * (redness), and b * (yellowness) using a colorimeter, and calculate the color difference value ΔE, $\Delta E = \sqrt{[(\Delta L^*)^2 + (\Delta a^*)^2 + (\Delta b^*)^2]}$.

2.5.2. Texture Characteristics

Using a texture analyzer (Stable Micro Systems, London, UK), the hardness, adhesiveness, resilience, cohesiveness, elasticity, gumminess, and chewiness of the samples were measured. Select 10 particles of similar size and full size for TPA extrusion testing, using a cylindrical P/36R probe with a pre-measurement speed of 5 mm/s, a mid-measurement speed of 1 mm/s, a post-measurement speed of 2 mm/s, and a compression of 75%.

2.5.3. Odor Based on GC-IMS

Measurement was performed using GC-IMS (G.A.S., Dortmund, Germany), following the method of Sun et al. [18] with slight modifications. Transfer 5.0 g of fresh fermented soybeans before and after fermentation into a 20 mL headspace vial.

Headspace injection conditions: incubation temperature of 60 °C, incubation time of 15 min, oscillation speed of 500 r/min, injection needle temperature of 85 °C, injection volume of 500 μL. After each injection, purge the injection needle with high-purity nitrogen gas (purity ≥ 99.999%) for 5 min.

GC analysis conditions: The chromatographic column is a WAXX quartz capillary column (15 m × 0.53 mm, 1 μm), the column temperature is 60 °C (constant temperature mode), the carrier gas is high-purity nitrogen gas (purity ≥ 99.999%), and the carrier gas flow rate gradient is: initial flow rate is 2 mL/min, maintained for 5 min; At 5–20 min,

https://doi.org/10.3390/foods15020237

linearly increase from 2 mL/min to 150 mL/min and maintain for 5 min, with a total running time of 25 min.

IMS analysis conditions: migration tube length of 98 mm, migration voltage of 500 V/cm, migration tube temperature of 45 °C, migration gas of high-purity nitrogen (purity ≥ 99.999%), flow rate of 150 mL/min, ionization source of tritium (3H), positive ion mode. Under the same chromatographic conditions, C4–C9 ketone (China National Pharmaceutical Group Chemical Reagent Beijing Co., Ltd., Beijing, China) was used as an external reference to calculate the retention indices (RIs) of the identified volatile compounds.

2.6. Data Processing

Each sample was set up in three parallel groups, and the results were expressed as mean ± standard deviation. The significance between samples was evaluated by IBM SPSS (24.0) using one-way analysis of variance (ANOVA) and the LSD test; $p < 0.05$ indicated significant differences. The results were visualized using Origin (2021) software. The VOCal (0.4.03) software was used to process the GC-IMS dataset. The Genescloud platform (https://www.genescloud.cn/) was used to generate the heatmap. The Metware Cloud platform (https://cloud.metware.cn/) was also utilized to draw a flavor wheel and a heatmap.

3. Results

3.1. Changes in Basic Nutritional Indicators During the Fermentation and Post-Ripening Process of Soybeans by Bacillus subtilis natto

3.1.1. Changes in Basic Nutritional Indicators

Basic nutritional indicators can reflect the nutrient sources utilized by *Bacillus subtilis natto* during fermentation. The specific changes are summarized in Table 1. During fermentation, the fat and reducing sugar content decrease. The fat content of cooked soybeans is 17.70 g/100 g, which decreases to 16.55 g/100 g after 24 h of fermentation. At 24 h post-ripening, the fat content decreases to 15.58 g/100 g, and the reducing sugar content decreases significantly during the fermentation post-ripening process ($p < 0.05$). The total protein content significantly increased during the 24 h fermentation stage ($p < 0.05$), rising from 35.85 g/100 g to 38.23 g/100 g. After post-ripening, there was a slight decline, but it remained significantly higher than in the unfermented group ($p < 0.05$). The content of water-soluble protein showed a significant increasing trend ($p < 0.05$), reaching 4.56 g/100 g before fermentation; After 24 h of fermentation, it rapidly increased to 8.54 g/100 g; After post-ripening, it continued to increase to 9.81 g/100 g.

Table 1. Changes in basic nutritional indicators of soybeans before and after fermentation by natto bacteria.

	Fat (g/100 g)	Reduced Sugar (g/100 g)	Protein (g/100 g)	Water Soluble Protein (g/100 g)
CS	17.7 ± 0.15^{a}	1.46 ± 0.09^{a}	35.85 ± 0.48^{a}	4.56 ± 0.21^{a}
FN	16.55 ± 0.31^{b}	1.24 ± 0.06^{b}	38.23 ± 0.31^{b}	8.54 ± 0.22^{b}
PRN	15.58 ± 0.28^{c}	0.92 ± 0.04^{c}	37.28 ± 0.35^{c}	9.81 ± 0.28^{c}

The values with different superscript letters in a column are significantly different ($p < 0.05$).

3.1.2. Changes in Free Amino Acids

Further analysis of the changes in free amino acid content showed that the total free amino acid content in unfermented soybeans was 151.0 mg/100 g; After 24 h of fermentation, it soared to 427.9 mg/100 g, and further increased to 615.7 mg/100 g after post-ripening. The content of 12 amino acids continued to significantly increase ($p < 0.05$), while the content of three amino acids continued to decrease(Figure 2A). Proline and

https://doi.org/10.3390/foods15020237

cysteine showed a trend of first increasing and then decreasing, and first decreasing and then increasing, respectively. A heatmap (Figure 2B) can more intuitively show the trend in the content of each amino acid, with 12 amino acids showing a continuous increase. In comparison, the other five amino acids exhibit different patterns of change.

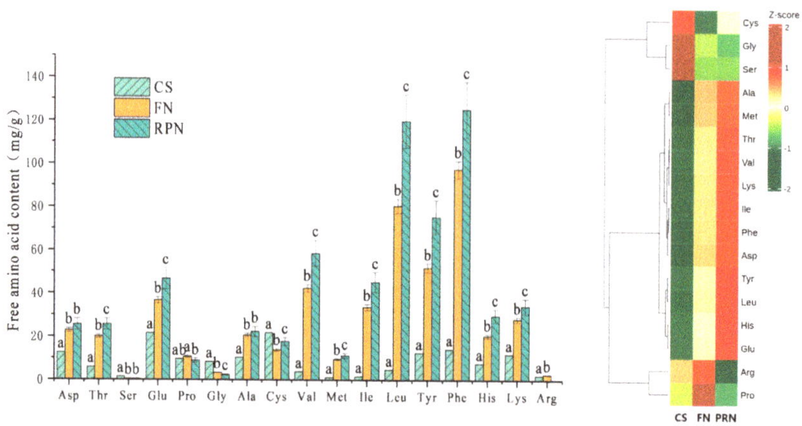

(A) (B)

Figure 2. Changes in free amino acid content during the fermentation process. (**A**) Bar chart of changes in free amino acid content. (**B**) Heat map of changes in free amino acid content. The experiment was set up with three biological replicates. The values with different superscript letters in a column are significantly different ($p < 0.05$).

3.2. Changes in Active Nutritional Indicators During the Fermentation and Post-Ripening Process of Soybeans by Bacillus subtilis natto

3.2.1. Changes in Effective Viable Bacterial Count and Nattokinase Activity

The number of viable *Bacillus subtilis natto* bacteria and the activity of nattokinase can reflect the growth characteristics of this strain in the natto production process (Figure 3). The results showed that after 24 h of fermentation, nattokinase activity reached 2434.22 IU/g, and the number of live bacteria in natto reached 8.24 log CFU/mL. After post-ripening, nattokinase activity increased significantly ($p < 0.05$), but the number of live bacteria showed a slight decrease during the post-ripening stage.

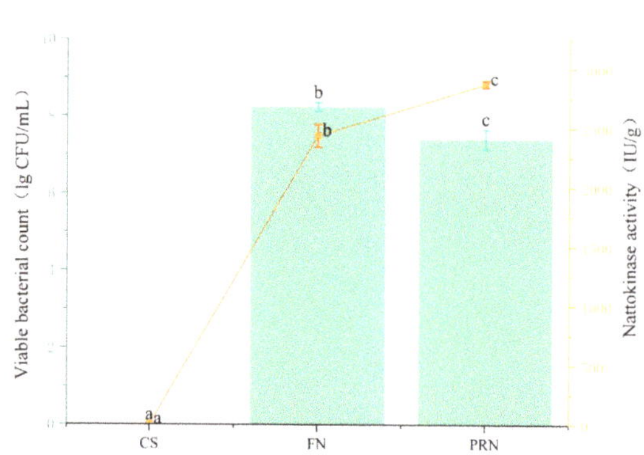

Figure 3. The effect of fermentation on the number of viable bacteria and nattokinase activity. The yellow line reflects the nattokinase activity across different treatments. The green columns reflects the vialbe bacteria across different treatments. The values with different superscript letters in a column are significantly different ($p < 0.05$).

https://doi.org/10.3390/foods15020237

3.2.2. Changes in Soy Isoflavones

The total content of soy isoflavones showed a slow decreasing trend throughout the fermentation post-ripening process ($p < 0.05$) (Figure 4). The cooked soybean group was 2.77 mg/g; after 24 h, it decreased to 2.52 mg/g; after post-ripening, it continued to decrease slightly to 2.44 mg/g. There is a significant fluctuation relationship between glycoside type isoflavones and aglycone type isoflavones. The content of glycosidic isoflavones decreased from 2.39 mg/g at the beginning to 1.87 mg/g after 24 h of fermentation, and then decreased to 1.61 mg/g after 24 h of post-ripening. During the post-ripening process, the content of aglycone-type isoflavones increased from 0.38 mg/g to 0.66 mg/g and finally reached 0.83 mg/g. The proportion of aglycones (aglycones/total isoflavones) increased from 13.7% to 34.0%.

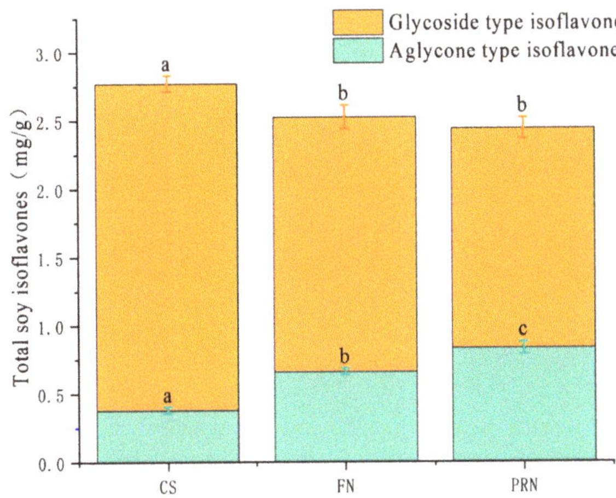

Figure 4. The total, glycosylated, and aglycone isoflavone contents across different treatments. The values with different superscript letters in a column are significantly different ($p < 0.05$).

3.2.3. Changes in Vitamin K2 and Gamma Aminobutyric Acid

The vitamin K2 content continues to increase during fermentation (Figure 5), reaching only 0.18 mg/g in the 0 h sample. After 24 h of fermentation, it rapidly and significantly increases ($p < 0.05$) to 5.58 mg/g, and further increases by 22.7% after 24 h of post-ripening. The gamma-aminobutyric acid content in the 0 h sample was only 5.66 mg/100 g. During the 24 h fermentation process, the bacterial cells secreted a large amount, increasing the content to 91.54 mg/100 g. After another 24 h of post-ripening, it continued to rise to 111.96 mg/100 g.

3.3. Changes in Sensory Indicators During the Fermentation and Post-Ripening Process of Soybeans by Bacillus subtilis natto

3.3.1. Changes in Color Difference

During fermentation and post-ripening, the surface of natto gradually changes from light yellow to off-white to a classic yellow-brown color. The L* value decreased from 58 after 0 h to 47, and then decreased to 45.59 after post-ripening; The value of a* increases continuously to 6.32, indicating that the sample has shifted from greenish to reddish; The overall change in b* value was not significant, but it also showed a significant decrease after fermentation ($p < 0.05$) (Figure 6). Compared with the post-ripening stage, the color difference changes significantly within 24 h of fermentation ($p < 0.05$), and the total color difference between cooked soybeans and post-ripe natto is $\Delta E = 13.1$. It has exceeded the threshold recognizable by the naked eye ($\Delta E \geq 3$), indicating that fermentation ripening has a significant impact on the appearance and color of natto.

https://doi.org/10.3390/foods15020237

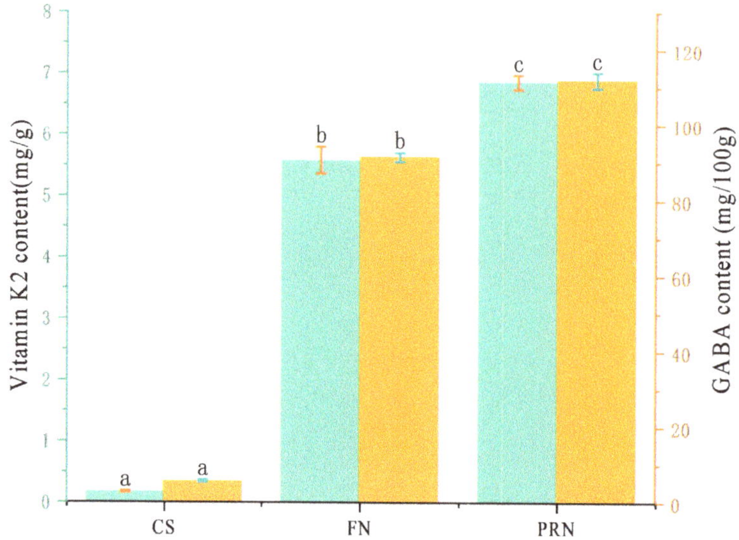

Figure 5. Total vitamin K2 and GABA content among different treatments. The green columns reflects the vitamin K2 content across different treatments. The yellow columns reflects the GABA content across different treatments. The values with different superscript letters in a column are significantly different ($p < 0.05$).

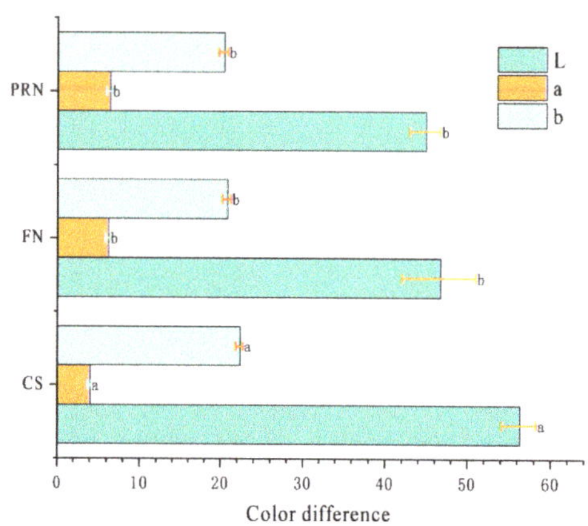

Figure 6. Color difference among different treatments. The values with different superscript letters in a column are significantly different ($p < 0.05$).

3.3.2. Changes in Texture Characteristics

Using the TPA method to determine the seven genetic characteristics of natto, the results are presented in Table 2. The decrease in natto hardness is mainly due to the hydrolysis of soybean storage protein into small molecule peptides by various proteases secreted by natto bacteria, which destroys the protein network structure; The enhancement of adhesinerness is related to the accumulation of γ-PGA and extracellular polysaccharides, ultimately forming the typical drawing characteristics of natto; The resilence showed no significant difference between 0 and 48 h and remained around 6.3; The cohesion sharply decreased from 0.255 to 0.052 ($p < 0.05$), indicating a transition from "tight brittle" to "loose viscous" protein network, consistent with the synergistic effect of protein degradation and polysaccharide lubrication; The springiness significantly increased from 40.2% to 78% ($p < 0.05$), reflecting

https://doi.org/10.3390/foods15020237

the enhancement of viscoelastic properties; The gumminess and chewiness decreased from 186.8 g and 75.1 g to 27.2 g and 20.9 g, respectively, with a decrease of more than 85%, indicating a shift in taste from "hard and crisp" to "soft and easy to chew", which highly corresponds to the sensory evaluation of "melt in the mouth".

Table 2. The effect of Fermentation on Texture Characteristics.

	Hardness	Adhesiveness	Resilence	Cohesion	Springiness	Gumminess	Chewiness
CS	732.82 ± 16.83 ^{a}	−14.95 ± 1.46 ^{a}	5.66 ± 0.96 ^{a}	0.26 ± 0.01 ^{a}	40.2 ± 2.56 ^{a}	186.85 ± 3.44 ^{a}	75.1 ± 4.81 ^{a}
FN	517.22 ± 32.2 ^{b}	−56.567 ± 3.16 ^{b}	6.35 ± 0.9 ^{a}	0.053 ± 0.02 ^{b}	78.31 ± 14.72 ^{b}	27.22 ± 8.54 ^{b}	20.86 ± 6.31 ^{b}
PRN	515.94 ± 43.06 ^{b}	−59.07 ± 2.47 ^{b}	6.32 ± 0.79 ^{a}	0.053 ± 0.01 ^{b}	78.18 ± 15.69 ^{b}	27.16 ± 8.65 ^{b}	18.86 ± 6.71 ^{b}

The values with different superscript letters in a column are significantly different ($p < 0.05$).

3.3.3. Changes in Volatile Aroma Components

Analysis of changes in volatile aroma compounds during fermentation and post-ripening was conducted using SPME-GC-IMS technology. The differential GC-IMS fingerprint spectrum (Figure 7B) and spectrum (Figure 7C) display a total of 41 compounds identified during fermentation, including 11 ketones, 9 alcohols, 7 esters, 4 acids, 4 pyrazines, 3 aldehydes, and 30 alkenes. The percentage of each type of VOC (Figure 7A) shows the proportion of each substance. Throughout the fermentation and post-ripening stages, the overall levels of alcohols, ketones, and esters increased significantly, driving a shift in aroma profile from "grassy-fruity" to "soy sauce–ester" notes. Alcohols (e.g., 3-methyl-1-butanol, 1-propanol) and ketones (e.g., 2-butanone, 2,3-pentanedione) served as precursors for esterification, synergizing with acids to enhance the complexity of the overall aroma. Esters (e.g., ethyl acetate, isoamyl acetate, ethyl isobutyrate) peaked at 24 h, contributing distinct fruity flavors. Aldehydes initially increased and then decreased, a pattern linked to the balance between lipid oxidation and reduction; during fermentation, lipoxygenase-generated aldehydes were reduced to corresponding alcohols by alcohol dehydrogenase/aldehyde reductase. Acids accumulated continuously, providing both sourness and substrates for esterification. Pyrazines (e.g., 2-acetylpyrazine, 2,3-dimethylpyrazine) increased steadily during post-ripening, imparting a "roasted-nutty" aftertaste. Sulfur-containing compounds (e.g., 1-propanethiol) stabilized at high levels after 24 h, contributing to the characteristic complex flavor of natto. We can see the aroma type of each compound through the flavor wheel (Figure 7D).

Radar chart analysis (Figure 7E) revealed distinct aroma profiles among unfermented soybeans, fermented natto, and post-ripened natto. Unfermented soybeans were dominated by creamy and nutty notes, while fermented natto exhibited a gradual transition toward fruity and sweet aromas. Compared to fermented natto, the post-ripened product showed a more pronounced aroma profile: the sensory intensity of banana and apple notes increased by 20%, whereas creamy notes slightly declined, potentially due to dynamic changes in esters such as isoamyl acetate.

The Sankey diagram (Figure 7F) traced the metabolic flow of volatile organic compounds (VOCs), identifying 2-butanone and 2-heptanone as key nodes connecting "fruity" and "nutty" aromas. Meanwhile, 2,3-dimethylpyrazine directly contributed to "caramel" notes. Methyl isovalerate and ethyl acetate, as primary contributors to "banana" and "apple" aromas, respectively, increased 1.8-fold during post-ripening ($p < 0.05$).

The network diagram (Figure 7G) illustrated correlations among flavor compounds, with proximity to green nodes indicating greater contributions to specific flavors. Compounds such as 2,3-pentanedione and 2,3-dimethylpyrazine were identified as major contributors to caramel notes. Others, like 2-acetylpyrazine and 2-methyltetrahydrofuran-3-one, exhibited significant nutty characteristics. Esters such as ethyl hexanoate and active ethyl 2-methylbutyrate were closely associated with fruity notes, while acetone and 1-

https://doi.org/10.3390/foods15020237

acetoxy-3-acetone were key components of creamy flavors. This analysis underscores how microbial fermentation transforms soybean substrates into a diverse aroma profile through coordinated biochemical pathways.

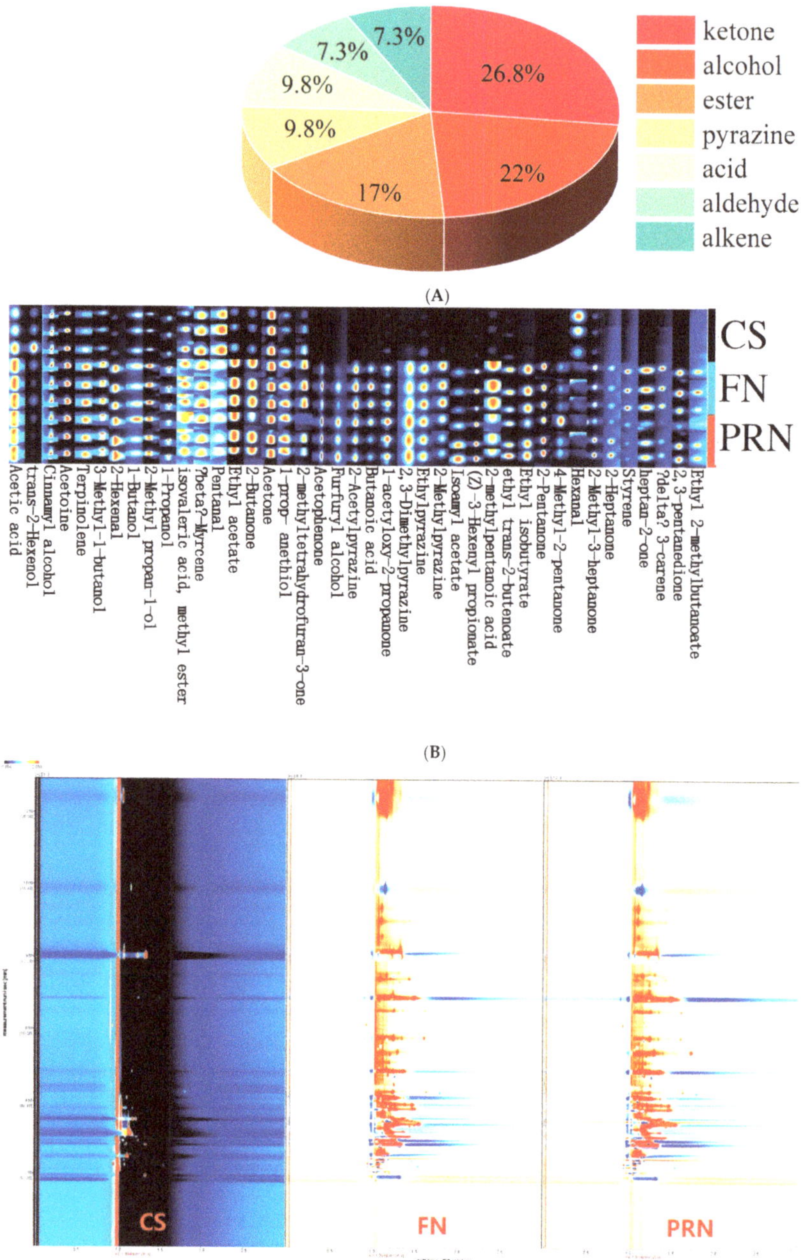

Figure 7. *Cont.*

https://doi.org/10.3390/foods15020237

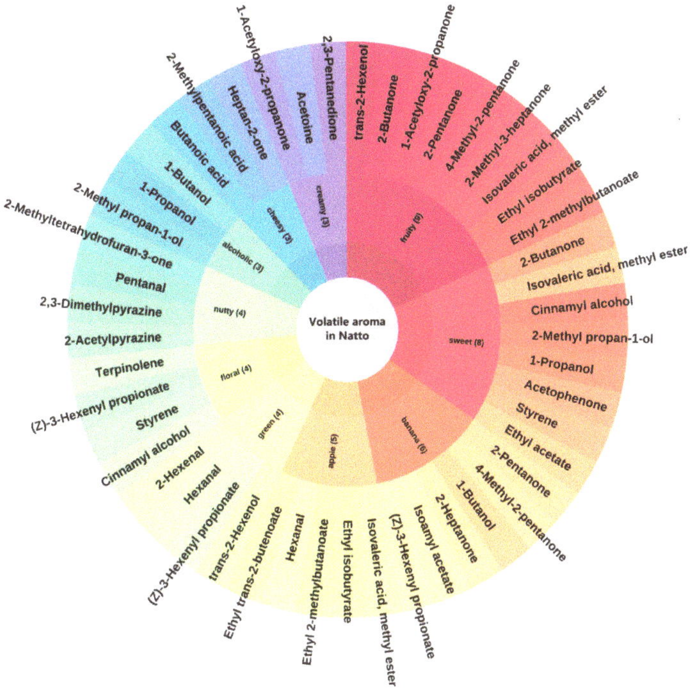

(D)

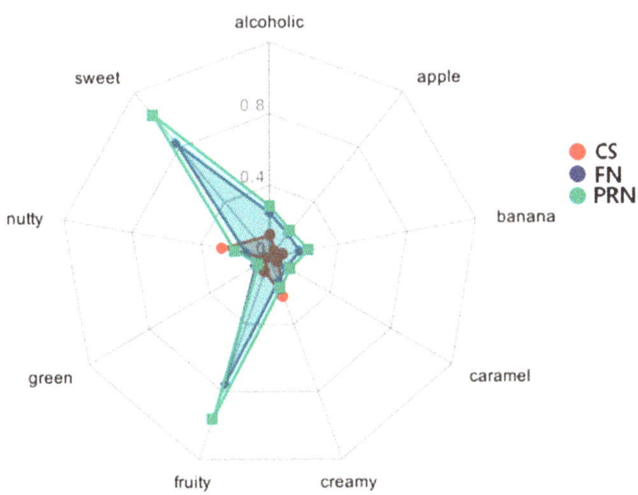

(E)

Figure 7. *Cont.*

https://doi.org/10.3390/foods15020237

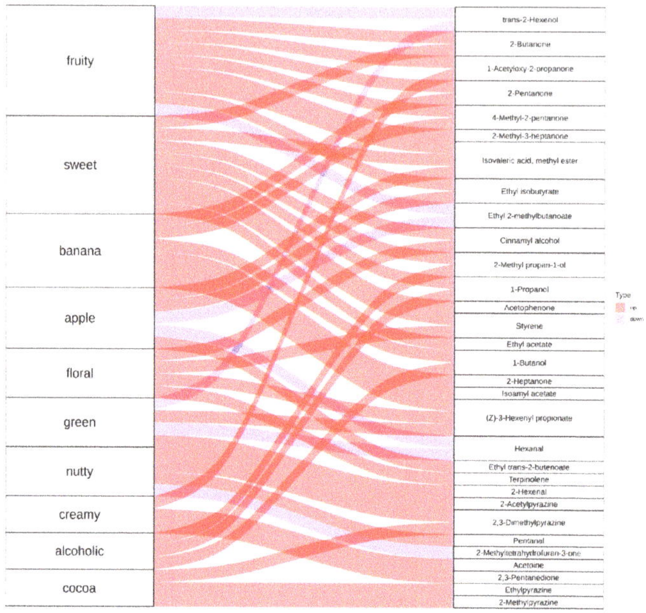

(F)

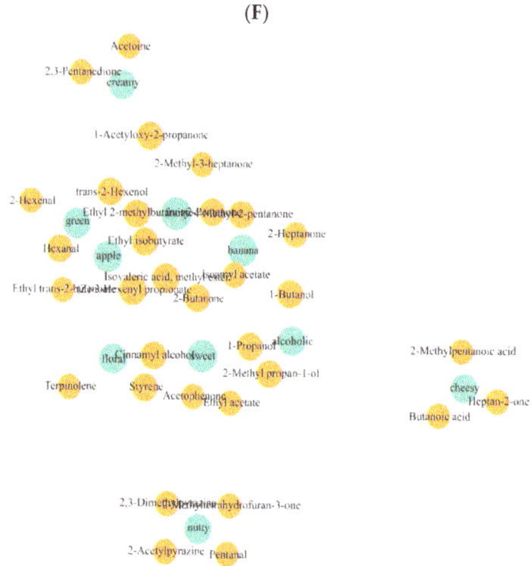

(G)

Figure 7. The effect of fermentation on color difference. (**A**) Percentage of each type of VOCs. (**B**) GC-IMS fingerprints. (**C**) GC-IMS spectrogram. Red indicates an increase in content, while blue indicates a decrease in content (**D**) Flavor wheel chart. (**E**) Radar chart. (**F**) Sankey diagram. (**G**) Network diagram. The experiment was set up with three biological replicates.

4. Discussion

This study systematically elucidates the multifaceted effects of the *Bacillus subtilis natto* (*B. natto*) strain JLCC513 on the nutritional, bioactive, and sensory attributes of soybeans during fermentation (0–24 h) and post-fermentation maturation (25–48 h). The results demonstrate that the impact of *B. natto* on soybeans is a complex biological process in-

https://doi.org/10.3390/foods15020237

volving the synergistic actions of microbial metabolism, enzymatic catalysis, and substrate transformation, ultimately representing an "omics effect" arising from the coordinated succession of the bacterial population, enzyme systems, and the soybean matrix.

Firstly, regarding the transformation of fundamental nutrients, *B. natto* continuously consumes reducing sugars and fats in soybeans throughout the fermentation process, providing ample carbon sources and energy for its growth, reproduction, and metabolic activities. Zhou et al. [19] suggested a strong correlation between *Bacillus* species and reducing sugars, indicating that the bacteria secrete enzymes to degrade polysaccharides. Their results showed an initial increase in reducing sugar content, followed by a continuous consumption and decrease during the mid to late fermentation stages. Chen et al. [20] also confirmed the decline in reducing sugar content during *B. natto* fermentation.

The significant increase ($p < 0.05$) in total protein content during the fermentation stage, concurrent with a gradual rise in viable count, indicates substantial synthesis of bacterial protein, signifying a period of vigorous microbial growth and metabolism. The sharp and continuous increase in water-soluble protein and total free amino acid content (reaching 615.7 mg/100 g) is attributed to the continuous hydrolysis of soybean storage proteins by proteases, including nattokinase, secreted by *B. natto*. Zhang et al. [21] concluded that insoluble proteins are hydrolyzed into water-soluble proteins, which are then broken down into dipeptides or amino acids, and are further metabolized into other metabolites under the action of *B. natto*. This process not only significantly enhances the digestibility and absorption rate of proteins but also lays the foundation for natto's umami flavor by releasing large amounts of flavor amino acids, such as aspartic acid and glutamic acid. However, the accumulation of bitter amino acids (like leucine and isoleucine) also contributes to the complexity of its flavor profile. The disappearance of arginine in the post-ripening stage suggests the activation of the arginine deiminase (ADI) pathway, which converts arginine into ornithine, citrulline, and polyamines. This may be related to the formation of amine odor precursors (such as putrescine and spermine) in natto. Furthermore, Zhou et al. [22] found that *Bacillus subtilis* efficiently utilizes arginine as an energy source for bacterial growth.

Regarding the transformation of bioactive components in soybeans, this study observed that their changes were not entirely synchronized with bacterial growth. During the post-fermentation maturation stage, the viable count decreased slightly, while levels of bioactive nutrients continued to rise. The nattokinase activity increased further in the post-fermentation stage, potentially due to the release of intracellular enzymes during bacterial autolysis or to the continued synthesis of proteases in the late fermentation phase, even though the viable count slightly declined. This indicates that using the viable count alone as an indicator of the probiotic function of natto might be insufficient, and the enzymatic activity of the matured product warrants close attention. The successful conversion of soy isoflavones from glycosidic forms to aglycones is a direct manifestation of the β-glucosidase activity of *B. natto*. Aglycone-type isoflavones can be directly absorbed through passive diffusion, increasing their bioavailability by 3–5 times, indicating that the fermentation post-ripening process significantly improves the bioavailability and physiological activity of isoflavones. Li et al. [23] similarly observed an increase in isoflavone content accompanied by elevated gene expression of β-glucosidase during natto fermentation. Vitamin K2 (MK-7) and γ-aminobutyric acid (GABA) showed a trend of continuous accumulation throughout the process. Vitamin K2 is mainly produced by the naphthoquinone synthesis pathway of natto bacteria, and its key enzymes MenA and MenD are continuously expressed in the logarithmic and stable phases. Therefore, the vitamin K2 content continues to increase during the fermentation period. Although the number of viable bacteria decreases in the post-maturation stage, the bacterial cells still maintain synthetic activity, and the intracellu-

https://doi.org/10.3390/foods15020237

lar storage form of vitamin K2 is released into the extracellular environment after autolysis, resulting in a continuous increase in content. Rocchi R. et al. [24] found that Vitamin K2 is synthesized by *Bacillus subtilis* during fermentation and is entirely of microbial origin; thus, it is hypothesized that the increase in Vitamin K2 is associated with the menaquinone synthesis pathway of the bacteria during the stationary phase and its release after cell autolysis. The enrichment of GABA (to 111.96 mg/100 g) is closely related to the increased content of the substrate glutamate and the activity of glutamate decarboxylase (GAD). Under the catalysis of glutamic acid decarboxylase, an alpha decarboxylation reaction occurs to generate gamma aminobutyric acid. From previous results, it is known that the free glutamic acid content during fermentation also promotes gamma-aminobutyric acid synthesis. Wang et al. [25] also confirmed that *B. natto* fermentation activates glutamate decarboxylase, catalyzing the conversion of glutamate to GABA. The increase in these two components equips natto with dual health benefits: promoting bone health (Vitamin K2) and calming nerves and regulating blood pressure (GABA).

The evolution of sensory qualities is a macroscopic manifestation of these biochemical transformations. In terms of color, the decrease in brightness (L* value) and increase in redness (a* value) resulted in a significant color difference ($\Delta E = 13.1$), primarily associated with the Maillard reaction, polyphenol oxidation, and the formation of bacterial colonies, collectively contributing to the characteristic yellowish-brown appearance of natto. The decrease in brightness is mainly due to the growth of bacterial cells and the formation of extracellular polysaccharides, which cause micro-wrinkles on the surface, increase light-scattering losses, and lead to the deposition of brown polymers due to polyphenol oxidation during fermentation. The increase in a* represents a shift in the color of natto from greenish to reddish, which is related to the oxidation of soy isoflavones and the polymerization of phenolic substances, and is the main contribution to the typical "amber" appearance of natto. The decrease in b* value indicates that carotenoid degradation and browning reactions gradually dominate, corresponding to the sensory stage of "yellow soybeans turning into dark natto". Kubo Y. et al. [26] reported similar findings, noting that natto hardness decreases with extended storage time. The textural results of this study similarly showed a sharp decrease in hardness, gumminess, and chewiness, alongside a significant increase in springiness and stickiness. This is presumably due to the degradation of the protein network and the production of viscous substances like γ-polyglutamic acid (γ-PGA), which collectively transform the product's texture from the hard and crisp" texture of raw beans to the characteristically soft, sticky, and smooth texture of natto, a key determinant of consumer acceptance.

The most intricate flavor-formation process is intuitively revealed through GC-IMS technology. In the initial fermentation stage, aldehydes and alcohols dominate, imparting a "grassy" note. Zhang et al. [27] noted that alcohols, with their low odor thresholds, significantly contribute to flavor development. As fermentation progresses, the aroma evolves into a complex profile shaped by esters (e.g., ethyl acetate, isoamyl acetate), pyrazines (e.g., 2,3-dimethylpyrazine), and ketones. Ma et al. [28] similarly observed increases in alcohols, ketones, and esters during co-fermentation with *Bacillus subtilis natto* and lactobacilli. The post-fermentation maturation stage is critical for flavor refinement: further ester generation enhances "fruity" notes (e.g., banana, apple), while Chen et al. [20] highlighted that ketones—often derived from amino acid degradation or unsaturated fatty acid oxidation—contribute to consumer-preferred flavors. Concurrently, the accumulation of pyrazines provides a robust "nutty, roasted" aftertaste. Sun et al. [29] mapped dynamic changes in key volatile organic compounds (VOCs) and free amino acids in natto to KEGG pathways, revealing interconversions via the tricarboxylic acid cycle and among certain amino acids, elucidating the pathway from primary metabolites to complex flavor compounds.

https://doi.org/10.3390/foods15020237

Comparing the process of the transformation from mature soybeans to post mature natto during the fermentation and post-ripening stages, it can be found that the changes in the fermentation stage are due to the metabolic activity of *Bacillus subtilis natto*, which consumes the original nutrient matrix in soybeans and generates a large amount of active nutrients under the action of enzymes, which is an accumulation process. The post-ripening stage is a process of improving the quality of food. In the post-ripening stage of natto, its active nutrient content further increases, and its yellow-brown appearance deepens, making the texture and taste softer and smoother. Lyu et al. [30] also confirmed, in their study of fermented egg and dairy beverages, that the post-ripening stage can enhance the nutritional quality of fermented foods. The change in aroma components during the post-ripening stage is the most crucial, shifting from the "green grass fruit aroma" at the end of fermentation to the "sauce ester aroma", making the flavor of natto more mellow. Wu et al. [31] found that the activity of microbial colonies during the post-ripening stage is closely related to the formation of product flavor. Deng et al. [32] also studied changes in the aroma of Jinhua ham during post-ripening. Weng et al. [33] also studied the evolution of aroma in several fermented teas during the post-ripening stage. This indicates that the post-ripening stage is an important stage for aroma transformation and for improving the flavor quality of fermented foods.

In summary, *Bacillus subtilis natto* fermentation systematically upgrades soybean quality. This process transforms nutritionally rich but hard-to-digest, bland-flavored soybeans into a health-promoting food with enhanced protein bioavailability, diverse functional components, and unique texture and flavor profiles. This study preliminarily elucidates the dynamics of nutritional and sensory quality changes during fermentation, laying a foundation for understanding the underlying mechanisms. Future research could integrate multi-omics approaches (e.g., transcriptomics, proteomics) with enzyme kinetics and metabolic flux analysis to dissect molecular-level transformation pathways and regulatory mechanisms. The findings provide a theoretical basis for selecting superior starter cultures, developing high-activity natto products, and establishing a more scientific, comprehensive quality evaluation system for natto.

5. Conclusions

This study demonstrates that *Bacillus subtilis natto* JLCC513 can systematically convert soybeans into a food product with high protein bioavailability, rich in various functional components, and a unique flavor during fermentation (0–24 h) and post-ripening (25–48 h).

During the fermentation stage, the mycelium rapidly consumes reducing sugars and fats as carbon and energy sources through metabolic activities, promoting mycelial protein synthesis and significantly increasing the total protein content. The secreted proteases, such as nattokinase, hydrolyze soybean storage proteins into water-soluble proteins and free amino acids, thereby greatly improving the digestion and absorption of proteins and releasing umami substances, such as glutamic acid, and bitter amino acids, such as leucine, enriching the flavor profile. The degradation of protein networks and the production of viscous substances, such as γ-polyglutamic acid, give it a soft, smooth texture. The flavor at this stage transitions from a single-bean aroma to a dominant, creamy, nutty aroma.

During the post-ripening stage, the number of viable bacteria decreases slightly, while the active components continue to accumulate. The activity of nattokinase is further. Vitamin K2 (MK-7) is continuously produced through the naphthoquinone synthesis pathway of the bacteria, and γ-aminobutyric acid (GABA) is enriched under the catalysis of glutamate decarboxylase. These components endow natto with various health potentials. During the post-ripening period, the color and taste changes are not significant, but the aroma transitions from "green grass fruit aroma" to "sauce ester aroma".

https://doi.org/10.3390/foods15020237

This study preliminarily elucidated the dynamic changes in nutrition and sensory quality during fermentation, providing a theoretical basis for subsequent analyses of molecular mechanisms, strain optimization, and the establishment of a comprehensive evaluation system for natto quality using multi-omics technology.

Author Contributions: Y.H.: Writing—original draft, Visualization, Methodology, Investigation. Y.J.: Writing—review and editing, Supervision, Methodology. D.L.: Visualization, Methodology, Investigation. X.O.: Methodology, Conceptualization. X.M.: Supervision, Investigation. M.S.: Visualization, Methodology. H.N.: Validation, Methodology, Conceptualization. M.H.: Methodology, Formal analysis, Conceptualization. Y.S.: Supervision, Resources, Project administration. J.W.: Writing—review and editing, Supervision, Resources, Project administration, Funding acquisition. Z.L.: Supervision, Resources, Funding acquisition, Formal analysis. All authors have read and agreed to the published version of the manuscript.

Funding: This work was financially supported by the Jilin Provincial Science and Technology Development Plan (20240404004ZP).

Institutional Review Board Statement: Not applicable.

Informed Consent Statement: Not applicable.

Data Availability Statement: The original contributions presented in this study are included in the article. Further inquiries can be directed to the corresponding authors.

Conflicts of Interest: The authors declare no conflicts of interest.

References

1. Leeuwendaal, N.K.; Stanton, C.; O'Toole, P.W.; Beresford, T.P. Fermented Foods, Health and the Gut Microbiome. *Nutrients* **2022**, *14*, 1527. [CrossRef]
2. Samuel, D.; Walker, H. Brewing and baking in ancient Egyptian art. In *Food in the Arts. Proceedings of the Oxford Symposium on Food and Cookery*; Oxford Symposium: London, UK, 1999; pp. 173–181.
3. Yang, Q.; Yao, H.; Liu, S.; Mao, J. Interaction and application of molds and yeasts in Chinese fermented foods. *Front. Microbiol.* **2022**, *12*, 664850. [CrossRef]
4. van Boekel, M.; Fogliano, V.; Pellegrini, N.; Stanton, C.; Scholz, G.; Lalljie, S.; Somoza, V.; Knorr, D.; Jasti, P.R.; Eisenbrand, G. A review on the beneficial aspects of food processing. *Mol. Nutr. Food Res.* **2010**, *54*, 1215–1247. [CrossRef]
5. Ayivi, R.D.; Gyawali, R.; Krastanov, A.; Aljaloud, S.O.; Worku, M.; Tahergorabi, R.; Da Silva, R.; Ibrahim, S.A. Lactic acid bacteria: Food safety and human health applications. *Dairy* **2020**, *1*, 202–232. [CrossRef]
6. Bourdichon, F.; Casaregola, S.; Farrokh, C.; Frisvad, J.C.; Gerds, M.L.; Hammes, W.P.; Harnett, J.; Huys, G.; Laulund, S.; Ouwehand, A. Food fermentations: Microorganisms with technological beneficial use. *Int. J. Food Microbiol.* **2012**, *154*, 87–97. [CrossRef] [PubMed]
7. Parvez, S.; Malik, K.A.; Ah Kang, S.; Kim, H.-Y. Probiotics and their fermented food products are beneficial for health. *J. Appl. Microbiol.* **2006**, *100*, 1171–1185. [CrossRef] [PubMed]
8. Hassan, S.M. Soybean, nutrition and health. *Soybean-Bio-Act. Compd.* **2013**, *20*, 453–473. [CrossRef]
9. Afzaal, M.; Saeed, F.; Islam, F.; Ateeq, H.; Asghar, A.; Shah, Y.A.; Ofoedu, C.E.; Chacha, J.S. Nutritional health perspective of natto: A critical review. *Biochem. Res. Int.* **2022**, *2022*, 5863887. [CrossRef]
10. Xu, X.; Cui, H.; Xu, J.; Yuan, Z.; Liu, X.; Fan, X.; Li, J.; Zhu, D.; Liu, H. Effects of different probiotic fermentations on the quality, soy isoflavone and equol content of soy protein yogurt made from soy whey and soy embryo powder. *LWT* **2022**, *157*, 113096. [CrossRef]
11. Ma, M.; Li, Y.; He, Y.; Li, D.; Niu, H.; Sun, M.; Miao, X.; Su, Y.; Zhang, H.; Hua, M.; et al. The Combination of Bacillus natto JLCC513 and Ginseng Soluble Dietary Fiber Attenuates Ulcerative Colitis by Modulating the LPS/TLR4/NF-κB Pathway and Gut Microbiota. *J. Microbiol. Biotechnol.* **2024**, *34*, 1287–1298. [CrossRef]
12. Sun, R.; Niu, H.; Li, Y.; Sun, M.; Hua, M.; Miao, X.; Su, Y.; Wang, J.; Li, D.; Wang, Y. Fermented natto powder alleviates obesity by regulating LXR pathway and gut microbiota in obesity rats. *J. Appl. Microbiol.* **2024**, *135*, lxae003. [CrossRef] [PubMed]
13. Sun, R.; Li, D.; Sun, M.; Miao, X.; Jin, X.; Xu, X.; Su, Y.; Xu, H.; Wang, J.; Niu, H. *Bacillus natto* ameliorates obesity by regulating PI3K/AKT pathways in rats. *Biochem. Biophys. Res. Commun.* **2022**, *603*, 160–166. [CrossRef]
14. Sun, R.; Niu, H.; Sun, M.; Miao, X.; Jin, X.; Xu, X.; Yanping, C.; Mei, H.; Wang, J.; Da, L.; et al. Effects of *Bacillus subtilis natto* JLCC513 on gut microbiota and intestinal barrier function in obese rats. *J. Appl. Microbiol.* **2022**, *133*, 3634–3644. [CrossRef]
15. Miao, X.; Niu, H.; Sun, M.; Dong, X.; Hua, M.; Su, Y.; Wang, J.; Li, D. A comparative study on the nutritional composition, protein structure and effects on gut microbiota of 5 fermented soybean products (FSPs). *Food Res. Int.* **2024**, *183*, 114199. [CrossRef]

16. Murphy, P.A.; Song, T.; Buseman, G.; Barua, K. Isoflavones in soy-based infant formulas. *J. Agric. Food Chem.* **1997**, *45*, 4635–4638. [CrossRef]
17. Xu, B.; Chang, S.K.C. Total phenolics, phenolic acids, isoflavones, and anthocyanins and antioxidant properties of yellow and black soybeans as affected by thermal processing. *J. Agric. Food Chem.* **2008**, *56*, 7165–7175. [CrossRef]
18. Sun, X.H.; Qi, X.; Han, Y.D.; Guo, Z.J.; Cui, C.B.; Lin, C.Q. Characteristics of changes in volatile organic compounds and microbial communities during the storage of pickles. *Food Chem.* **2023**, *409*, 135285. [CrossRef] [PubMed]
19. Zhou, X.; Guo, T.; Lu, Y.; Hadiatullah, H.; Li, P.; Ding, K.; Zhao, G. Effects ofamino acid composition of yeast extract on the microbiota and aroma quality offermented soy sauce. *Food Chem.* **2022**, *393*, 133289. [CrossRef] [PubMed]
20. Chen, X.; Lu, Y.; Zhao, A.; Wu, Y.; Zhang, Y.; Yang, X. Quantitative analyses for several nutrients and volatile components during fermentation of soybean by *Bacillus subtilis natto*. *Food Chem.* **2022**, *374*, 131725. [CrossRef] [PubMed]
21. Zhang, L.; Wu, J.L.; Xu, P.; Guo, S.; Zhou, T.; Li, N. Soy protein degradation drives diversity of amino-containing compounds via *Bacillus subtilis natto* fermentation. *Food Chem.* **2022**, *388*, 133034. [CrossRef]
22. Zhou, B.; Semanjski, M.; Orlovetskie, N.; Bhattacharya, S.; Alon, S.; Argaman, L.; Jarrous, N.; Zhang, Y.; Macek, B.; Sinai, L.; et al. Arginine dephosphorylation propels spore germination in bacteria. *Proc. Natl. Acad. Sci. USA* **2019**, *116*, 14228–14237. [CrossRef]
23. Li, C.; Xu, T.; Liu, X.W.; Wang, X.; Xia, T. The expression of β-glucosidase during natto fermentation increased the active isoflavone content. *Food Biosci.* **2021**, *43*, 101286. [CrossRef]
24. Rocchi, R.; Zwinkels, J.; Kooijman, M.; Garre, A.; Smid, E.J. Development of novel natto using legumes produced in Europe. *Heliyon* **2021**, *10*, e26849. [CrossRef]
25. Wang, H.J.; Chang, L.; Lin, Y.S. Changes in functionality of germinated and non-germinated brown rice fermented by *Bacillus natto*. *Foods* **2021**, *10*, 2779. [CrossRef]
26. Kubo, Y.; Noguchi, T.; Kimura, K. Storage temperature and quality changes of natto. *Food Sci. Technol. Res.* **2021**, *27*, 497–504. [CrossRef]
27. Zhang, K.; Zhang, T.T.; Guo, R.R.; Ye, Q.; Zhao, H.L.; Huang, X.H. The regulation of key flavor of traditional fermented food by microbial metabolism: A review. *Food Chem. X* **2023**, *19*, 100871. [CrossRef]
28. Ma, Y.N.; Xu, X.; Chen, L.F.; Zhou, J.P.; Cao, Z.H.; Lin, Q.Y. Process optimization and evaluation of quality properties of natto with co-culture of *Bacillus subtilis natto* and *Limosilactobacillus* fermentum. *Curr. Res. Microb. Sci.* **2025**, *8*, 100347. [CrossRef]
29. Sun, M.; Yang, N.; He, Y.; Miao, X.; Niu, H.; Hua, M.; Wang, J.; Li, D. Characterisation of dynamic changes in nutritional components and aroma compound formation in natto during fermentation and post-ripening. *Food Res. Int.* **2025**, *219*, 116961. [CrossRef]
30. Lyu, S.; Yang, Q.; Duan, X.; Liu, J.; Yu, Y.; Pan, F.; Zhang, T. Impact of different cooling times during post-maturation on physicochemical and texture properties of fermented egg-milk beverage. *Food Biosci.* **2023**, *54*, 102906. [CrossRef]
31. Wu, Q.; Xu, Z.; Feng, S.; Shi, X.; Qin, L.; Zeng, H. Correlation analysis between microbial communities and flavor compounds during the post-ripening fermentation of traditional chili bean paste. *Foods* **2024**, *13*, 1209. [CrossRef] [PubMed]
32. Deng, J.; Xu, H.; Li, X.; Wu, Y.; Xu, B. Correlation of characteristic flavor and microbial community in Jinhua ham during the post-ripening stage. *LWT* **2022**, *171*, 114067. [CrossRef]
33. Weng, Y.; Chen, L.; Kun, J.; He, S.; Tong, H.; Chen, Y. The unique aroma of ripened Pu-erh tea, Liupao tea and Tietban tea: Associated post-fermentation condition and dominant microorganism with key aroma-active compound. *Food Chem.* **2025**, *464*, 141788. [CrossRef] [PubMed]

https://doi.org/10.3390/foods15020237

foods

MDPI

Article

Screening of Lactic Acid Bacteria and RSM-Based Optimization for Enhancing γ-Aminobutyric Acid (GABA) Accumulation in Orange Juice

Shufeng Yin, Yiyao Wang, RuiXue Zhao, Ning Zhao, Hao Liu, Yining Tang, Ningbo Qin, Yiwei Dai * and Xinping Lin *

State Key Laboratory of Marine Food Processing & Safety Control, National Engineering Research Center of Seafood, School of Food Science and Technology, Dalian Polytechnic University, Dalian 116034, China; yinshufeng1220@163.com (S.Y.); wangyiyao0506@163.com (Y.W.); zhaorx04@outlook.com (R.Z.); ningzhao0221@163.com (N.Z.); 18763666325@163.com (H.L.); 18904084306@163.com (Y.T.); qinningbo6902765@163.com (N.Q.)

* Correspondence: ywdai6228@126.com (Y.D.); yingchaer@163.com (X.L.); Tel.: +86-15242684612 (Y.D.); +86-15898160352 (X.L.)

Abstract

Inoculated fermentation can enhance the flavor, nutrition, and functionality of juice. The lactic acid bacteria (LAB) are commonly used as starter cultures. This study screened LAB for orange juice fermentation and optimized fermentation factors using response surface methodology (RSM) to improve GABA content in orange juice. A total of 52 LAB strains were screened, and *Lacticaseibacillus paracasei* ZY (*Lcb. paracasei* ZY) and *Lacticaseibacillus rhamnosus* SN12 (*Lcb. rhamnosus* SN12) presented higher GABA yields and adaptability to substrates. The optimized fermentation factors for GABA enhancement in orange juice were as follows: initial pH of 5.5, fermentation temperature of 37 °C, soluble solids content of 12.0 °Bx, inoculum ratio of *Lcb. paracasei* ZY to *Lcb. rhamnosus* SN12 as 1:1, inoculum size of 6 Log CFU/mL, and fermentation time of 96 h. Under these optimized conditions, the GABA content reached 0.89 g/L, representing a 39.06% increase compared to uninoculated orange juice. This indicates that RSM-based optimization is conducive to increasing GABA content in orange juice and provides a scientific basis for the development of GABA-enriched functional fermented juices.

Keywords: γ-aminobutyric acid (GABA); orange juice; *Lacticaseibacillus paracasei*; *Lacticaseibacillus rhamnosus*; single-factor experiment; response surface methodology (RSM); functional beverage

1. Introduction

γ-aminobutyric acid (GABA) is a non-protein amino acid that is widely present in plants and animals. As a major inhibitory neurotransmitter in the central nervous system, GABA has multiple physiological functions, including improving sleep [1], alleviating anxiety [2], lowering blood pressure [3], and regulating blood sugar [4]. Compared with chemical synthesis and plant extraction methods, microbial fermentation is safer and more sustainable for GABA production [5]. Lactic acid bacteria (LAB), as generally recognized as safe (GRAS) probiotics, can irreversibly decarboxylate glutamic acid or monosodium glutamate (MSG) to convert GABA through glutamic acid decarboxylase (GAD). In recent years, they have received increasing attention in the food industry [6]. The reported species of GABA-producing LAB include *Levilactobacillus brevis* (*L. brevis*) [7], *Lactiplantibacillus plantarum* (*Lpb. plantarum*) [8], *Lacticaseibacillus rhamnosus* (*Lcb. rhamnosus*) [9], *Lacticaseibacillus paracasei* (*Lcb. paracasei*) [10], and *Limosilactobacillus fermentum* (*Lim. fermentum*) [11].

Oranges are rich in nutrients, containing carbohydrates, vitamin C, and polyphenols [12]. Their high juice content makes them highly attractive to consumers. At present, the main processed products of orange juice are concentrated juice and freshly squeezed juice. However, traditional processing methods are difficult to meet consumers' growing demands for nutritional value. Fermentation, as an effective means to enhance flavor, nutritional value, and bioactive substances, has been widely applied in the production of juice. Due to their safety and probiotic properties, LAB are often used in the preparation of non-alcoholic fermented drinks. Wu et al. [13] reported that fermenting citrus juice with *Lpb. plantarum* A72 could significantly increase the contents of total phenols, flavonoids, and antioxidant activity. Quan et al. [14] evaluated the effect of six strains of LAB on orange juice fermentation and found that the orange juice fermented by *Lcb. paracasei* had the best flavor and overall sensory acceptance. Huang et al. [15] confirmed that fermentation with *Lactiplantibacillus paraplantarum* M23 could significantly enhance the antioxidant and antibacterial properties of orange juice. However, current research on fermented orange juice mainly focuses on its antioxidant activity, while discussions on other functional properties remain relatively limited.

GABA content found in natural foods tends to be low and may fail to meet biological requirements. While the direct addition of GABA to food products is possible as a food additive, its stability and safety cannot be guaranteed. GABA-producing LAB are effective in developing GABA-enriched fermented foods. Wang et al. [16] reported that, after litchi juice was fermented by *Lpb. plantarum* HU-C2W for 40 h, the GABA content increased by approximately 40 mg/100 mL. Zheng et al. [17] found that fermenting *Moringa oleifera leaves* with *Lpb. plantarum* LK-1 could increase the content of GABA, polyphenols, and organic acids. Kanklai et al. [18] demonstrated that, when fermenting mulberry juice with *L. brevis* F064A, the final GABA content reached 3.31 mg/mL. Jin et al. [19] reported that fermenting litchi juice supplemented with 0.5% MSG using *L. brevis* LBG-29 increased the GABA content to 3.07 g/L. In contrast, Zhou et al. [20] achieved only 0.22 g/L of GABA in persimmon juice through co-fermentation with *Lpb. plantarum* C17 and *L. brevis* Lp-B at a 1:1 ratio for 48 h at 30 °C. GABA production in fermented juices depends not only on the substrate but also on the interplay of key process variables such as temperature and pH. Thus, rational optimization of fermentation conditions is essential for enhancing GABA yield. The traditional single-factor optimization is difficult to accurately reflect the mutual influence among these factors. In contrast, the response surface methodology (RSM) can quantitatively analyze the interactions among the factors and establish a more reliable and efficient fermentation model. Wang et al. [21] optimized the fermentation conditions of soymilk using RSM, and the results showed that, by using *Lpb. plantarum* Lp3 and *Streptococcus thermophilus* in a ratio of 2:1 as the starter, adding 0.4% MSG at 35 °C for 10 h, the highest GABA content (0.55 mg/mL) could be obtained. It is significantly superior to a single strain or other co-fermentation combinations. Shan et al. [8] optimized the process of fermenting yogurt with *Lpb. plantarum* NDC75017 using RSM and found that an MSG concentration of 80 mM, pyridoxal phosphate at 18 μM, and a fermentation temperature of 36 °C produced a maximum GABA content of 314.56 mg/100 g.

In this study, two high GABA-producing LAB strains were identified through the Berthelot colorimetric method and high-performance liquid chromatography (HPLC) screening. Their fermentation performance and probiotic properties were systematically evaluated, and the suitability of orange juice as a fermentation substrate for LAB growth was assessed. A single-factor experiment combined with RSM was employed to optimize initial

https://doi.org/10.3390/foods15010071

pH, fermentation temperature, soluble solids content, inoculum ratio, inoculum size, and fermentation time, aiming to produce GABA-enriched orange juice. These findings provide a theoretical foundation and process guidance for the development of functional fermented juices.

2. Materials and Methods

2.1. Starter Preparation

In the initial stage of the experiment, 52 strains of LAB were isolated from fermented foods to construct a GABA-producing screening library (Table A1). The strains were streaked onto MRS agar (Haibo, Qingdao, China) and then incubated at 37 °C for 48 h. Subsequently, single colonies were incubated in MRS broth (Haibo, Qingdao, China) at 37 °C for 16–20 h to reach a cell density of 6–8 log CFU/mL. Bacterial cells were collected by centrifugation (6000× *g*, 5 min, 4 °C) and resuspended in sterile saline. The suspension was inoculated into MRS broth containing 10 g/L MSG (99% purity; Meihua, Tongliao, Inner Mongolia, China) to achieve an initial cell density of approximately 6 log CFU/mL, followed by incubation at 37 °C for 48 h. The culture medium without LAB served as the control. All experimental equipment was sterilized by autoclaving at 121 °C for 15 min, and all procedures were performed under aseptic conditions.

2.2. GABA-Producing LAB Screening

2.2.1. Rapid Screening of GABA by Berthelot Colorimetric Method

The Berthelot colorimetric method was slightly revised based on the method reported by Zhang et al. [22]. LAB fermentation cultures prepared in Section 2.1 were subjected to centrifugation at 6000× *g* for 10 min at 4 °C, and 0.5 mL of the obtained supernatant was aliquoted. Subsequently, 1 mL of 6.0% (*w*/*v*) phenol (Macklin, Shanghai, China), 0.4 mL of 7.0% (*w*/*v*) sodium hypochlorite (Sangon Biotech, Shanghai, China), and 0.2 mL of 0.2 mol/L borate buffer (pH 9.0; Damao, Tianjin, China) were added and thoroughly mixed. The reaction mixtures were incubated in a boiling water bath for 7 min and then immediately cooled in an ice bath for 10 min. Subsequently, 2 mL of 60% (*v*/*v*) ethanol (Fuyu, Tianjin, China) was added, and the absorbance was measured at 630 nm. The top 15 LAB strains with the highest GABA production were further quantified by HPLC.

2.2.2. Quantitative Screening of GABA by HPLC

The GABA content in the fermented orange juice and the top 15 LAB fermentation cultures (Section 2.1) were determined using an Essentia LC-16 HPLC system (Shimadzu Ltd., Tokyo, Japan) according to Wu et al. [13]. After centrifugation at 6000× *g* for 10 min at 4 °C, an aliquot of 200 µL from the resulting supernatant was subjected to derivatization with ortho-phthaldialdehyde (OPA; Aladdin, Shanghai, China) for 5 min, followed by filtration through a 0.22 µm membrane filter. Subsequently, 10 µL of the filtrate was injected into a HPLC system for analysis. Chromatographic separation was carried out on an Agilent Extend-C18 column (4.6 mm × 250 mm, 5 µm) with a mobile phase composed of acetonitrile (phase A) and 50 mM sodium acetate solution (phase B) under an isocratic elution of 35:65 (*v*/*v*). The column temperature was kept at 30 °C, the mobile phase flow rate was set to 0.5 mL/min, and detection was performed at 334 nm.

2.3. Evaluation of LAB Fermentation Performance and Probiotic Properties

High GABA-producing LAB were cultured in MRS broth, centrifuged, and resuspended as described in Section 2.1. LAB were inoculated into 96-well plates containing 200 µL of MRS broth to an initial OD_{600} of 0.2, and growth at 37 °C was monitored every 2 h using a Bioscreen C automated growth curve analyzer (Oy Growth Curves Ab Ltd., Turku,

https://doi.org/10.3390/foods15010071

Finland). Acid tolerance (pH 3.0, 2 h) and bile salt tolerance (0.03% w/v, 3 h) were assessed following Meena et al. [23]. Simulated Gastric Fluid (SGF) and Simulated Intestinal Fluid (SIF) tolerance were evaluated as described by Huang et al. [24], with results expressed as survival rates (%).

2.4. *Evaluation of Substrate Adaptation*

High GABA-producing LAB were cultured in MRS broth, and centrifuged as described in Section 2.1. After being washed twice with sterile saline, the LAB were resuspended in saline and inoculated into orange juice to achieve an initial cell density of approximately 6 Log CFU/mL. The orange juice was fermented at 37 °C for 48 h. Viable cells were counted on MRS agar using the standard plate count method and expressed as log CFU/mL.

2.5. *Preparation of Fermented Orange Juice*

Oranges were purchased from the Qianhe Market (Dalian, Liaoning, China), and fermented orange juice was prepared following the method of Wu et al. [13]. After peeling and deseeding, the fruits were homogenized with an equal volume of deionized water and filtered through three layers of cheesecloth to obtain clarified juice. Before fermentation, 8 g/L MSG was added. The juice was pasteurized at 97 °C for 3 min using a vertical pressure steam sterilizer (DSX-30L-I, Shanghai SHENAN Medical Device Co., Ltd., Shanghai, China). Bacterial cells were prepared for inoculation as described in Section 2.4. When cooled to room temperature (25 °C), LAB were inoculated into the orange juice. Orange juice fermentation was conducted under static conditions, with the orange juice without LAB set as the control group. Fermentation conditions were adjusted according to the experimental design, and collected samples were stored at −80 °C for later analysis within 2 weeks.

2.6. *Design of Single-Factor Experiment*

The influence of initial pH, fermentation temperature, soluble solids content, inoculum ratio, inoculum size, and fermentation time on GABA production by GABA-producing LAB in orange juice was evaluated through single-factor experiments (Table A2). The aim was to determine the optimal range of each factor in RSM. The initial pH of orange juice was adjusted with sodium bicarbonate and measured using a pH meter (FE28 Standard, Mettler Toledo, Greifensee, Switzerland). The content of soluble solids content was adjusted with glucose and measured using a handheld refractometer (LYT-380, ATAGO, Tokyo, Japan), and the results were expressed as °Bx.

2.7. *Design of RSM*

2.7.1. Plackett–Burman Design

Based on the findings from the single-factor experiments, the Plackett–Burman design was utilized to determine significant factors for GABA yields from fermented orange juice [25]. In this design, the GABA content was taken as the response value (Y), while the initial pH (A), fermentation temperature (B), soluble solids content (C), inoculum ratio (D), inoculum size (E), and fermentation time (F) were selected as the Plackett–Burman design factors (Table A3).

2.7.2. Box–Behnken Design

Based on the significant factors identified by the Plackett–Burman design, a Box–Behnken design (Table A5) was employed to investigate the interactive effects among each factor [26]. Non-significant factors were fixed at their optimal levels established by single-factor experiments.

https://doi.org/10.3390/foods15010071

2.8. Model Validation

Triplicate experiments were performed under the optimized fermentation conditions to verify the model derived from RSM.

2.9. Statistical Analysis

Experiments were carried out in three separate replicates, with outcomes reported as the mean ± standard deviation (mean ± SD). Statistical evaluation was performed using SPSS (version 24.0; IBM Corp., Armonk, NY, USA). Data visualization was accomplished with Origin (version 2021b; OriginLab, Northampton, MA, USA), while RSM modeling and analysis were conducted through Design-Expert software (version 13.0; Stat-Ease, Minneapolis, MN, USA). The experimental runs of the Plackett–Burman design (Table A4) and Box–Behnken design (Table A6) were randomly generated using Design-Expert software (version 13.0; Stat-Ease, Inc., Minneapolis, MN, USA).

3. Results and Discussion

3.1. Analysis of GABA-Producing LAB

As shown in Figure 1A, 52 LAB isolates were rapidly screened via the Berthelot colorimetric method, among which forty-three strains exhibited GABA-producing ability. The top five LAB strains were identified as *Lcb. paracasei* ZY (1.00 g/L), *Lcb. rhamnosus* SN12 (0.82 g/L), *Lpb. plantarum* M35 (0.75 g/L), *Lpb. plantarum* 2-22-LJ (0.71 g/L), and *Lpb. plantarum* Z5-24 (0.68 g/L). To further validate these results, the top 15 GABA-producing LAB strains identified by the Berthelot colorimetric method were recultured and quantified using HPLC. Figure 1B shows the chromatograms of the medium, the LAB strains (*Lcb. paracasei* ZY and *Lcb. rhamnosus* SN12), and the 1 g/L GABA standard, while the HPLC results are shown in Figure 1C. HPLC analysis confirmed that the highest GABA producers were *Lcb. paracasei* ZY (0.65 g/L), *Lcb. rhamnosus* SN12 (0.70 g/L), *Lpb. plantarum* 2-22-LJ (0.58 g/L), *Lpb. plantarum* LS3-1 (0.52 g/L), and *Lpb. plantarum* Z5-24 (0.50 g/L). We found that GABA content measured by the Berthelot colorimetric method was generally higher than that obtained by HPLC, consistent with the findings of Le et al. [27]. This is because the Berthelot colorimetric method lacks specificity, and other amino-containing substances in the sample react with it, leading to an overestimation of GABA content. Considering the GABA production ability of the strains and their stability during fermentation, *Lcb. paracasei* ZY and *Lcb. rhamnosus* SN12 were ultimately selected as the subjects for subsequent research.

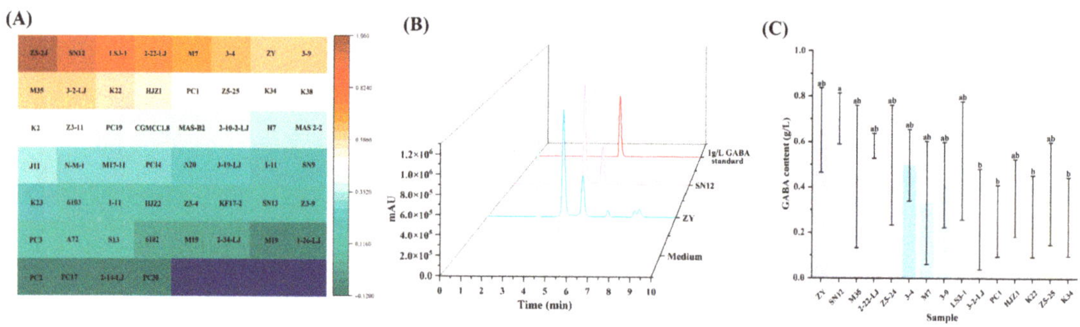

Figure 1. GABA determination via the Berthelot colorimetric method (**A**), HPLC chromatogram of GABA (**B**), and GABA quantification by HPLC (**C**). Different letters indicate significant differences among different treatments ($p < 0.05$).

https://doi.org/10.3390/foods15010071

3.2. Analysis of Fermentation Performers and Probiotic Properties

Both *Lcb. paracasei* ZY and *Lcb. rhamnosus* SN12 have shown classic 'S' curves for growth (Figure 2A). In the first 6 h of incubation, growth was slow for both strains, followed by log phase growth. *Lcb. paracasei* ZY and *Lcb. rhamnosus* SN12 entered the stationary phase at 20 and 24 h of culture, respectively. At the end of 72 h culture, neither strain showed a significant decrease in absorbance, thus making them good fermentation performers.

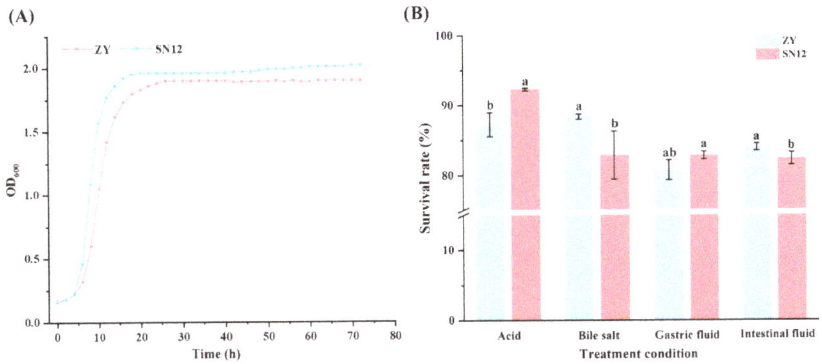

Figure 2. Fermentation performance (**A**) and probiotic properties (**B**) of *Lcb. paracasei* ZY and *Lcb. rhamnosus* SN12. Different letters indicate significant differences among different treatments ($p < 0.05$).

Figure 2B shows the tolerance of the strains to acids, bile salts, SGF, and SIF. At an acidic pH 3.0, both strains were able to sustain a survival rate of more than 87.00%, while *Lcb. rhamnosus* SN12 showed 92.23% survival at that pH, thereby displaying high-level tolerance to highly acidic environments. Since 0.3% bile salt concentration is equivalent to its approximate concentration found in the human small intestine [28], this concentration was tested. In MRS both containing 0.3% bile salt, both strains maintained survival rates over 82.00%, demonstrating strong resistance to bile salts. Both strains exhibited survival rates exceeding 80.00% in SGF and SIF, which was higher than that of *L. brevis* DSM 32386 and DSM 20054 [7], displaying strong resistance to digestive enzymes. Both strains showed excellent probiotic properties and hold potential as starter cultures for food fermentation.

3.3. LAB Adaptability Evaluation in Orange Juice

Adaptability is also very important for applying strains to orange juice. Based on Figure A1, it is shown that *Lcb. paracasei* ZY and *Lcb. rhamnosus* SN12 have an upward trend in orange juice. At the initial growth stage of fermentation, the viable cell counts for both LAB strains were 6.68 and 6.19 log CFU/mL, respectively. In the initial 12 h of fermentation, both LAB strains grew very quickly and then entered a period of relatively slow growth. After 48 h, the viable cell counts of *Lcb. paracasei* ZY and *Lcb. rhamnosus* SN12 reached approximately 8.74 and 8.53 log CFU/mL, respectively, and tended to stabilize. These results are consistent with the findings of Jin et al. [19], who reported a significant increase in the population of *L. brevis* within the first 16 h of litchi fermentation ($p < 0.05$), followed by a stable phase. Therefore, the two LAB strains can maintain excellent adaptability in orange juice.

3.4. Analysis of Single-Factor Experiment

There have been studies exploring GABA synthesis by LAB influenced by fermentation processes [6]. In this study, we selected initial pH, fermentation temperature, soluble solids content, inoculum ratio, inoculum size, and fermentation time as optimization factors.

https://doi.org/10.3390/foods15010071

3.4.1. pH

pH is very significant for GAD activity for GABA synthesis. Previous studies have shown that most LAB exhibit enhanced metabolic activity under mildly acidic conditions [29]. In our experiments (Figure 3A), GABA production increased as pH increased to reach its peak at 1.36 g/L at pH 5.0 and started to reduce gradually. This optimal pH likely reflects the maximum GAD activity at pH 5.0, facilitating the efficient conversion of Glutamic acid and MSG to GABA. Similarly, Komatsuzaki et al. [10] also found that *Lcb. paracasei* NFRI 7415 produced 210 mM GABA at an initial pH of 5.0, which was markedly higher than production at pH 4.0 or 6.0 ($p < 0.05$).

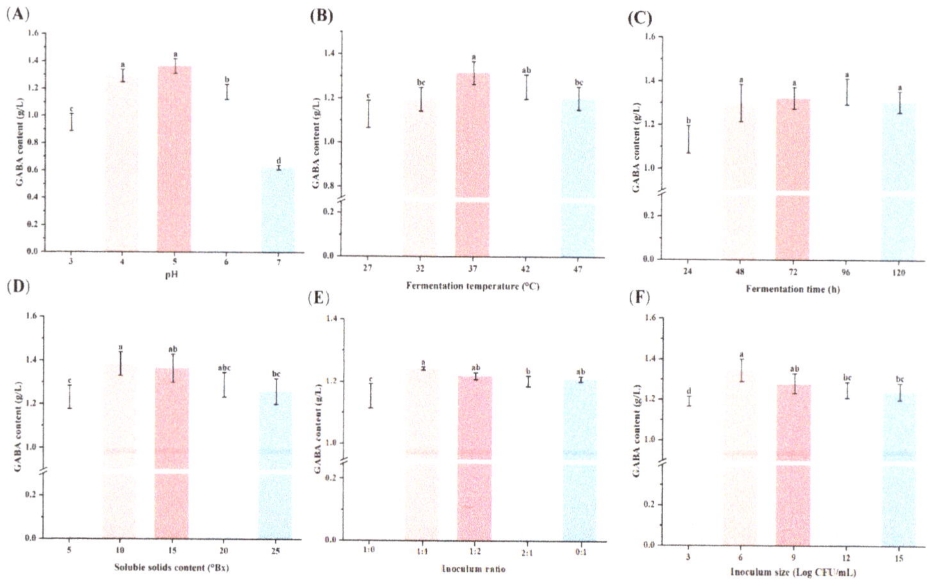

Figure 3. Effects of single factors on GABA production in fermented orange juice, including initial pH (**A**), fermentation temperature (**B**), soluble solids content (**C**), inoculum ratio (**D**), inoculum size (**E**), and fermentation time (**F**). Different letters indicate significant difference among different treatments ($p < 0.05$).

3.4.2. Fermentation Temperature

Temperature is another important factor to consider during fermentation. As illustrated in Figure 3B, GABA content in orange juice rose with increasing temperature up to 37 °C, but began to decrease when the temperature exceeded this point. Based on single-factor optimization experiments focusing on temperature, 37 °C is likely the optimal fermentation temperature, given that GAD enzyme activity is significantly enhanced at this temperature. This result is also identical to Komatsuzaki et al. [10], who observed that *Lcb. paracasei* NFRI 7415 produced the highest content of GABA (302 mmol/L) at 37 °C, with a notable drop at 43 °C. Rayavarapu et al. [30] also applied RSM to optimize GABA production in soymilk fermented with *Lim. fermentum* and found that the maximum GABA content (4.2 g/L) was also achieved at 37 °C.

3.4.3. Soluble Solids Content

The soluble solids content had a significant effect on the accumulation of GABA in fermented orange juice. As shown in Figure 3C, the GABA content reached its maximum when the soluble solids content was 10.0 °Bx. This result was attributed to the balance between carbon source supply and osmotic regulation. An adequate carbon source not

https://doi.org/10.3390/foods15010071

only provides the necessary substrates and energy for cell growth, but also maintains the permeability and metabolic activity of the cell membrane [31].

3.4.4. Inoculum Ratio

Figure 3D shows the effect of the inoculation ratio on the accumulation of GABA in fermented orange juice. When the inoculum ratio of *Lcb. paracasei* ZY to *Lcb. rhamnosus* SN12 was 1:1, the GABA content was the highest (1.36 g/L). This is because at this ratio, the combination of the two LAB strains' metabolism resulted in the effective utilization of substrates for higher GABA concentration.

3.4.5. Inoculum Size

Figure 3E shows the influence of inoculum size on GABA content in orange juice. It was found that 6 Log CFU/mL inoculum size resulted in a higher GABA content than any other inoculum size. This suggests that a moderate initial cell density helps the strain establish dominance more quickly, boosting metabolic activity and ultimately increasing GABA synthesis. A similar trend was observed by Rayavarapu et al. [30], who studied soymilk fermentation with *Lim. fermentum* and found that a 5.8 Log CFU/mL inoculum size led to the highest GABA content among the tested levels.

3.4.6. Fermentation Time

LAB start to produce GABA during the logarithmic growth phase and continue through to the stationary growth phase [32]. Figure 3F presents the effect of fermentation time on GABA content in orange juice. GABA content showed significant increments from 0 to 96 h ($p < 0.05$), while, from 96 to 120 h, the increment was very slight ($p > 0.05$). This shows that LAB start to produce GABA between 24 and 96 h. After 96 h, the accumulation of GABA stopped due to the consumption of substrates (such as glutamic acid) and the production of lactic acid by LAB, which led to a decrease in pH value.

Based on single-factor experiments, the most suitable factor levels for GABA accumulation in orange juice are the following: initial pH of 5.0, fermentation temperature of 37 °C, soluble solids content of 10.0 °Bx, inoculum ratio of *Lcb. paracasei* ZY to *Lcb. rhamnosus* SN12 as 1:1, inoculum size of 6 Log CFU/mL and fermentation time of 96 h. These results not only offer proper levels for each factor but also bring ideal guidance for RSM.

3.5. Optimization of GABA Content in Orange Juice Using RSM

3.5.1. Analysis of Plackett–Burman Design

The results obtained by the Plackett–Burman design are shown in Table A4, and the multiple linear regression equations between each factor and the response values are as follows:

$$Y = 1.12 - 0.0586A - 0.0491B - 0.0618C + 0.0181D + 0.0137E - 0.0269F$$

As shown in Table 1, the analysis results of ANOVA suggest that the overall model is significant ($p < 0.05$). The coefficient of determination (R^2) of the model is 0.9140, demonstrating a good match between the model and the data. The adjusted coefficient of determination (adjusted R^2) is 0.8109, and this value implies that the model explains 81.09% of the data variance. The coefficient of variation (CV) is 4.41%, further proof of its repeatability and reliability [33]. The p-values of initial pH (A), fermentation temperature (B), and soluble solids content (C) are 0.0094, 0.0186, and 0.0076, respectively ($p < 0.05$), indicating that these three factors all have a significant effect on the response value. This result represents the fact that initial pH, fermentation temperature, and soluble solids content are the significant factors influencing GABA accumulation in fermented orange juice.

https://doi.org/10.3390/foods15010071

Table 1. Analysis of variance (ANOVA) for the Plackett–Burman design.

Source	Sum of Squares	Df	Mean Square	F-Value	p-Value	Significance
Model	0.1310	6	0.0218	8.86	0.0150	Significant
A-Initial pH	0.0413	1	0.0413	16.74	0.0094	**
B-Fermentation temperature	0.0290	1	0.0290	11.76	0.0186	*
C-Soluble solids content	0.0459	1	0.0459	18.62	0.0076	*
D-Starter ratio	0.0039	1	0.0039	1.60	0.2623	
E-Inoculum size	0.0023	1	0.0023	0.91	0.3830	
F-Fermentation time	0.0087	1	0.0087	3.54	0.1188	
Residual	0.0123	5	0.0025			
Cor Total	0.1433	11				
R^2	0.9140					
Adjusted R^2	0.8109					
Predicted R^2	0.5049					
C.V.%	4.41					

Note: * indicates a significant difference at $p < 0.05$, and ** indicates a highly significant difference at $p < 0.01$.

3.5.2. Analysis of Box–Behnken Design

The results obtained by the Plackett–Burman design are shown in Table A6. A total of 17 runs for experimentation have been conducted to estimate the GABA content of fermented orange juice as response values (Y). The variables include initial pH (A), fermentation temperatures (B), and soluble solids contents (C). The following nonlinear regression equation was fitted:

$$Y = 0.7765 + 0.0240A + 0.0109B + 0.0085C - 0.0197AB + 0.0128AC - 0.0082BC - 0.0323A^2 - 0.0276B^2 - 0.0188C^2$$

Through ANOVA analysis of the Box–Behnken design (Table 2), the model had an F-value of 9.85 and a p-value of 0.0032, which indicates that the model is extremely significant. The lack of fit was not significant ($p > 0.05$), suggesting that the model fits the data well and is reliable. The model's R^2 is 0.9268, and its adjusted R^2 (0.8327) is close to this value, indicating the model is robust and can explain 83.27% of the variation in the response value. The predictive coefficient of determination (predictive R^2) is 0.7656, further validating the model's predictive performance. The CV value is 1.96%, which is below 5%, showing that the experimental data exhibit good precision and repeatability. It is well-suited for application in the process optimization of GABA production from orange juice. Significance analysis revealed that the initial pH (A), the quadratic term of initial pH (A^2), and the quadratic term of fermentation temperature (B^2) exert extremely significant effects on GABA production ($p < 0.01$), while the interaction between initial pH and fermentation temperature (AB) and the quadratic term of soluble solids content (C^2) exert significant effects ($p < 0.05$). Based on the F-values of each factor for GABA production, the degree of influence of each factor on GABA production is as follows: initial pH > fermentation temperature > soluble solids content.

To clarify the effects of significant factors' interactions on GABA production, response surface and contour plots are presented (Figure 4). The steeper the slope on the response surface plot, the greater the effect of that factor on GABA content, thereby aiding in understanding the interactions among various factors [33]. Analysis results indicated that mutual interactive effects existed between each pair of factors. The interactive effect of initial pH and fermentation temperature is illustrated in Figure 4A,D, revealing that the GABA yield reached the highest level under weakly acidic conditions (pH 5.0–6.0) and within the temperature range of 30–38 °C. As shown in Figure 4B,E, the interactive effect of

https://doi.org/10.3390/foods15010071

initial pH and soluble solids content demonstrated that the effect of soluble solids content on GABA yield gradually enhanced as the pH value approached neutrality. As presented in Figure 4C,F, the contour lines between fermentation temperature and soluble solids content tend to be circular, indicating a relatively weak interactive effect between these two factors. The analysis of response surface and contour plots indicates that fermentation temperature has the largest influence on GABA production, followed by initial pH, and finally soluble solids content. This order differs slightly from the ranking based on *F*-values because the response surface plots better reflect the actual strength of combined effects under real experimental conditions.

Table 2. Analysis of variance (ANOVA) for the quadratic model of Box–Behnken design.

Source	Sum of Squares	Df	Mean Square	*F*-Value	*p*-Value	Significance
Model	0.0187	9	0.0021	9.85	0.0032	Significant
A-Initial pH	0.0046	1	0.0046	21.88	0.0023	**
B-Fermentation temperature	0.0009	1	0.0009	4.48	0.0721	
C-Soluble solids content	0.0006	1	0.0006	2.72	0.1433	
AB	0.0015	1	0.0015	7.33	0.0303	*
AC	0.0007	1	0.0007	3.09	0.1220	
BC	0.0003	1	0.0003	1.28	0.2959	
A^2	0.0044	1	0.0044	20.82	0.0026	**
B^2	0.0032	1	0.0032	15.23	0.0059	**
C^2	0.0015	1	0.0015	7.07	0.0325	*
Residual	0.0015	7	0.0002			
Lack of fit	0.0002	3	0.0001	0.17	0.9108	Not significant
Pure error	0.0013	4	0.0003			
Cor toal	0.0202	16				
R^2	0.9268					
Adjusted R^2	0.8327					
Predicted R^2	0.7656					
C.V.%	1.96					

Note: * indicates a significant difference at $p < 0.05$, and ** indicates a highly significant difference at $p < 0.01$.

The optimal fermentation conditions, determined by combining single-factor experiments with RSM, were initial pH 5.46, fermentation temperature 36.89 °C, soluble solids content 11.92 °Bx, inoculum ratio of *Lcb. paracasei* ZY to *Lcb. rhamnosus* SN12 1:1, inoculum size 6 Log CFU/mL, and fermentation time 96 h. Under these conditions, the model estimated that the maximum GABA production would reach 0.78 g/L.

3.6. Effect of Optimal Fermentation Conditions on GABA Content in Orange Juice

To assess the reliability and practical usefulness of the RSM, fermentation experiments were conducted under the optimized conditions. These included initial pH 5.5, fermentation temperature 37 °C, soluble solids content 12.0 °Bx, inoculum ratio of *Lcb. paracasei* ZY to *Lcb. rhamnosus* SN12 1:1, inoculum size 6 Log CFU/mL, and fermentation time 96 h. Uninoculated orange juice was used as the control. The results showed that the GABA content in the fermented orange juice reached 0.89 g/L, representing an increase of 39.06% compared with the uninoculated orange juice (0.65 g/L GABA) (Figure 5).

https://doi.org/10.3390/foods15010071

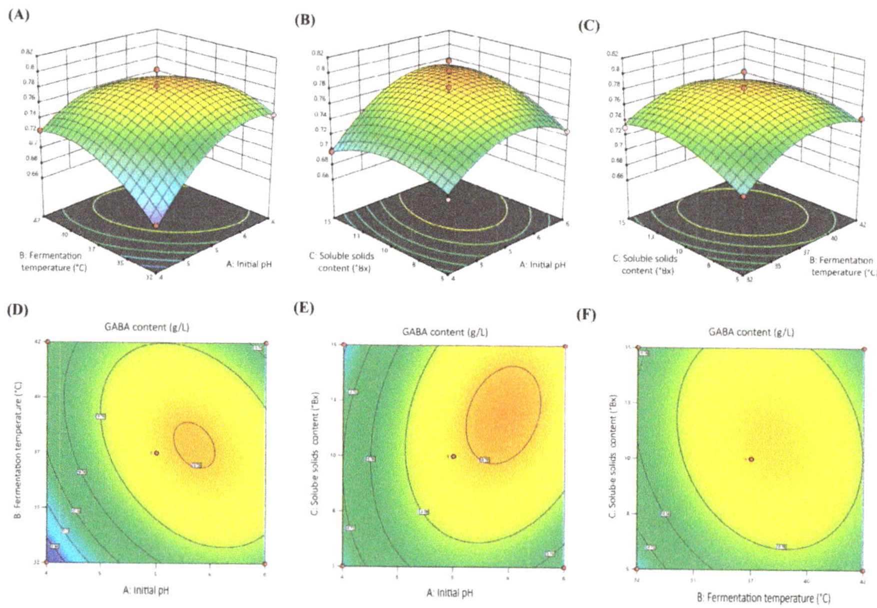

Figure 4. Response surface and contour plots showing interactive effects of initial pH and fermentation temperature (**A**,**D**), initial pH and soluble solids content (**B**,**E**), and fermentation temperature and soluble solids content (**C**,**F**) on GABA production. Red and pink dots in the response surface plots (**A**–**C**) denote observed values above and below the predicted surface, respectively; red dots in the contour plots (**D**–F) correspond to experimental design points. The color transition from blue to yellow indicates increasing GABA content.

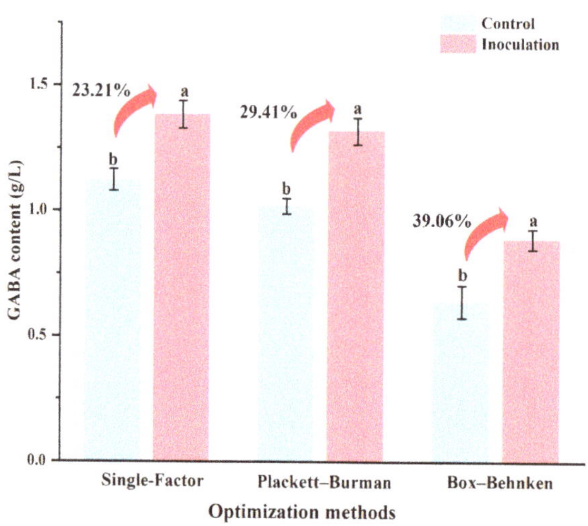

Figure 5. Enhancement of GABA content in fermented orange juice by different optimization methods. Different letters indicate significant differences before and after the same optimization methods ($p < 0.05$). "Control" represents orange juice without bacterial inoculation, while "inoculation" represents orange juice inoculated with LAB and subjected to optimized fermentation.

Different optimization methods had different effects on GABA production. As shown in Figure 5, the highest GABA content (1.38 g/L) was achieved at a soluble solids content of 10.0 °Bx in the single-factor experiments, representing a 23.21% increase compared to uninoculated orange juice (1.12 g/L). The Plackett–Burman design yielded a GABA content of 1.33 g/L, which corresponds to a 29.41% increase. Although the initial GABA content prior to Box–Behnken optimization was relatively low (0.65 g/L), this method achieved the most significant enhancement, with a 39.06% increase. Wang et al. [21] optimized the fermentation process of soymilk using *Lpb. plantarum* Lp 3, yielding a final GABA concentration of 0.55 g/L, which represents an approximately 20.00% improvement compared to the unfermented control. In another study, Jin et al. [19] fermented litchi juice with *L. brevis* LBJ-29 for 48 h and reported a 10-fold increase in GABA content relative to the unfermented counterpart. The high GABA yields observed in these studies may be attributed to the strains' robust metabolic activity and efficient substrate utilization. However, since *L. brevis* is not generally recognized as a food-grade microorganism, its application in the food industry remains limited. While *Lpb. plantarum* Lp 3, *Lcb. paracasei* ZY, and *Lcb. rhamnosus* SN12 exhibit a lower GABA conversion capacity than *L. brevis* LBJ-29, process optimization remains a viable and effective strategy to enhance GABA yield. While single-factor optimization facilitates the rapid identification of reasonable ranges and levels for individual factors, RSM is more effective than the former in examining the interactions between various factors.

4. Conclusions

This study increased the content of GABA in fermented orange juice by optimizing the fermentation process. A total of fifty-two LAB strains were screened for GABA production using the Berthelot colorimetric method, and forty-three strains showed GABA-producing ability. The top fifteen strains were further analyzed by HPLC. *Lcb. paracasei* ZY and *Lcb. rhamnosus* SN12 exhibited high and stable GABA-producing ability and were selected for orange juice fermentation. Single-factor experiments were conducted to evaluate the effects of initial pH, fermentation temperature, soluble solids content, inoculum ratio, inoculum size, and fermentation time on GABA production in fermented orange juice. Through Plackett–Burman design analysis, initial pH, fermentation temperature, and soluble solids content were identified as the significant factors affecting GABA accumulation. The further optimization of these three factors was performed by Box–Behnken designs. The final optimized fermentation conditions are as follows: initial pH of 5.5, fermentation temperature of 37 °C, soluble solids content of 12.0 °Bx, inoculum ratio of *Lcb. paracasei* ZY to *Lcb. rhamnosus* SN12 as 1:1, inoculum size of 6 Log CFU/mL, and fermentation time of 96 h. Under these conditions, the GABA content in fermented orange juice increased to 0.89 g/L, which was 39.06% higher than that in uninoculated orange juice. This research not only screened out excellent strains for the development of functional juices, but also provided process references and technical support for the industrial production of functional products by enterprises.

Author Contributions: S.Y. (Shufeng Yin): Writing—Original Draft, Writing—Review and Editing, Visualization. Y.W.: Methodology, Investigation; R.Z.: Validation, Investigation; N.Z.: Methodology, Visualization; H.L.: Visualization; Y.T.: Validation; N.Q.: Methodology, Resources; Y.D.: Conceptualization, Supervision, Project Administration, Writing—Review and Editing; X.L.: Conceptualization, Supervision, Project Administration, Writing—Review and Editing. All authors have read and agreed to the published version of the manuscript.

Funding: This research was funded by the National Key Research and Development Program of China (2023YFD2100304) and the High-Quality Innovation and Entrepreneurship Project of Dalian Polytechnic University (8250011002).

Institutional Review Board Statement: Not applicable.

Informed Consent Statement: Not applicable.

https://doi.org/10.3390/foods15010071

Data Availability Statement: The original contributions presented in the study are included in the article, further inquiries can be directed to the corresponding authors.

Conflicts of Interest: The authors declare no conflicts of interest.

Appendix A

Table A1. Information on the strains used in this study.

Number	Strain Name	Isolation Source	Acquisition Method
1	*Lacticaseibacillus paracasei* ZY	Fermented milk	Laboratory isolation
2	*Lacticaseibacillus rhamnosus* SN12	Fermented milk	Laboratory isolation
3	*Lactiplantibacillus plantarum* M35	Fermented meat	Laboratory isolation
4	*Lactiplantibacillus plantarum* 2-22-LJ	Fermented meat	Laboratory isolation
5	*Lactiplantibacillus plantarum* Z5-24	Fermented meat	Laboratory isolation
6	*Lacticaseibacillus paracasei* 3-4	Fermented meat	Laboratory isolation
7	*Lactiplantibacillus plantarum* M7	Fermented meat	Laboratory isolation
8	*Lacticaseibacillus paracasei* 3-9	Fermented meat	Laboratory isolation
9	*Lactiplantibacillus plantarum* LS3-1	Fermented pickles	Laboratory isolation
10	*Lactiplantibacillus plantarum* 3-2-LJ	Fermented meat	Laboratory isolation
11	*Limosilactobacillus reuteri* PC1	Fermented pickles	Laboratory isolation
12	*Lactiplantibacillus plantarum* HJZ1	Distiller's grains	Laboratory isolation
13	*Lacticaseibacillus paracasei* K22	Fermented milk	Laboratory isolation
14	*Lactiplantibacillus plantarum* Z5-25	Fermented meat	Laboratory isolation
15	*Lacticaseibacillus casei* K34	Fermented milk	Laboratory isolation
16	*Lacticaseibacillus paracasei* K38	Fermented milk	Laboratory isolation
17	*Pediococcus acidilactici* K2	Fermented meat	Laboratory isolation
18	*Lactiplantibacillus plantarum* 3-19-LJ	Fermented meat	Laboratory isolation
19	*Lacticaseibacillus paracasei* MAS-B2	Fermented milk	Laboratory isolation
20	*Pediococcus pentosaceus* 2-10-2-LJ	Fermented meat	Laboratory isolation
21	*Limosilactobacillus fermentum* CGMCC1.8	——	Obtained from CGMCC
22	*Lactiplantibacillus plantarum* J11	Fermented meat	Laboratory isolation
23	*Lacticaseibacillus rhamnosus* H7	Fermented meat	Laboratory isolation
24	*Lactococcus lactis subsp. lactis* MAS2-2	Fermented meat	Laboratory isolation
25	*Lactiplantibacillus plantarum* 1-26-LJ	Fermented meat	Laboratory isolation
26	*Lactiplantibacillus plantarum* N-M-1	Fermented milk	Laboratory isolation
27	*Latilactobacillus sakei* PC19	Fermented pickles	Laboratory isolation
28	*Lactiplantibacillus plantarum* PC14	Fermented pickles	Laboratory isolation
29	*Lactobacillus acidophilus* A20	Fermented meat	Laboratory isolation
30	*Lacticaseibacillus paracasei* Z3-11	Fermented meat	Laboratory isolation
31	*Lactiplantibacillus plantarum* 1-1	Fermented milk	Laboratory isolation
32	*Lacticaseibacillus rhamnosus* SN9	Fermented pickles	Laboratory isolation
33	*Streptococcus salivarius subsp.thermophilus* M17-11	Fermented milk	Laboratory isolation
34	*Lactiplantibacillus plantarum* 2-14-LJ	Fermented meat	Laboratory isolation
35	*Lactiplantibacillus plantarum* 1-11	Fermented milk	Laboratory isolation
36	*Pediococcus pentosaceus* HJZ2	Distiller's grains	Laboratory isolation
37	*Lactiplantibacillus plantarum* M19	Fermented meat	Laboratory isolation
38	*Streptococcus salivarius subsp. thermophilus* KF17-2	——	Obtained from CICC
39	*Latilactobacillus sakei* S13	Fermented milk	Laboratory isolation
40	*Lacticaseibacillus paracasei* Z3-9	Fermented meat	Laboratory isolation
41	*Lacticaseibacillus paracasei* K23	Fermented milk	Laboratory isolation
42	*Lactobacillus delbrueckii subsp.bulgaricus* 6103	——	Obtained from CICC
43	*Lacticaseibacillus rhamnosus* SN13	Fermented pickles	Laboratory isolation
44	*Lactobacillus helveticus* 6102	——	Obtained from CICC
45	*Lacticaseibacillus paracasei* Z3-4	Fermented meat	Laboratory isolation
46	*Limosilactobacillus fermentum* 2-34-LJ	Fermented meat	Laboratory isolation
47	*Lactobacillus acidophilus* LS2-3	Fermented pickles	Laboratory isolation

https://doi.org/10.3390/foods15010071

Table A1. *Cont.*

Number	Strain Name	Isolation Source	Acquisition Method
48	*Lactiplantibacillus plantarum* A72	Fermented pickles	Laboratory isolation
49	*Latilactobacillus sakei* PC2	Fermented pickles	Laboratory isolation
50	*Latilactobacillus sakei* PC3	Fermented pickles	Laboratory isolation
51	*Latilactobacillus sakei* PC17	Fermented pickles	Laboratory isolation
52	*Latilactobacillus sakei* PC30	Fermented pickles	Laboratory isolation

Note: CGMCC refers to the China General Microbiological Culture Collection Center; CICC refers to the China Industrial Microbial Culture Collection Center.

Table A2. Experimental factors and levels in the single-factor design.

Factors	Unit	Levels				
Initial pH	—	3.0	4.0	5.0	6.0	7.0
Fermentation temperature	°C	27	32	37	42	47
Soluble solids content	°Bx	5.0	10.0	15.0	20.0	25.0
Inoculum ratio	—	1:0	1:1	1:2	2:1	0:1
Inoculum size	Log CFU/mL	3	6	9	12	15
Fermentation time	h	48	72	96	108	120

Table A3. Experimental factors and coded levels in the Plackett–Burman design.

Symbols	Factors	Unit	Coded Levels	
A	Initial pH	—	5.0	7.0
B	Fermentation temperature	°C	37	47
C	Soluble solids content	°Bx	10.0	25.0
D	Inoculum ratio	—	1:0	1:1
E	Inoculum size	Log CFU/mL	6	15
F	Fermentation time	h	96	120

Table A4. Plackett–Burman design and experimental results.

Run	A	B	C	D	E	F	GABA Content (g/L)
1	1	−1	−1	−1	1	−1	1.18 ± 0.02
2	−1	1	−1	1	1	1	1.2 ± 0.05
3	−1	1	1	−1	1	−1	1.07 ± 0.10
4	1	−1	1	1	1	−1	1.14 ± 0.06
5	1	1	−1	−1	−1	1	1.06 ± 0.04
6	−1	−1	−1	1	1	1	1.29 ± 0.06
7	−1	−1	−1	−1	−1	−1	1.32 ± 0.10
8	−1	−1	1	−1	−1	1	1.06 ± 0.01
9	1	1	1	−1	1	1	0.92 ± 0.07
10	−1	1	1	1	−1	−1	1.13 ± 0.08
11	1	−1	1	1	−1	1	1.03 ± 0.10
12	1	1	−1	1	−1	−1	1.04 ± 0.14

Table A5. Experimental factors and coded levels in the Box–Behnken design.

Symbols	Factors	Unit	Coded Levels		
			−1	0	1
A	Initial pH	—	4.0	5.0	6.0
B	Fermentation temperature	°C	32	37	42
C	Soluble solids content	°Bx	5.0	10.0	15.0

https://doi.org/10.3390/foods15010071

Table A6. Box–Behnken design and experimental results.

Run	Initial pH	Fermentation Temperature (°C)	Soluble Solids Content (°Bx)	GABA Content (g/L)
1	−1	−1	0	0.67 ± 0.04
2	1	0	1	0.78 ± 0.12
3	−1	1	0	0.72 ± 0.08
4	0	0	0	0.80 ± 0.08
5	1	0	−1	0.73 ± 0.03
6	0	1	1	0.74 ± 0.05
7	1	1	0	0.73 ± 0.10
8	0	1	−1	0.75 ± 0.04
9	0	0	0	0.77 ± 0.05
10	0	0	0	0.76 ± 0.07
11	−1	0	1	0.70 ± 0.11
12	0	−1	−1	0.70 ± 0.13
13	0	0	0	0.78 ± 0.07
14	0	−1	1	0.73 ± 0.09
15	−1	0	−1	0.70 ± 0.06
16	0	0	0	0.77 ± 0.10
17	1	−1	0	0.75 ± 0.07

Appendix B

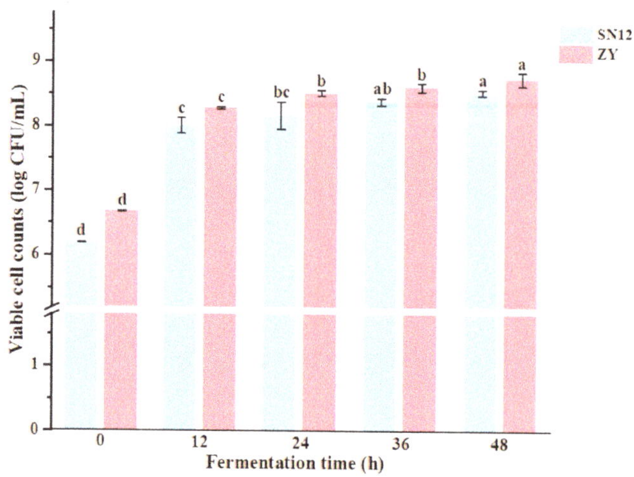

Figure A1. Viable cell counts of *Lcb. paracasei* ZY and *Lcb. rhamnosus* SN12 in orange juice during fermentation.

References

1. Varinthra, P.; Anwar, S.; Shih, S.C.; Liu, I.Y. The role of the GABAergic system on insomnia. *Tzu Chi Med. J.* **2024**, *36*, 103–109. [CrossRef]
2. Jiang, C.J.; Chen, Y.; Sun, T. From the gut to the brain, mechanisms and clinical applications of γ-aminobutyric acid (GABA) on the treatment of anxiety and insomnia. *Front. Neurosci.* **2025**, *19*, 1570173. [CrossRef] [PubMed]
3. Inoue, K.; Shirai, T.; Ochiai, H.; Kasao, M.; Hayakawa, K.; Kimura, M.; Sansawa, H. Blood-pressure-lowering effect of a novel fermented milk containing γ-aminobutyric acid (GABA) in mild hypertensives. *Eur. J. Clin. Nutr.* **2003**, *57*, 490–495. [CrossRef] [PubMed]
4. Dastgerdi, A.H.; Sharifi, M.; Soltani, N. GABA administration improves liver function and insulin resistance in offspring of type 2 diabetic rats. *Sci. Rep.* **2021**, *11*, 23155. [CrossRef] [PubMed]

https://doi.org/10.3390/foods15010071

5. Han, J.R.; Zhao, X.L.; Zhao, X.; Wang, Q.; Li, P.; Gu, Q. Microbial-Derived γ-Aminobutyric Acid: Synthesis, Purification, Physiological Function, and Applications. *J. Agric. Food Chem.* **2023**, *71*, 14931–14946. [CrossRef]
6. Pannerchelvan, S.; Rios-Solis, L.; Wong, F.W.F.; Zaidan, U.H.; Wasoh, H.; Mohamed, M.S.; Tan, J.S.; Mohamad, R.; Halim, M. Strategies for improvement of gamma-aminobutyric acid (GABA) biosynthesis via lactic acid bacteria (LAB) fermentation. *Food Funct.* **2023**, *14*, 3929–3948. [CrossRef]
7. Mancini, A.; Carafa, I.; Franciosi, E.; Nardin, T.; Bottari, B.; Larcher, R.; Tuohy, K.M. In vitro probiotic characterization of high GABA producing strain Lactobacilluas brevis DSM 32386 isolated from traditional "wild" Alpine cheese. *Ann. Microbiol.* **2019**, *69*, 1435–1443. [CrossRef]
8. Shan, Y.; Man, C.X.; Han, X.; Li, L.; Guo, Y.; Deng, Y.; Li, T.; Zhang, L.W.; Jiang, Y.J. Evaluation of improved γ-aminobutyric acid production in yogurt using Lactobacillus plantarum NDC75017. *J. Dairy Sci.* **2015**, *98*, 2138–2149. [CrossRef]
9. Lin, Q. Submerged fermentation of Lactobacillus rhamnosus YS9 for γ-aminobutyric acid (GABA) production. *Braz. J. Microbiol.* **2013**, *44*, 183–187. [CrossRef]
10. Komatsuzaki, N.; Shima, J.; Kawamoto, S.; Momose, H.; Kimura, T. Production of γ-aminobutyric acid (GABA) by Lactobacillus paracasei isolated from traditional fermented foods. *Food Microbiol.* **2005**, *22*, 497–504. [CrossRef]
11. Thuy, D.T.B.; An, N.T.; Jayasena, V.; Vandamme, P. A comprehensive investigation into the production of gamma-aminobutyric acid by Limosilactobacillus fermentum NG16, a tuna gut isolate. *Acta Aliment.* **2022**, *51*, 302–311. [CrossRef]
12. Miles, E.A.; Calder, P.C. Effects of Citrus Fruit Juices and Their Bioactive Components on Inflammation and Immunity: A Narrative Review. *Front. Immunol.* **2021**, *12*, 712608. [CrossRef]
13. Wu, Q.; Zou, S.B.; Yan, X.; Yue, Y.; Zhang, S.F.; Ji, C.F.; Chen, Y.X.; Dai, Y.W.; Lin, X.P. Lactiplantibacillus plantarum A72 inoculation enhances Citrus juice quality: A systematic analysis of the strain's fermentation properties, antioxidant capacities, and transcriptome pathways. *Food Biosci.* **2025**, *64*, 105878. [CrossRef]
14. Quan, Q.; Liu, W.; Guo, J.J.; Ye, M.L.; Zhang, J.H. Effect of Six Lactic Acid Bacteria Strains on Physicochemical Characteristics, Antioxidant Activities and Sensory Properties of Fermented Orange Juices. *Foods* **2022**, *11*, 1920. [CrossRef]
15. Huang, M.L.; Lai, C.L.; Liang, Y.; Xiong, Q.; Chen, C.Y.; Ju, Z.R.; Jiang, Y.M.; Zhang, J. Improving the functional components and biological activities of navel orange juice through fermentation with an autochthonous strain Lactiplantibacillus paraplantarum M23. *Food Bioprod. Process.* **2025**, *149*, 249–260. [CrossRef]
16. Wang, D.W.; Wang, Y.; Lan, H.B.; Wang, K.; Zhao, L.; Hu, Z.Y. Enhanced production of γ-aminobutyric acid in litchi juice fermented by Lactobacillus plantarum HU-C2W. *Food Biosci.* **2021**, *42*, 101155. [CrossRef]
17. Zheng, L.; Lu, X.L.; Yang, S.T.; Zou, Y.; Zeng, F.K.; Xiong, S.H.; Cao, Y.P.; Zhou, W. The anti-inflammatory activity of GABA-enriched Moringa oleifera leaves produced by fermentation with Lactobacillus plantarum LK-1. *Front. Nutr.* **2023**, *10*, 1093036. [CrossRef] [PubMed]
18. Kanklai, J.; Somwong, T.C.; Rungsirivanich, P.; Thongwai, N. Screening of GABA-Producing Lactic Acid Bacteria from Thai Fermented Foods and Probiotic Potential of *Levilactobacillus brevis* F064A for GABA-Fermented Mulberry Juice Production. *Microorganisms* **2021**, *9*, 33. [CrossRef]
19. Jin, Y.W.; Wu, J.Y.; Hu, D.; Li, J.; Zhu, W.W.; Yuan, L.X.; Chen, X.S.; Yao, J.M. Gamma-Aminobutyric Acid-Producing *Levilactobacillus brevis* Strains as Probiotics in Litchi Juice Fermentation. *Foods* **2023**, *12*, 302. [CrossRef]
20. Zhou, C.; Li, J.; Mao, K.M.; Gao, J.; Li, X.Y.; Zhi, T.X.; Sang, Y.X. Anti-hangover and anti-hypertensive effects in vitro of fermented persimmon juice. *CYTA-J. Food* **2019**, *17*, 960–966. [CrossRef]
21. Wang, X.J.; Wang, Y.; Nan, B.; Cao, Y.; Piao, C.H.; Li, X.; Wang, Y.H. Optimization of fermentation for gamma-aminobutyric acid (GABA) production by Lactiplantibacillus plantarum Lp3 and the development of fermented soymilk. *LWT-Food Sci. Technol.* **2024**, *195*, 115841. [CrossRef]
22. Zhang, Y.C.; Zhang, M.; Li, T.; Zhang, X.X.; Wang, L. Enhance Production of γ-Aminobutyric Acid (GABA) and Improve the Function of Fermented Quinoa by Cold Stress. *Foods* **2022**, *11*, 3908. [CrossRef]
23. Meena, K.K.; Taneja, N.K.; Jain, D.; Ojha, A.; Kumawat, D.; Mishra, V. In Vitro Assessment of Probiotic and Technological Properties of Lactic Acid Bacteria Isolated from Indigenously Fermented Cereal-Based Food Products. *Fermentation* **2022**, *8*, 529. [CrossRef]
24. Huang, Y.; Adams, M.C. In vitro assessment of the upper gastrointestinal tolerance of potential probiotic dairy propionibacteria. *Int. J. Food Microbiol.* **2004**, *91*, 253–260. [CrossRef]
25. Ghashghaei, T.; Soudi, M.R.; Hoseinkhani, S. Optimization of Xanthan Gum Production from Grape Juice Concentrate Using Plackett-Burman Design and Response Surface Methodology. *Appl. Food Biotechnol.* **2016**, *3*, 15–23.
26. Sharma, D.; Sharma, N. Optimization of fermentation of himalayan black tea beverage using response surface methodology. *LWT-Food Sci. Technol.* **2025**, *229*, 118107. [CrossRef]
27. Le, P.H.; Verscheure, L.; Le, T.T.; Verheust, Y.; Raes, K. Implementation of HPLC Analysis for γ-Aminobutyric Acid (GABA) in Fermented Food Matrices. *Food Anal. Methods* **2020**, *13*, 1190–1201. [CrossRef]

28. Hofmann, A.F. The continuing importance of bile acids in liver and intestinal disease. *Arch. Intern. Med.* **1999**, *159*, 2647–2658. [CrossRef] [PubMed]
29. Cui, Y.H.; Miao, K.; Niyaphorn, S.; Qu, X.J. Production of Gamma-Aminobutyric Acid from Lactic Acid Bacteria: A Systematic Review. *Int. J. Mol. Sci.* **2020**, *21*, 995. [CrossRef] [PubMed]
30. Rayavarapu, B.; Tallapragada, P.; Usha, M.S. Optimization and comparison of γ-aminobutyric acid (GABA) production by LAB in soymilk using RSM and ANN models. *Beni-Suef Univ. J. Basic Appl. Sci.* **2021**, *10*, 14. [CrossRef]
31. Nguyen, H.; Perrier, J.M.; Gervais, P. Effect of the osmotic conditions during sporulation on the subsequent resistance of bacterial spores. *Appl. Microbiol. Biotechnol.* **2008**, *80*, 107–114. [CrossRef] [PubMed]
32. Kook, M.C.; Cho, S.C. Production of GABA (gamma amino butyric acid) by Lactic Acid Bacteria. *Korean J. Food Sci. Anim. Resour.* **2013**, *33*, 377–389. [CrossRef]
33. Zhao, L.; Maimaitiyiming, R.; Hong, J.Y.; Wang, L.; Mu, Y.; Liu, B.Z.; Zhang, H.M.; Chen, K.P.; Aihaiti, A. Optimization of tomato (*Solanum lycopersicum* L.) juice fermentation process and analysis of its metabolites during fermentation. *Front. Nutr.* **2024**, *11*, 1344117, Corrigendum in *Front. Nutr.* **2024**, *11*, 1423909. https://doi.org/10.3389/fnut.2024.1423909. [CrossRef] [PubMed]

https://doi.org/10.3390/foods15010071

foods

MDPI

Article

Two-Dimensional GC–ToFMS Analysis of Volatile Organic Compounds in Fermented Camel Milk (Shubat)

Sagyman Zhadyra [1,2], Fei Tao [1,*] and Ping Xu [1]

[1] State Key Laboratory of Microbial Metabolism, School of Life Sciences and Biotechnology, Shanghai Jiao Tong University, Shanghai 200240, China; sagyman@alumni.sjtu.edu.cn (S.Z.); pingxu@sjtu.edu.cn (P.X.)

[2] Laboratory of Biotechnology, Research Institute for Biotechnology and Ecology, Zhetysu University, Taldykorgan 040009, Kazakhstan

* Correspondence: taofei@sjtu.edu.cn

Abstract: Shubat, a traditional fermented camel milk from Kazakhstan, is renowned for its unique flavor and nutritional properties, though its volatile compound profile remains poorly characterized. In this study, headspace solid-phase microextraction coupled with comprehensive two-dimensional gas chromatography–time-of-flight mass spectrometry (HS-SPME-GC×GC–ToFMS) was employed to qualitatively identify and semi-quantitatively analyze volatile metabolites in seven Shubat samples collected from four regions of Kazakhstan. Of the 372 volatile organic compounds initially detected, 202 were retained after screening, predominantly comprising esters, acids, alcohols, ketones, and aldehydes. Esters, acids, and alcohol were found to be the most abundant categories. Diversity analyses (α and β) revealed substantial variation across regions, likely influenced by Shubat's rich and region-specific microbiome. An UpSet analysis demonstrated that 75 volatile compounds were shared among all samples, accounting for over 87% of the total volatile content, indicating a chemically stable core. These findings underscore the chemical complexity of Shubat and provide novel insights into its metabolite composition, thereby establishing a foundation for future sensory, microbial, and quality-related research.

Keywords: Shubat fermented camel milk; volatile organic compounds; GC×GC–TOF-MS; HS-SPME; metabolomics

1. Introduction

Shubat is traditionally produced by fermenting fresh raw camel milk with a small amount of previously soured milk starter culture (~50 mL/L) in goat skin bags or wooden containers for 1–2 days [1]. During fermentation, the product is continuously or intermittently stirred to ensure sufficient aeration and uniformity. Owing to its long history of consumption among Kazakh nomads, Shubat is widely recognized for its high nutritional value and health-promoting properties. Shubat shares similarities with other fermented camel milk products, which are increasingly recognized for their functional properties [2,3]. For instance, fermented camel milk has been reported to possess anti-inflammatory, antidiabetic, lipid-lowering, and antimicrobial activities, attributed to its unique bioactive components, including organic acids and peptides. Moreover, short-chain fatty acids (SCFAs) such as acetic, propionic, and butyric acids, commonly produced through microbial fermentation, are known to play key roles in gut health, the modulation of immune function, and intestinal barrier integrity [4]. These properties suggest that the volatile organic compounds detected in Shubat may contribute to its health benefits and support its potential as a functional beverage.

Metabolism plays a central role in shaping the phenotypic characteristics of all living organisms, where low-molecular-weight biochemical compounds (metabolites) serve as intermediate and end products of enzymatically catalyzed reactions within cells [5]. These metabolites, often modulated by genetic and environmental factors, provide critical insights into cellular functions and physiological states. Following the emergence of genomics, transcriptomics, and proteomics, "metabolomics" has recently become an essential high throughput "omics" platform that enables the systematic study of the metabolic profile of organisms. It is particularly valuable for revealing biochemical mechanisms underlying food composition, fermentation, and flavor development [6,7].

In the context of fermented dairy products, microbial diversity and its associated metabolic pathways have a profound impact on final flavor, texture, and quality [8]. The characteristic flavor of fermented milk products is primarily attributed to volatile organic compounds (VOCs), such as esters, alcohols, acids, aldehydes, and ketones, many of which are generated through microbial fermentation processes [9]. Given their sensory importance, the identification and quantification of VOCs in complex food matrices has become a central focus in food science, as VOCs are often responsible for desirable aromas and product differentiation across cultures and regions.

Various analytical strategies have been developed for VOC profiling, among which headspace solid-phase microextraction (HS-SPME) has proven to be an efficient, solvent-free, and highly sensitive sample preparation method. First introduced by Pawliszyn et al. in 1989 [10], HS-SPME relies on the partitioning of analytes between a coated fiber and the sample matrix under equilibrium conditions. It can be coupled with gas chromatography (GC) for the separation and detection of volatile compounds. Fiber selection, extraction conditions, and analyte properties significantly affect the efficiency of this technique [11].

However, due to the complexity and diversity of food metabolites, conventional one-dimensional (1D) GC-MS methods often suffer from co-elution and limited resolution, restricting comprehensive VOC characterization. To overcome these challenges, the comprehensive two-dimensional gas chromatography–time-of-flight mass spectrometry (GC×GC–ToFMS) technique has emerged as a powerful tool, offering improved separation efficiency, structured chromatographic space, enhanced sensitivity, and the ability to capture low-abundance compounds that contribute significantly to flavor perception [12,13]. This technique has shown excellent performance in analyzing VOCs in complex matrices such as environmental samples, oils, and fermented foods [14].

From a molecular perspective, fermented dairy products, such as Shubat, represent highly dynamic chemical systems composed of proteins, lipids, amino acids, organic acids, carbohydrates, and aromatic volatiles [15]. Accurately profiling the volatile metabolome of Shubat requires integrative analytical platforms capable of capturing this diversity. GC×GC–ToFMS, when combined with optimized sample preparation such as HS-SPME, enables the high-resolution detection of a wide spectrum of volatile compounds, providing a comprehensive metabolic fingerprint of Shubat.

Recent research, including our previous metagenomic analysis of Shubat [16], has demonstrated that the rich microbial diversity in fermented camel milk plays a vital role in modulating its flavor-associated metabolic profile. These microbial communities, composed of diverse lactic acid bacteria (LAB) and yeast, participate in the biosynthesis of esters, alcohols, and acids that define the sensory identity of the product. Yet, the volatile metabolomic composition of Shubat remains underexplored.

Therefore, in this study, we applied headspace solid-phase microextraction coupled with comprehensive two-dimensional gas chromatography–time-of-flight mass spectrometry (HS-SPME-GC×GC–ToFMS) to comprehensively profile and classify volatile organic compounds (VOCs) in Shubat samples collected from different locations in Kazakhstan.

This high-resolution technique offers enhanced separation efficiency and sensitivity compared to conventional GC–MS, enabling the accurate characterization of trace-level compounds in complex dairy matrices. The aim of this pilot study was to characterize the major classes of VOCs present in Shubat and to explore compositional differences among samples from distinct collection sites.

2. Materials and Methods

2.1. Sample Collection

Seven fermented camel milk (Shubat) samples were obtained from rural households located across four provinces of Kazakhstan: Almaty (S1, S2), Kyzylorda (S3), Zhetisu (S4, S7), and Turkestan (S5, S6). Each representative sample was formed by pooling three individual subsamples, which were separately sourced from different families residing in the same village. Although certain samples were sourced from the same administrative region, they were derived from geographically isolated villages, thus maintaining the independence of the samples.

After collection, the subsamples were carefully homogenized, and 50 mL of each pooled sample was transferred into sterile polypropylene tubes using aseptic c techniques. The samples were then pre-cooled, packed in insulated containers with ice packs, and immediately transported to the laboratory, where they were stored at −20 °C until subsequent analysis.

2.2. Volatile Metabolite Analysis

2.2.1. Headspace Solid-Phase Microextraction (HS-SPME)

Frozen Shubat samples were thawed at room temperature, and 5 mL of the samples were transferred into a 15 mL glass vial. The vial was sealed with a PTFE/silicone septum and magnetic screw cap (Agilent Technologies, Santa Clara, CA, USA), followed by equilibration at 55 °C for 30 min. A 1 cm DVB/CAR/PDMS fiber (Supelco) (Merck KGaA, Darmstadt, Germany) was then inserted into the vial headspace for extraction. The sample was pre-incubated at 55 °C for 10 min, followed by extraction for 50 min. After extraction, the fiber was desorbed in the GC injection port for 5 min, and then thermally cleaned at 290 °C for 10 min to avoid cross-contamination.

2.2.2. GC×GC–ToFMS Analysis

Volatile compounds were analyzed using a GC×GC–ToFMS system, consisting of an Agilent 7890 gas chromatograph (Agilent Technologies, Santa Clara, CA, USA) coupled with a Pegasus 4D time-of-flight mass spectrometer (LECO Corporation, St. Joseph, MI, USA). The analysis parameters were as follows: Primary column: DB-WAX (30 m × 250 μm × 0.25 μm; Agilent Technologies, Santa Clara, CA, USA). Secondary column: DB-17MS (2 m × 100 μm × 0.10 μm; Agilent Technologies, Santa Clara, CA, USA), Injection temperature: 250 °C. Carrier gas: Helium (99.99%) at 1.0 mL/min (constant flow, splitless mode). Oven temperature program: 40 °C for 5 min; ramped at 5 °C/min to 140 °C (held 2 min); then 10 °C/min to 250 °C (held 12 min). Secondary column oven: 15 °C higher than the primary column. Modulator temperature: 15 °C higher than the secondary column. Modulation period: 5.0 s. Transfer line temperature: 270 °C. Ion source temperature: 230 °C. Ionization mode: Electron impact (EI), 70 eV. Detector voltage: 1670 V. Mass range: m/z 33–550. Acquisition rate: 50 spectra/s.

2.3. Data Preprocessing, Compound Identification, and Visualization

Raw chromatographic data were processed using LECO ChromaTOF® software (version 4.72.0.0), which performed automated baseline correction, peak deconvolution, alignment, and noise reduction. Primary and secondary dimension peak widths were set to 16 s and 0.2 s, respectively.

To eliminate non-sample-related signals, procedural blanks were analyzed under identical conditions, and all peaks detected in blanks were excluded from the dataset. Retention indices (RIs) were determined using a C5–C40 n-alkane standard mixture and calculated via the Kovats index method. Mass spectral identification was conducted using the NIST 17 Mass Spectral Library, and only peaks with forward and reverse match scores above 700 were retained [17]. The calculated RI values were compared to literature data (e.g., NIST Chemistry WebBook), and compounds with deviations exceeding ±5% were excluded. Identification confidence corresponds to Level 2 of the Metabolomics Standards Initiative (MSI), indicating putative annotation based on spectral and RI matching without validation by authentic standards. To ensure data robustness, peaks containing halogen or silicon atoms were removed to avoid potential artifacts from column bleed or contamination. Additionally, low-frequency compounds, defined as those detected in fewer than four of the seven Shubat samples, were excluded prior to statistical analysis.

Volatile compound abundance matrices were generated after data preprocessing, which included removing contaminants, compounds detected in blank samples, and non-organic compounds. Peak areas of each compound were converted to relative abundances within each sample. To further focus on the major contributing compounds and enhance the interpretability of the results, the normalized dataset was ranked based on the total abundance of each compound across all samples. The top 50 most abundant volatile compounds were selected for subsequent statistical analysis and visualized in tabular form as percentage values relative to the total volatile profile.

To investigate the overlap and uniqueness of volatile compounds across different Shubat samples, a presence/absence matrix was constructed based on the qualitative identification results. In this binary matrix, the presence of a compound in each sample was encoded as 1 and absence was encoded as 0. The binary data were then analyzed using the upsetplot (v0.8.0). An UpSet plot was generated to visualize the number of shared and unique compounds among the seven Shubat samples.

Alpha diversity indices were calculated using Python (v3.12) using the pandas and scikit-bio packages. Beta diversity was assessed using Bray–Curtis dissimilarity based on relative abundance data, and principal coordinate analysis (PCoA) was performed using scikit-bio. Data visualization, including heatmaps and PCoA plots, was carried out using matplotlib and seaborn. Due to logistical and field constraints, only one representative Shubat sample was collected from each village across four regions. Replicates or composite samples were not available for each region. Therefore, the diversity analyses presented in this study are exploratory in nature and do not allow for the statistical validation of intra-regional variability.

3. Results

3.1. Major Volatile Compound Categories in Shubat Samples

To characterize the VOCs, present in the seven Shubat samples we performed qualitative identification and relative quantification using headspace solid-phase microextraction coupled with GC×GC–TOF-MS. A total of 372 volatile organic compounds (VOCs) were detected in Shubat samples collected from four regions of Kazakhstan, of which 202 were retained for compositional analysis. These VOCs were primarily classified into six major categories: esters, acids, alcohols, ketones, aldehydes, and aromatic compounds. The relative abundance of each category varied across the seven samples (Figure 1), with detailed 3D chromatograms for individual samples provided in Supplementary Figure S1. Esters and acids were generally the dominant compound classes, though the proportion of alcohol and aldehydes showed notable variation among samples. Aromatic compounds were present in lower abundance but showed marked enrichment in S4 and S6.

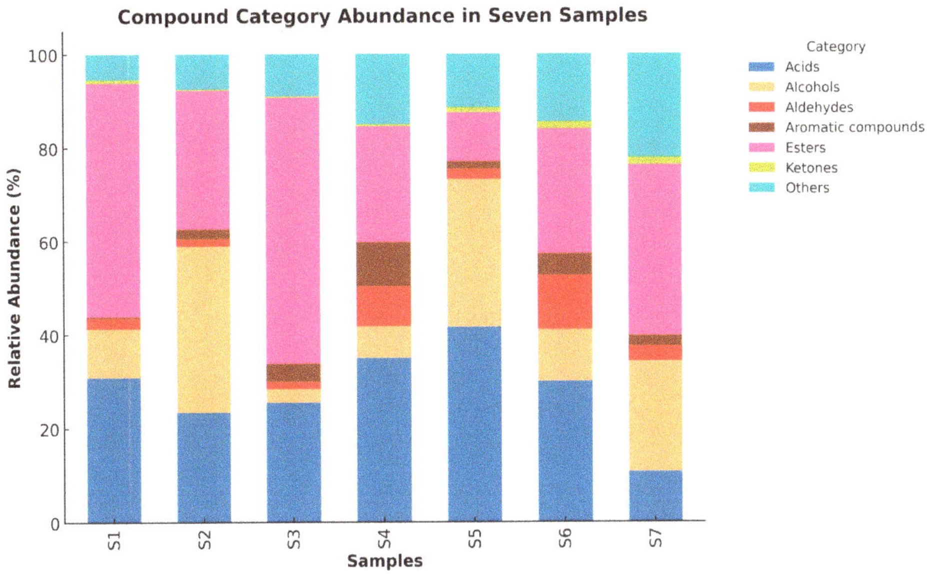

Figure 1. Stacked bar plot illustrating the relative abundance of major volatile compound categories across seven Shubat samples (S1–S7).

Overall, the compositional differences in VOC categories across samples highlight the influence of regional origin, household fermentation practices, and microbiota composition on the volatile metabolome of Shubat.

A total of 50 volatile organic compounds (VOCs) with the highest average relative abundance were identified across the seven Shubat samples and are listed in Table 1. The detailed information for each sample is provided in Supplementary Table S1 (S1_top50–S7_top50), which includes retention indices, CAS numbers, and spectral match scores. Most of these VOCs belonged to esters, organic acids, and alcohols. Among them, acetic acid and ethyl acetate were the most abundant overall, with mean relative abundances of 19.60% and 15.27%, respectively. Acetic acid showed particularly high levels in samples S4, S5, and S6, whereas ethyl acetate was dominant in S1 and S6.

Ethanol, another major fermentation product, exhibited large variation among samples, with concentrations ranging from 0.00% to 36.28%. Medium-chain fatty acid ethyl esters such as ethyl decanoate, ethyl dodecanoate, and ethyl octanoate were also prominent, with the highest levels observed in samples S2 and S3. These esters are typically associated with fruity and creamy flavor characteristics in fermented dairy products.

Several volatile organic acids, including hexanoic acid, octanoic acid, and dodecanoic acid, were consistently detected across all samples. These acids contribute sour, pungent, or goaty notes depending on concentration and matrix interaction. Additionally, aromatic compounds such as benzaldehyde and 2-phenylethyl acetate were identified in moderate abundance, with the latter reaching 14.05% in sample S7, indicating sample-specific enrichment.

Table 1. Top 50 Most Abundant Volatile Compounds Detected in Shubat Samples (Based on Total Relative Abundance).

Compounds	Relative Abundance (%)						
	S1	S2	S3	S4	S5	S6	S7
Acetic acid	24.27	14.01	15.13	25.62	31.17	27.01	0.01
Ethanol	9.27	36.28	0.00	0.00	26.02	0.60	14.77
Ethyl Acetate	37.87	2.52	3.99	12.52	7.00	22.73	20.26
Decanoic acid, ethyl ester	0.45	6.52	32.85	1.64	0.17	0.59	0.50
Hexanoic acid	3.46	4.48	1.25	4.94	6.24	3.10	5.05
Dodecanoic acid, ethyl ester	3.42	10.09	2.47	1.91	0.25	0.81	0.73
Benzaldehyde	0.17	1.91	3.53	9.56	1.31	4.72	1.80
1-Butanol, 3-methyl-	2.02	1.40	1.29	7.99	0.73	7.42	2.12
Acetic acid, 2-phenylethyl ester	1.92	1.49	2.63	1.16	0.94	1.51	14.05
Octanoic acid, ethyl ester	2.48	3.47	3.01	4.47	0.29	0.84	0.79
ETHYL (S)-(-)-LACTATE	0.00	0.00	0.00	6.31	1.37	5.42	10.07
Phenylethyl Alcohol	0.88	0.51	2.19	4.64	1.27	3.49	4.01
Tetradecanoic acid, ethyl ester	1.45	1.55	2.65	2.29	0.36	0.79	0.76
Pentadecanoic acid, ethyl ester	1.15	1.62	4.26	0.13	0.00	0.00	0.00
Octanoic acid	0.92	1.04	0.49	1.85	2.08	1.16	1.64
Dodecanoic acid	1.00	1.25	2.02	1.47	0.27	0.19	0.27
Ethyl tridecanoate	0.76	1.79	2.76	0.09	0.00	0.00	0.00
Tetradecanoic acid	1.09	0.58	2.34	1.41	0.28	0.19	0.28
1-Hexanol	0.14	0.00	0.00	0.26	0.93	3.08	4.58
Hexadecanoic acid, ethyl ester	1.45	0.16	2.59	0.70	0.00	0.11	0.00
1,3,5,7-Cyclooctatetraene	0.00	1.57	2.38	0.00	0.06	0.07	0.00
n-Decanoic acid	1.18	0.60	1.01	0.95	0.41	0.29	0.36
Butanoic acid	0.25	0.88	1.58	0.39	0.79	0.24	0.44
2-Nonenal, (E)-	0.03	0.02	0.04	0.40	1.87	2.52	1.71
Hexanal	0.18	0.06	0.08	0.78	0.58	3.52	0.26
Furan, 2-ethyl-	0.04	0.02	0.05	0.30	2.43	0.88	1.83
Ethyl 9-decenoate	0.40	0.80	1.47	0.56	0.02	0.05	0.04
Heptanoic acid	0.07	0.08	0.15	0.14	2.21	0.57	1.36
Hexanoic acid, ethyl ester	0.39	0.21	1.04	1.74	0.04	0.18	0.15
(E)-9-Octadecenoic acid ethyl ester	0.54	0.77	1.07	0.23	0.00	0.07	0.03
Benzoic acid	0.09	0.62	1.00	0.87	0.12	0.09	0.13
Acetoin	0.02	0.01	0.01	0.12	1.76	0.18	2.10
Butanoic acid, ethyl ester	0.12	0.64	1.25	0.25	0.00	0.05	0.05
Furan, 2-pentyl-	0.06	0.04	0.04	0.29	0.93	1.12	1.36
Nonanal	0.05	0.04	0.06	0.32	1.04	0.79	0.92
E-11-Hexadecenoic acid, ethyl ester	0.08	1.01	0.37	0.02	0.03	0.14	0.13
Propanoic acid, 2-methyl-	0.26	0.04	0.07	0.04	1.54	0.13	0.54
1-Heptanol	0.00	0.00	0.01	0.14	0.45	0.98	1.79
n-Hexadecanoic acid	0.16	0.48	0.65	0.35	0.08	0.07	0.10
9-Decenoic acid	0.17	0.36	0.62	0.69	0.06	0.04	0.05
Benzene	0.14	0.15	0.27	0.45	0.38	0.47	0.64
Ethyl 9-hexadecenoate	0.75	0.10	0.17	0.72	0.03	0.13	0.11
2-Decenal, (Z)-	0.04	0.02	0.04	0.18	0.84	0.77	0.90
Butanoic acid, 3-methyl-	0.22	0.04	0.00	0.11	0.85	0.08	1.31
1-Penten-3-one	0.01	0.00	0.01	0.10	1.63	0.22	0.23
1-Propanol, 2-methyl-	0.23	0.16	0.05	0.05	0.09	1.08	0.47
Z-7-Tetradecenoic acid	0.17	0.37	0.60	0.20	0.00	0.00	0.00
2-Octenal, (E)-	0.02	0.01	0.02	0.31	0.49	0.96	0.31
2-Undecenal	0.04	0.03	0.05	0.14	0.55	0.34	0.75
1-Propanol	0.11	0.18	0.41	0.19	0.03	0.20	0.24

3.2. *Core and Unique Volatile Compounds Across Shubat Samples*

To explore the overlap and uniqueness of volatile compounds among the seven Shubat samples, an UpSet plot was constructed. As shown in Figure 2, a total of 75 compounds were found to be shared by all seven samples. Although these shared compounds only account for a small fraction of all detected compounds, their abundance accounts for about

87% of the total abundance of the overall volatile components. This indicates that a core set of volatile metabolites dominates the flavor composition of Shubat, suggesting that despite differences in origin, there is a high degree of consistency between samples at the metabolic level.

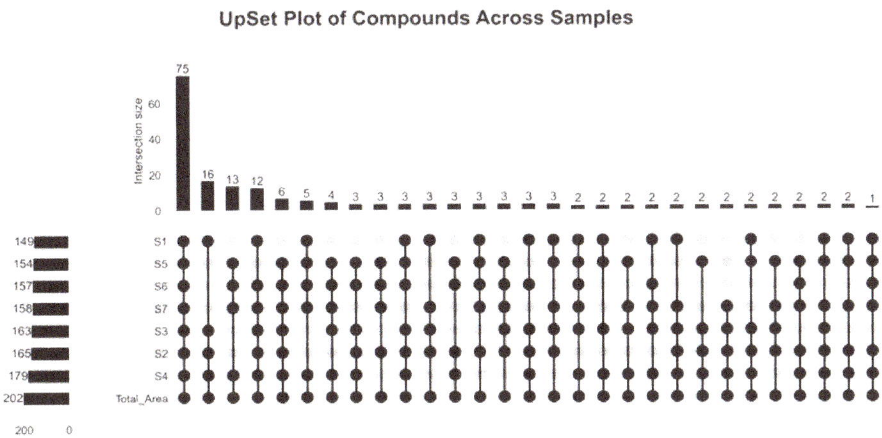

Figure 2. Visualization of Core and Sample-Specific Volatile Compounds in Shubat.

3.3. Alpha and Beta Diversity Analysis of Volatile Compounds

The diversity of volatile organic compounds (VOCs) in the seven Shubat samples was evaluated using three alpha diversity indices: richness, Shannon, and Simpson. As shown in Figure 3a, richness ranged from 149 (S1) to 179 (S4), indicating moderate variation in the number of detected VOCs across samples. The Shannon index values ranged from 2.43 to 3.29, with the highest diversity observed in S7 and the lowest in S1, suggesting differences in both the number and evenness of VOCs. Simpson indices were relatively stable across the samples, ranging from 0.80 to 0.92, indicating a relatively even distribution of compounds with no single compound overwhelmingly dominant in any sample.

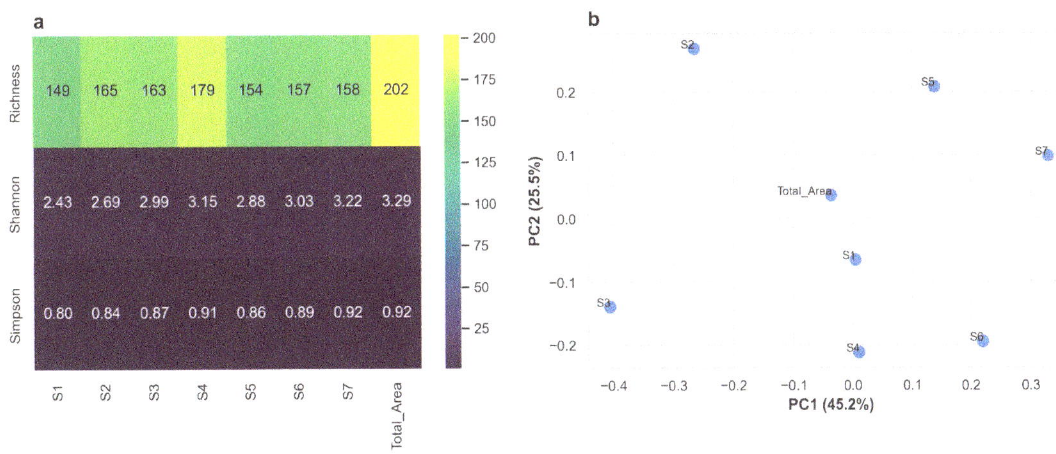

Figure 3. Diversity Analysis of Volatile Compounds in Shubat Samples. (**a**) Alpha diversity indices (Richness, Shannon, and Simpson) of volatile compounds across seven samples. (**b**) Beta diversity analysis based on Bray–Curtis dissimilarity visualized by PCoA.

Beta diversity was assessed using Principal Component Analysis (PCA) based on VOC composition. The first two principal components explained 45.2% and 25.5% of the total variance, respectively (Figure 3b). The samples were clearly separated into the PCA space, reflecting substantial variation in volatile profiles among the different Shubat samples. No clear clustering pattern was observed in the PCA plot, indicating high inter-sample variability in volatile compound composition.

4. Discussion

In this study, we applied HS-SPME–GC×GC–TOF-MS to profile volatile organic compounds (VOCs) in Shubat samples, detecting a total of 372 compounds, which significantly exceeds the counts typically reported using conventional one-dimensional GC–MS methods [3,18]. This improved coverage reflects the enhanced separation performance and greater sensitivity afforded by the GC×GC platform.

Despite advances in deconvolution algorithms for one-dimensional GC–MS, co-elution remains a major challenge in complex matrices such as fermented dairy. In contrast, comprehensive two-dimensional GC delivers markedly improved peak capacity and selectivity, enabling better resolution of co-eluting compounds across two chromatographic dimensions (volatility and polarity) [19–21]. The modulation process in GC×GC refocuses analytes into sharp pulses, leading to substantially improved signal-to-noise ratios (S/N) and facilitates the detection of low-abundance metabolites that often go undetected in 1D GC–MS [22]. For Shubat, a highly complex fermented dairy matrix containing diverse proteins, lipids, and microbial metabolites, the use of GC×GC–TOFMS was particularly advantageous. This approach enabled the resolution of trace, aroma-relevant VOCs that would likely remain masked in conventional 1D analysis, thereby yielding a more reliable metabolic fingerprint specific to Shubat fermentation.Moreover, this study emphasizes the contribution of Shubat's rich microbiome to the observed diversity of volatile compounds. According to our recent metagenomic investigation of Shubat [16], the product harbors a diverse and metabolically versatile microbial community, particularly LAB such as *Lactobacillus*, *Lactococcus*, and *Streptococcus*. These microbes are involved in amino acid metabolism, carbohydrate utilization, and secondary metabolite biosynthesis, which may collectively contribute to the unique and complex volatile profiles observed in Shubat. This microbial influence provides an important biochemical basis for Shubat's distinctive flavor characteristics and potential functional properties.

Compared to previous findings, the volatile compound profile obtained in this study shares both commonalities and differences. For instance, Ning et al. [18] identified esters, alcohols, and acids as the dominant volatile classes in fermented milk products, which aligns with our results, where esters and acids exhibited the highest relative abundance. Similarly, Bilal et al. [3] reported the presence of key volatiles such as ethyl acetate, acetic acid, and 2,3-butanediol in traditional fermented dairy products compounds that were also found at notably high levels in the Shubat samples examined here.

In the present study, the volatile profiles of Shubat were predominantly composed of acids, esters, and alcohols, followed by aldehydes, aromatic compounds, ketones, and minor volatile classes. Such a composition aligns with previous findings that fermented dairy products, including camel milk derivatives, typically contain these compound classes as key aroma contributors [23,24]. Notably, the dominance of esters in Shubat underscores their important role in flavor development; their formation via microbial esterification and lipid metabolism during dairy fermentation is well documented [25].

The relative stability of acids across samples suggests that they form a consistent chemical backbone for Shubat's aroma, whereas variation in esters and alcohols likely reflects microbial metabolic differences linked to fermentation conditions or raw milk characteristics. Lower-abundance volatiles such as aldehydes and ketones, though present in smaller quantities, may still impart distinctive sensory nuances. This pattern echoes broader reports in fermented dairy research where compound diversity contributes to flavor complexity [26].

Esters are among the most prevalent volatile constituents in dairy and fermented milk products [27,28]. At low concentrations, these compounds contribute pleasant fruity aromas to milk and its derivatives [29]. However, elevated levels of esters may result in undesirable fruity off-flavors in dairy products [30]. In the present study, over 90 distinct ester compounds were identified, representing the most abundant chemical class among all detected volatiles. It is hypothesized that the biosynthesis of ethyl acetate and ethyl lactate may primarily originate from the esterification of microbial fermentation metabolites, lactic acid and hexanoic acid, with ethanol. These short-chain ethyl esters are considered favorable contributors to the fruity flavor observed in Shubat samples [31,32].

Two major biochemical pathways, esterification and alcoholysis, are proposed to play key roles in the formation of ester compounds in dairy matrices. Esterification refers to the formation of esters from the reaction of free fatty acids (FFAs) (released during the lipolysis of triglycerides by lipases or esterases) with alcohols. Alcoholysis, in contrast, involves the transesterification of fatty acyl moieties from mono-, di-, or triacylglycerols or fatty acyl-CoA derivatives directly to alcohols generated through amino acid or carbohydrate metabolism. Both pathways are enzymatically catalyzed by esterases or lipases [33]. LAB has been shown to exhibit both esterase and lipase activities, contributing significantly to the biosynthesis of ethyl esters in fermented dairy products. Several studies have reported esterase activity in LAB strains [34–36]. These enzymes not only catalyze the hydrolysis of milk fat triglycerides to release FFAs but also participate in transesterification reactions, leading to the synthesis of esters from glycerides and alcohols [37].

One of the most abundant acids, acetic acid, may originate from the heterofermentation of glucose, a hydrolysis product of lactose. In addition, high-abundance acetic acid bacteria are capable of oxidizing ethanol to produce acetic acid. FFAs with carbon chains of four or more atoms are often derived from the lipolysis of milk fat, which is a key factor influencing the flavour profile of fermented dairy products [38]. Triglycerides, comprising a glycerol backbone esterified with three fatty acid chains, represent the primary form of fat in milk. FFAs are fatty acids not bound to glycerol molecules, and they play a central role in the flavour development of dairy products, particularly cheeses and fermented milk [39].

Free fatty acids (FFAs) in dairy systems can be released enzymatically by lipases and esterases, which are either endogenous to milk or produced by microorganisms during fermentation. However, acid generation is not limited to lipid breakdown. Streptococcus spp. can convert lactic acid into formate, acetaldehyde, ethanol, and acetic acid [32,40,41], whereas *Propionibacterium* spp. utilize lactate to form propionic acid, acetic acid, and CO_2. Additionally, certain lactic acid bacteria (LAB) strains can synthesize volatile fatty acids through amino acid catabolism [33,42]. These organic acids not only significantly influence the sensory properties of fermented dairy products but also serve as essential precursors for the formation of other volatiles, such as esters, ketones, and alcohols.

Alcohols represented the fourth most abundant category among the volatile constituents identified in Shubat. The most prominent alcohol detected was ethanol, which consistently showed the highest relative abundance across all samples. Ethanol is a well-known volatile compound commonly found in fermented dairy products, typically produced as a byproduct of microbial metabolism, especially by yeasts and certain lactic acid bacteria [31,33]. Its presence is indicative of active fermentation processes and contributes to the characteristic aroma of fermented milk.

Alcohols such as phenylethyl alcohol (2-phenylethanol), 1-butanol, 3-methyl-, and other higher alcohols contribute floral and green sensory nuances, enhancing the aromatic complexity of Shubat. 2-Phenylethanol, present at relatively high levels, is well known for its floral and sweet aroma notes and makes an important contribution to the sensory complexity of Shubat. Aromatic alcohol 2-phenylethanol, with a distinct rose-like fragrance [43], is widely used in the food, cosmetic, and fragrance industries [44]. Microbial synthesis is considered an important route for producing natural 2-phenylethanol. Various microorganisms such as *Aspergillus niger*, *Saccharomyces vini*, *Torulopsis utilis*, *Kluyveromyces marxianus*, and *Microbacterium* spp. have been reported to generate 2-phenylethanol during fermentation [45–47]. In our previous study [16], metagenomic sequencing revealed a high relative abundance of *Kluyveromyces marxianus* in the Shubat samples, which may partially explain the presence of 2-phenylethanol.

Other identified alcohols such as 1-hexanol are also commonly observed volatile constituents in dairy products [24,48]. These alcohols can arise through several biochemical pathways, including the reduction of ketones or aldehydes, and the microbial catabolism of amino acids, lactose, lipids, and citric acid [31,49].

Diversity analysis revealed distinct regional differences in the volatile metabolite profiles of Shubat. Both α- and β-diversity metrics indicated considerable chemical variation among the seven samples, suggesting strong region-specific influences. However, due to the absence of replicate or composite samples within each region, intra-regional variation cannot be statistically assessed. As a result, diversity comparisons should be regarded as preliminary, and future studies incorporating biological replicates are needed to validate these findings.

To assess beta diversity, we employed Principal Coordinates Analysis (PCoA) based on Bray–Curtis dissimilarity derived from relative volatile metabolite abundances. PCoA is particularly advantageous for exploratory metabolomics studies with limited or non-existent replicates, as it does not assume data normality or homogeneity of variance and is capable of handling non-Euclidean distance measures. This method is widely used for visualizing compositional dissimilarities among samples in metabolomics and microbial profiling (e.g., Bray–Curtis PCoA) [50,51]. Importantly, studies have demonstrated that β-diversity metrics such as Bray–Curtis are highly sensitive in detecting differences, even with small sample sizes, whereas α-diversity metrics may not reveal these distinctions under the same conditions [52]. In this context, PCoA provided a robust, informative view of regional variation in Shubat volatile profiles, despite the lack of replication.

5. Conclusions

This study presents the first in-depth profiling of volatile organic compounds in traditionally fermented Shubat using HS-SPME-GC×GC–ToFMS. After rigorous data processing, contaminant removal, and retention index filtering, 202 high-confidence compounds were retained from an initial 372. Esters, acids, and alcohol were the most abundant classes, reflecting typical fermentation-derived metabolic activity.

Alpha diversity analyses revealed substantial chemical complexity across the samples, and β-diversity assessments indicated regional variation in volatile profiles, likely shaped by distinct household fermentation practices and microbial communities. Despite this heterogeneity, 75 core volatile compounds were shared across all samples and accounted for over 87% of the total volatile abundance, suggesting a chemically conserved backbone associated with traditional Shubat fermentation.

Several key metabolites such as acetic acid, ethanol, and ethyl acetate were abundant across most samples, underscoring their critical role in shaping sourness, aroma, and effervescence. The detection of medium-chain ethyl esters and higher alcohols further suggests active microbial esterification and amino acid metabolism during fermentation. These compounds are central to Shubat's distinctive flavor and may serve as potential markers for quality and authenticity.

Limitations include the relatively small number of samples and the lack of direct sensory or functional validation. Future studies should combine chemical, microbial, and sensory data to better link volatile profiles with flavor and health-related properties.

Supplementary Materials: The following supporting information can be downloaded at: https://www.mdpi.com/article/10.3390/foods14172995/s1, Figure S1: Representative 3D chromatograms of volatile compounds detected in Shubat samples S1–S7 using GC×GC–ToFMS. Table S1: Top 50 volatile organic compounds (VOCs) across all samples (S1–S7). Data are provided in an Excel file containing seven sheets (S1_top50–S7_top50), where each sheet corresponds to one Shubat sample.

Author Contributions: Conceptualization, P.X.; data curation, F.T.; formal analysis, S.Z. and F.T.; investigation, S.Z.; methodology, F.T.; software, S.Z.; supervision, F.T. and P.X.; validation, F.T. and P.X.; visualization, S.Z.; writing—original draft, S.Z.; writing—review and editing, S.Z., F.T. and P.X. All authors have read and agreed to the published version of the manuscript.

Funding: The article processing charge (APC) was funded by the Science Committee of the Ministry of Science and Higher Education of the Republic of Kazakhstan, grant number AP25793809.

Institutional Review Board Statement: Not applicable.

Informed Consent Statement: Not applicable.

Data Availability Statement: The original contributions presented in this study are included in the article/Supplementary Material, further inquiries can be directed to the corresponding author.

Acknowledgments: The first author wishes to express gratitude to Shanghai Jiao Tong University for their invaluable support. Additionally, he is deeply grateful to his advisors, Ping Xu and Fei Tao, for their invaluable guidance and support throughout this study.

Conflicts of Interest: The authors declare no conflict of interest.

References

1. Zhadyra, S.; Han, X.; Anapiyayev, B.B.; Tao, F.; Xu, P. Bacterial Diversity Analysis in Kazakh Fermented Milks Shubat and Ayran by Combining Culture-Dependent and Culture-Independent Methods. *LWT Food Sci. Technol.* **2021**, *141*, 110877. [CrossRef]
2. Sepehr, A.; Miri, S.T.; Aghamohammad, S.; Rahimirad, N.; Milani, M.; Pourshafie, M.-R.; Rohani, M. Health Benefits, Antimicrobial Activities, and Potential Applications of Probiotics: A Review. *Medicine* **2024**, *103*, e32412. [CrossRef]
3. Bilal, Z.; Kondybayev, A.; Ospanova, A.; Tormo, H.; Akhmetsadykova, S.; Amutova, F.; Faye, B.; Konuspayeva, G. Volatile Organic Compounds of Camel Milk and Shubat across Kazakhstan's Regions, Seasons, and Breeds. *Heliyon* **2024**, *10*, e35365. [CrossRef] [PubMed]
4. Ragavan, M.L.; Hemalatha, S. The Functional Roles of Short Chain Fatty Acids as Postbiotics in Human Gut: Future Perspectives. *Food Sci. Biotechnol.* **2024**, *33*, 275–285. [CrossRef] [PubMed]
5. Malviya, J.; Alameri, A.A.; Al-Janabi, S.S.; Fawzi, O.F.; Azzawi, A.L.; Obaid, R.F.; Alsudani, A.A.; Alkhayyat, A.S.; Gupta, J.; Mustafa, Y.F. Metabolomic Profiling of Bacterial Biofilm: Trends, Challenges, and an Emerging Antibiofilm Target. *World J. Microbiol. Biotechnol.* **2023**, *39*, 212. [CrossRef]
6. Viant, M.R.; Sommer, U. Mass Spectrometry Based Environmental Metabolomics: A Primer and Review. *Metabolomics* **2013**, *9*, 144–158. [CrossRef]
7. Fraga-Corral, M.; Carpena, M.; Garcia-Oliveira, P.; Pereira, A.G.; Prieto, M.A.; Simal-Gandara, J. Analytical Metabolomics and Applications in Health, Environmental and Food Science. *Crit. Rev. Anal. Chem.* **2022**, *52*, 712–734. [CrossRef] [PubMed]
8. An, F.; Wu, J.; Feng, Y.; Pan, G.; Ma, Y.; Jiang, J.; Yang, X.; Xue, R.; Wu, R.; Zhao, M. A Systematic Review on the Flavor of Soy-based Fermented Foods: Core Fermentation Microbiome, Multisensory Flavor Substances, Key Enzymes, and Metabolic Pathways. *Compr. Rev. Food Sci. Food Saf.* **2023**, *22*, 2773–2801. [CrossRef]
9. Lin, J.; Hua, B.; Xu, Z.; Li, S.; Ma, C. The Impact of Proteolytic Pork Hydrolysate on Microbial, Flavor and Free Amino Acids Compounds of Yogurt. *Korean J. Food Sci. Anim. Resour.* **2016**, *36*, 558. [CrossRef]

10. Belardi, R.P.; Pawliszyn, J.B. The Application of Chemically Modified Fused Silica Fibers in the Extraction of Organics from Water Matrix Samples and Their Rapid Transfer to Capillary Columns. *Water Qual. Res. J.* **1989**, *24*, 179–191. [CrossRef]
11. Starowicz, M. Analysis of Volatiles in Food Products. *Separations* **2021**, *8*, 157. [CrossRef]
12. Zhang, Y.; Cui, Z.; Li, J.; Wei, M.; Wang, Y.; Jiang, W.; Fang, Y.; Sun, X.; Ge, Q. Aroma Identification and Traceability of the Core Sub-Producing Area in the Helan Mountain Eastern Foothills Using Two-Dimensional Gas Chromatography and Time-of-Flight Mass Spectrometry and Chemometrics. *Foods* **2024**, *13*, 3644. [CrossRef] [PubMed]
13. Scarlett, A.G.; Despaigne-Diaz, A.I.; Wilde, S.A.; Grice, K. An Examination by GC×GC-TOFMS of Organic Molecules Present in Highly Degraded Oils Emerging from Caribbean Terrestrial Seeps of Cretaceous Age. *Geosci. Front.* **2019**, *10*, 5–15. [CrossRef]
14. Feng, Y.; Liu, S.; Fang, Y.; Li, J.; Ma, M.; Yang, Z.; Shang, L.; Guo, X.; Hua, R.; Sun, D. Volatile Flavor of Tricholoma Matsutake from the Different Regions of China by Using GC×GC-TOF MS. *Foods* **2025**, *14*, 1824. [CrossRef]
15. Zhang, K.; Zhang, T.-T.; Guo, R.-R.; Ye, Q.; Zhao, H.-L.; Huang, X.-H. The Regulation of Key Flavor of Traditional Fermented Food by Microbial Metabolism: A Review. *Food Chem. X* **2023**, *19*, 100871. [CrossRef]
16. Zhadyra, S.; Tao, F.; Xu, P. Exploring the Microbiome and Functional Metabolism of Fermented Camel Milk (Shubat) Using Metagenomics. *Foods* **2025**, *14*, 1102. [CrossRef]
17. Shao, D.; Cheng, W.; Jiang, C.; Pan, T.; Li, N.; Li, M.; Li, R.; Lan, W.; Du, X. Machine-Learning-Assisted Aroma Profile Prediction in Five Different Quality Grades of Nongxiangxing Baijiu Fermented During Summer Using Sensory Evaluation Combined with GC×GC–TOF-MS. *Foods* **2025**, *14*, 1714. [CrossRef]
18. Li, N.; Zheng, F.-P.; Chen, H.-T.; Liu, S.-Y.; Gu, C.; Song, Z.-Y.; Sun, B.-G. Identification of Volatile Components in Chinese Sinkiang Fermented Camel Milk Using SAFE, SDE, and HS-SPME-GC/MS. *Food Chem.* **2011**, *129*, 1242–1252. [CrossRef]
19. Amaral, M.S.S.; Marriott, P.J. The Blossoming of Technology for the Analysis of Complex Aroma Bouquets—A Review on Flavour and Odorant Multidimensional and Comprehensive Gas Chromatography Applications. *Molecules* **2019**, *24*, 2080. [CrossRef]
20. Szeitz, A.; Sutton, A.G.; Hallam, S.J. A Matrix-Centered View of Mass Spectrometry Platform Innovation for Volatilome Research. *Front. Mol. Biosci.* **2024**, *11*, 1421330. [CrossRef]
21. Fisher, K.; Perrault, K. *Detangling the Complex Web of GC×GC Method Development to Support New Users*; LCGC International: Iselin, NJ, USA, 2024.
22. Dallüge, J.; Beens, J.; Udo, A. Comprehensive Two-Dimensional Gas Chromatography: A Powerful and Versatile Analytical Tool. *J. Chromatogr. A* **2003**, *1000*, 69–108. [CrossRef]
23. Kondybayev, A.; Zhakupbekova, A.A.; Amutova, F.B.; Omarova, A.S.; Nurseitova, M.; Akhmetsadykova, S.; Akhmetsadykov, N.; Konuspayeva, G.; Faye, B. Volatile Organic Compounds Profiles in Milk Fermented by Lactic Bacteria. *Int. J. Biol. Chem.* **2018**, *11*, 57. [CrossRef]
24. Dan, T.; Wang, D.; Wu, S.; Jin, R.; Ren, W.; Sun, T. Profiles of Volatile Flavor Compounds in Milk Fermented with Different Proportional Combinations of *Lactobacillus delbrueckii* Subsp. *Bulgaricus and Streptococcus thermophilus*. *Molecules* **2017**, *22*, 1633.
25. Li, N.; Huang, G.; Zhang, Y.; Zheng, N.; Zhao, S.; Wang, J. Diversity of Volatile Compounds in Raw Milk with Different N-6 to n-3 Fatty Acid Ratio. *Animals* **2022**, *12*, 252. [CrossRef]
26. Dan, T.; Ren, W.; Liu, Y.; Tian, J.; Chen, H.; Li, T.; Liu, W. Volatile Flavor Compounds Profile and Fermentation Characteristics of Milk Fermented by *Lactobacillus delbrueckii* Subsp. *Bulgaricus*. *Front. Microbiol.* **2019**, *10*, 2183. [CrossRef]
27. Mukdsi, M.C.A.; Medina, R.B.; Alvarez, M.d.F.; González, S.N. Ester Synthesis by Lactic Acid Bacteria Isolated from Goat's and Ewe's Milk and Cheeses. *Food Chem.* **2009**, *117*, 241–247. [CrossRef]
28. Dan, T.; Jin, R.; Ren, W.; Li, T.; Chen, H.; Sun, T. Characteristics of Milk Fermented by Streptococcus Thermophilus MGA45-4 and the Profiles of Associated Volatile Compounds during Fermentation and Storage. *Molecules* **2018**, *23*, 878. [CrossRef]
29. Friedrich, J.E.; Acree, T.E. Gas Chromatography Olfactometry (GC/O) of Dairy Products. *Int. Dairy J.* **1998**, *8*, 235–241. [CrossRef]
30. Bills, D.D.; Day, E.A. Determination of the Major Free Fatty Acids of Cheddar Cheese. *J. Dairy Sci.* **1964**, *47*, 733–738. [CrossRef]
31. Cheng, H. Volatile Flavor Compounds in Yogurt: A Review. *Crit. Rev. Food Sci. Nutr.* **2010**, *50*, 938–950. [CrossRef]
32. Cadwallader, K.R.; Singh, T.K. Flavours and Off-Flavours in Milk and Dairy Products. In *Advanced Dairy Chemistry: Volume 3: Lactose, Water, Salts and Minor Constituents*; Springer: New York, NY, USA, 2009; pp. 631–690.
33. Aghlara, A.; Mustafa, S.; Manap, Y.A.; Mohamad, R. Characterization of Headspace Volatile Flavor Compounds Formed during Kefir Production: Application of Solid Phase Microextraction. *Int. J. Food Prop.* **2009**, *12*, 808–818. [CrossRef]
34. Fenster, K.M.; Rankin, S.A.; Steele, J.L. Accumulation of Short N-Chain Ethyl Esters by Esterases of Lactic Acid Bacteria under Conditions Simulating Ripening Parmesan Cheese. *J. Dairy Sci.* **2003**, *86*, 2818–2825. [CrossRef] [PubMed]
35. Gobbetti, M.; Fox, P.F.; Smacchi, E.; Stepaniak, L.; Damiani, P. Purification and Characterization of a Lipase from Lactobacillus Plantarum 2739. *J. Food Biochem.* **1996**, *20*, 227–246. [CrossRef]
36. Holland, R.; Liu, S.-Q.; Crow, V.L.; Delabre, M.-L.; Lubbers, M.; Bennett, M.; Norris, G. Esterases of Lactic Acid Bacteria and Cheese Flavour: Milk Fat Hydrolysis, Alcoholysis and Esterification. *Int. Dairy J.* **2005**, *15*, 711–718. [CrossRef]
37. Tsakalidou, E.; Kalantzopoulos, G. Purification and Partial Characterization of an Esterase from *Lactococcus lactis* Ssp *lactis* Strain ACA-DC 127. *Le Lait* **1992**, *72*, 533–543. [CrossRef]

38. Liu, S.-Q.; Holland, R.; Crow, V.L. Purification and Properties of Intracellular Esterases from Streptococcus Thermophilus. *Int. Dairy J.* **2001**, *11*, 27–35. [CrossRef]
39. Chen, C.; Zhao, S.; Hao, G.; Yu, H.; Tian, H.; Zhao, G. Role of Lactic Acid Bacteria on the Yogurt Flavour: A Review. *Int. J. Food Prop.* **2017**, *20*, S316–S330. [CrossRef]
40. Liu, G.; Qiao, Y.; Zhang, Y.; Leng, C.; Chen, H.; Sun, J.; Fan, X.; Li, A.; Feng, Z. Metabolic Profiles of Carbohydrates in Streptococcus Thermophilus during PH-Controlled Batch Fermentation. *Front. Microbiol.* **2020**, *11*, 1131. [CrossRef]
41. Price, C.E.; Zeyniyev, A.; Kuipers, O.P.; Kok, J. From Meadows to Milk to Mucosa–Adaptation of Streptococcus and Lactococcus Species to Their Nutritional Environments. *FEMS Microbiol. Rev.* **2012**, *36*, 949–971. [CrossRef]
42. Thierry, A.; Valence, F.; Deutsch, S.M.; Even, S.; Falentin, H.; Le Loir, Y.; Jan, G.; Gagnaire, V. Strain-to-Strain Differences within Lactic and Propionic Acid Bacteria Species Strongly Impact the Properties of Cheese—A Review. *Dairy Sci. Technol.* **2015**, *95*, 895–918. [CrossRef]
43. Etschmann, M.; Bluemke, W.; Sell, D.; Schrader, J. Biotechnological Production of 2-Phenylethanol. *Appl. Microbiol. Biotechnol.* **2002**, *59*, 1–8. [CrossRef] [PubMed]
44. Chung Jr, H.; Lee, S.L.; Chou, C.C. Production and Molar Yield of 2-Phenylethanol by Pichia Fermentans L-5 as Affected by Some Medium Components. *J. Biosci. Bioeng.* **2000**, *90*, 142–147. [CrossRef] [PubMed]
45. Jollivet, N.; Bézenger, M.-C.; Vayssier, Y.; Belin, J.-M. Production of Volatile Compounds in Liquid Cultures by Six Strains of Coryneform Bacteria. *Appl. Microbiol. Biotechnol.* **1992**, *36*, 790–794. [CrossRef]
46. Lomascolo, A.; Lesage-Meessen, L.; Haon, M.; Navarro, D.; Antona, C.; Faulds, C.; Marcel, A. Evaluation of the Potential of Aspergillus Niger Species for the Bioconversion of L-Phenylalanine into 2-Phenylethanol. *World J. Microbiol. Biotechnol.* **2001**, *17*, 99–102. [CrossRef]
47. Alonso-Vargas, M.; Téllez-Jurado, A.; Gómez-Aldapa, C.A.; Ramírez-Vargas, M.d.R.; Conde-Báez, L.; Castro-Rosas, J.; Cadena-Ramírez, A. Optimization of 2-Phenylethanol Production from Sweet Whey Fermentation Using Kluyveromyces Marxianus. *Fermentation* **2022**, *8*, 39. [CrossRef]
48. Ianni, A.; Bennato, F.; Martino, C.; Grotta, L.; Martino, G. Volatile Flavor Compounds in Cheese as Affected by Ruminant Diet. *Molecules* **2020**, *25*, 461. [CrossRef]
49. Beale, D.J.; Pinu, F.R.; Kouremenos, K.A.; Poojary, M.M.; Narayana, V.K.; Boughton, B.A.; Kanojia, K.; Dayalan, S.; Jones, O.A.H.; Dias, D.A. Review of Recent Developments in GC–MS Approaches to Metabolomics-Based Research. *Metabolomics* **2018**, *14*, 1–31. [CrossRef]
50. Quinn, R.A.; Navas-Molina, J.A.; Hyde, E.R.; Song, S.J.; Vázquez-Baeza, Y.; Humphrey, G.; Gaffney, J.; Minich, J.J.; Melnik, A.V.; Herschend, J. From Sample to Multi-Omics Conclusions in under 48 Hours. *mSystems* **2016**, *1*, e00038-16. [CrossRef]
51. Ceccarani, C.; Foschi, C.; Parolin, C.; D'Antuono, A.; Gaspari, V.; Consolandi, C.; Laghi, L.; Camboni, T.; Vitali, B.; Severgnini, M. Diversity of Vaginal Microbiome and Metabolome during Genital Infections. *Sci. Rep.* **2019**, *9*, 14095. [CrossRef]
52. Kers, J.G.; Saccenti, E. The Power of Microbiome Studies: Some Considerations on Which Alpha and Beta Metrics to Use and How to Report Results. *Front. Microbiol.* **2022**, *12*, 796025. [CrossRef]

foods

MDPI

Article

Evaluation of Nutritional Quality and Oxidation Stability of Fermented Edible Insects

Anja Vehar [1,2], Doris Potočnik [1], Marjeta Mencin [1], Mojca Korošec [3], Blaž Ferjančič [3], Marta Jagodic Hudobivnik [1], Polona Jamnik [3,†], Ajda Ota [4], Lenka Kouřimská [5], Martin Kulma [6], David John Heath [1] and Nives Ogrinc [1,2,*]

[1] Department of Environmental Sciences, Jožef Stefan Institute, 1000 Ljubljana, Slovenia; anja.vehar@ijs.si (A.V.); doris.potocnik@ijs.si (D.P.); marjeta.mencin@ijs.si (M.M.); marta.jagodic@ijs.si (M.J.H.); david.heath@ijs.si (D.J.H.)
[2] Jožef Stefan International Postgraduate School, 1000 Ljubljana, Slovenia
[3] Department of Food Science and Technology, Biotechnical Faculty, University of Ljubljana, 1000 Ljubljana, Slovenia; mojca.korosec@bf.uni-lj.si (M.K.); blaz.ferjancic@bf.uni-lj.si (B.F.); polona.jamnik@bf.uni-lj.si (P.J.)
[4] Emona Nutrition Research & Development Department, Jata Emona, 1000 Ljubljana, Slovenia; ajda.ota@e-rcp.si
[5] Department of Microbiology, Nutrition and Dietetics, Czech University of Life Sciences, 165 00 Praha-Suchdol, Czech Republic; kourimska@af.czu.cz
[6] Department of Zoology and Fisheries, Czech University of Life Sciences, 165 00 Praha-Suchdol, Czech Republic; kulma@af.czu.cz
* Correspondence: nives.ogrinc@ijs.si
† Deceased.

Abstract: Fermentation, a traditional method for enhancing nutritional value and functionality, has significant potential for improving the quality, safety and acceptability of farmed insect products. In this study, yellow mealworm, house cricket and migratory locust were fermented using *Lactobacillus plantarum* and a commercial starter culture for 48 h. Samples were analyzed for proximate composition, amino and fatty acid profiles, elemental composition and oxidation stability. Fermentation reduced total dietary fiber in yellow mealworm (33%) and house cricket (12%), and increased non-protein nitrogen (38% and 16%), while total and protein nitrogen remained unaffected. Fatty acid profiles also remained unchanged, whereas the amino acid composition varied depending on the species and fermentation culture. Essential mineral concentrations varied depending on species and fermentation culture Fe (19–23%), K (25%), Mg (12–23%), Mn (36–378%), Na (20–49%) and P (22%) increased, levels of Se (15%), and Cu (16%) decreased, while Zn levels showed inconsistent trends among treatments. Oxidation stability of yellow mealworm (41–42%) and migratory locust (21–29%) decreased, but improved for house cricket (153–167%). Overall, fermentation enhanced the nutritional value of edible insects, although the extent of improvement varied by species and fermentation culture.

Keywords: edible insects; fermentation; nutritional value; amino acids; fatty acids; elemental composition; oxidation stability

1. Introduction

Edible insects represent a diverse and nutrient-rich source of macro- and micronutrients, increasingly recognized as a sustainable alternative to traditional animal-based protein [1]. Compared to conventional livestock, farmed insects have a significantly lower environmental footprint. For example, they require 75–95% less water and 80–90% less land. They also emit up to 80% fewer greenhouse gases while providing high-quality protein with a complete amino acid profile, essential fatty acids, and key micronutrients such as copper, iron, magnesium, zinc, and B vitamins [2]. In addition, insect farming is adaptable to vertical systems, generates minimal noise, and produces almost no waste, as insect frass can be used as an organic fertilizer [3,4]. Moreover, utilizing agricultural by-products and food waste as insect feed can further reduce production costs, addressing a major barrier to market entry [5–9], while contributing to food waste reduction in regions where such practices are allowed under local regulations [10–14].

The European Commission also recognizes the potential of insects as sustainable sources of food and feed by approving several species as novel foods under various regulations. Approved species include yellow mealworm (*Tenebrio molitor*; Commission Regulation 2021/882, 2022/169, 2025/89) [15–17], migratory locust (*Locusta migratoria*; Commission Regulation 2021/1975) [18], house cricket (*Acheta domesticus*; Commission Regulations 2022/188, 2023/5) [19,20], and lesser mealworm (*Alphitobius diaperinus*; Commission Regulation 2023/58) [21]. Furthermore, Commission Regulation 2021/1372 authorizes eight insect species for feed in poultry, pig, and fish farming [22]. These include tropical house cricket (*Gryllodes sigillatus*), black soldier fly (*Hermetia illucens*), Jamaican field cricket (*Gryllus assimilis*), house fly (*Musca domestica*), and silkworm moth (*Bombyx mori*), in addition to the aforementioned species.

Although insects have a long history of widespread consumption in several countries [23], they are met with significant skepticism, safety concerns, and unfamiliarity in Western societies [24,25]. One of the main factors influencing consumer perception is related to product characteristics, such as taste, price, texture, appearance, smell, the degree of insect visibility, and the proportion of insect-based ingredients [24]. However, these characteristics can be changed and improved using differing processing techniques, such as blanching, boiling, curing, drying, fermentation, freeze-drying, deep frying, marination, roasting, smoking and stewing, which can also reduce microbial loads and extend insect product shelf life [8,26–29].

Among these methods, fermentation is a widely recognized technique for improving the nutritional, functional, and sensory qualities of various food products [30]. The process involves spontaneous or controlled (starter culture-assisted) microbial activity that breaks down large molecules into smaller ones through various oxidation–reduction reactions. It can serve multiple purposes, including (i) enhancing sensory qualities such as flavor, aroma, and texture; (ii) increasing nutritional value by improving nutrient bioavailability; (iii) eliminating anti-nutritional factors; (iv) increasing levels of bioactive compounds, such as amino acids, polyphenols, organic acids, (v) promoting intestinal health by incorporating beneficial microorganism; and (vi) extending shelf life and reducing microbiological risks by inhibiting pathogenic microorganisms [30–33].

Recent studies have begun to explore the potential of fermentation for improving the nutritional and functional properties of edible insects, particularly species such as yellow mealworm [34–37], black soldier fly [38–41], house cricket [42,43], desert locust [36] and Mexican grasshopper [37]. Various microbial cultures, primarily lactic acid bacteria, *Bacillus* spp., and *Aspergillus* spp., have been used, with reported improvements in protein digestibility, amino acid profiles, fatty acid composition, antioxidant activity, shelf life and microbiological safety.

Changes in proximate composition and nutrient metabolism appear consistently across species, although the underlying mechanisms differ. For example, lactic acid fermentation of black soldier fly larvae and related products has been shown to modify proximate composition, amino acid and fatty acid profiles, reduce chitin content, and show antimicrobial activity against pathogens [38]. Also, similar studies using *Bacillus subtilis*, *Aspergillus niger*, *Lactobacillus crispatus* and *Pichia kudriavzevii*, report shifts in carbon, amino acid, and fatty acid metabolism as well as changes in bioactive components and microbial community structure [39–41]. In house cricket, *Lactobacillus plantarum* fermentation has been reported to modify proximate composition, organic acids, alcohols, fatty and amino acid profiles, volatile compounds, mineral content, and the composition of the microbiome [43,44]. Other approaches, including the use of methods that mimic Thai fermented shrimp paste, have highlighted effects on peptide content, mineral composition, amino acids, and in vitro protein digestibility [42]. In yellow mealworm, fermentation with lactic acid bacteria, commercial meat starter cultures, and *Aspergillus oryzae* has been reported to modify proximate composition, amino acid and fatty acid profiles, antioxidant activity, and microbiome composition [44], along with microbial counts, sugar consumption, organic acid production, and sensory properties [26,34–37].

While these studies demonstrate the potential of fermentation to improve the nutritional and functional quality of edible insects, variations in insect species, fermentation conditions, and analytical approaches limit direct cross-study comparisons. Moreover, there remains a substantial knowledge gap regarding how different fermentation cultures and conditions influence the nutritional, functional, and safety characteristics of various insect species.

The present study investigates the effects of fermentation using *Lactobacillus plantarum* and a commercial starter culture on three farmed edible insect species: yellow mealworm, migratory locust, and house cricket, in terms of proximate composition, amino acid and fatty acid profiles, elemental composition, and oxidation stability. This work addresses an existing knowledge gap, as data on yellow mealworm and house cricket under such fermentation conditions remain limited, and, to the authors' knowledge, no studies on fermented migratory locust have yet been reported.

2. Materials and Methods

2.1. Chemicals and Reagents

Dichloromethane (CH_2Cl_2), methanol (MeOH), and hexane (C_6H_{14}) for organic residue analysis were sourced from J. T. Baker B.V. (Deventer, The Netherlands). Sodium hydroxide (NaOH) (puriss. p.a., ACS reagent, ≥98%) and 14% boron trifluoride in methanol were obtained from Sigma-Aldrich (St. Louis, MO, USA). Trichloroacetic acid (≥99.5%) was obtained from Merck, Germany. Ultrapure water (18.2 MΩ) was prepared using the Milli-Q System (Merck Millipore, Watertown, MA, USA). A Supelco 37 Component FAME Mix in dichloromethane, containing methyl esters of 37 fatty acids, was obtained from Supelco (Bellefonte, PA, USA). Amino acid standards (≥98%) used for analysis included L-valine, L-leucine, L-isoleucine, L-threonine, L-serine, L-proline, L-methionine, L-glutamic acid, L-phenylalanine, L-lysine, L-histidine, and L-tyrosine, sourced from Sigma Aldrich. Standards for alanine, glycine, and aspartic acid were purchased from Fluorochem (Derbyshire, UK). The internal standard, Stable Isotope Labelled Amino Acid Mix Solution 1, containing labelled variants of 17 amino acids, was obtained from Supelco (Buchs, Switzerland). Hydrochloric acid (37%) was sourced from VWR Chemicals (Fontenay-sous-Bois, France), while thioglycolic acid (≥99%, France) and phenol (≥99%, USA) were purchased from Sigma Aldrich. All other reagents for amino acid analysis were provided in the EZ Amino Acid Analysis Kit from Phenomenex (Torrance, CA, USA). For elemental composition analysis, nitric acid (65% Suprapur® for trace analysis, HNO_3) and hydrogen peroxide (30% Suprapur, for trace analysis, H_2O_2) were obtained from Supelco. For insect fermentation, De Man, Rogosa and Sharpe (MRS) broth was obtained from Merck (Darmstadt, Germany), and sodium chloride (NaCl) was obtained from Sigma-Aldrich (Steinheim, Germany).

2.2. Insect Rearing

Three edible insect species, the yellow mealworm, the house cricket, and the migratory locust, were selected based on their approval as novel foods by the EU (Commission Regulations 2021/882, 2021/1975, 2022/169, 2022/188, 2025/89). All species were sourced from the insect-rearing facility at the Faculty of Agrobiology, Food, and Natural Resources, Czech University of Life Sciences, Prague and were maintained under controlled conditions.

House crickets and mealworms were reared in a rack system under controlled conditions at 27 ± 1 °C, 40–50% relative humidity, and a 12:12 h light–dark photoperiod. Migratory locusts were maintained at the same temperature and photoperiod, but at a lower relative humidity of 30%. The humidity was controlled using a DeLonghi dehumidifier (Treviso, Italy) and measured daily with a Comet C3120 thermohygrometer (Comet system, Rožnov pod Radhoštěm, Czech Republic). No additional heating was used for house crickets and yellow mealworms, while migratory locusts were provided with heating foils and 40 W bulbs to create warm zones. House crickets and migratory locusts were reared in plastic boxes (56 × 39 × 28 cm, IKEA, Prague, Czech Republic), fitted with egg trays (Schubert Partner, Prague, Czech Republic) and secured with 80% aluminum anti-insect mesh lids. Yellow mealworms were reared in 60 × 40 × 12.5 cm breeding trays (Beekenkamp, Maasdijk, The Netherlands).

Insects were fed *ad libitum*. House crickets received chicken feed (77.9% wheat, 17.6% soybean meal, 1.8% rapeseed oil, 2.7% mineral and vitamin premix; particle size < 1 mm), while migratory locusts were provided wheat bran, grass hay, and fresh Poaceae grass. Mealworms were fed on a 4:1 mixture of wheat bran and chicken feed. Water gel (Oslavan, Náměšť nad Oslavou, Czech Republic) was supplied daily for house crickets and migratory locusts. House crickets and locusts were maintained in the same containers from hatching until harvest at the adult stage, while mealworms were harvested upon first pupal emergence.

All insects were starved for 24 h prior to euthanasia by freeze-killing (−20 °C for 24 h). Samples were lyophilized (Trigon Plus, Čestlice, Czech Republic), homogenized (A10; IKA Werke GmbH & Co. KG, Staufen, Germany), and stored at −20 °C until further analysis. The dry matter content (%) of fresh insects for further calculations was determined by calculating the mass difference before and after lyophilization and was 30%.

2.3. Insect Fermentation

The insect samples (yellow mealworm, house cricket and migratory locust) were subjected to lactic acid fermentation using *L. plantarum* and a commercial starter culture for meat processing. *Lactobacillus plantarum* IM 527 was obtained from the Institute of Dairy Science and Probiotics, Department of Animal Science, Biotechnical Faculty, University of Ljubljana, Slovenia. An aliquot of *L. plantarum* stock culture (100 µL) was inoculated into 20 mL of sterilized liquid MRS medium and incubated on a rotary shaker (150 rpm, 30 °C, 12 h) until the late exponential phase was reached. The obtained broth (2 mL) was then centrifuged (10,000× *g*, 5 min) and washed twice with a saline solution to obtain the inoculum suspension.

A stock suspension of a commercial starter culture SM-194 Bactoferm® (Chr Hansen A/S, Hørsholm, Denmark) was prepared according to the manufacturer's instructions by dissolving 0.125 g of the culture in 5 mL of sterile saline solution. The starter culture comprised *Staphylococcus xylosus, Debaryomyces hansenii, Lactobacillus sakei, Staphylococcus carnosus*, and *Pediococcus pentosaceus*.

Both fermentation types were performed using 8 g of lyophilized and milled insect material, placed in 50 mL polyethylene centrifuge tubes and then capped. For house cricket and migratory locust, a substrate-to-sterile double distilled H_2O ratio of 1:2 (8 g:16 mL) was used, whereas a ratio of 1:1.5 (8 g:12 mL) was applied for yellow mealworm. Samples were then inoculated with the *L. plantarum* or commercial starter culture suspension (240 μL for house cricket and migratory locust; 200 μL for yellow mealworm) and incubated under static conditions at 30 °C for 48 h.

Fermentation was performed in triplicate. To monitor the fermentation process, samples were collected at 0, 24, and 48 h, and the pH and colony-forming units (CFU) were measured. A reduction in pH (3–30%), accompanied by an increase in CFU count (29–47%) over the fermentation period, indicated active microbial growth and metabolic activity, confirming successful fermentation. Following fermentation, all samples were lyophilized (72 h, −53 °C, 0.280 mbar) and stored at −20 °C until further analysis.

2.4. Determination of Proximate Composition

Proximate composition of yellow mealworm and house cricket, both unfermented and fermented with *L. plantarum*, was determined using official AOAC methods for macronutrient analysis, though due to insufficient sample amounts, the analysis was limited to these two species. This decision was based on several practical considerations: (1) proximate composition requires relatively large sample quantities compared to other analyses in this study; (2) initial difficulties in rearing migratory locust resulted in limited biomass; and (3) the insectarium's capacity to produce large-scale insect biomass was constrained. Consequently, yellow mealworm and house cricket were selected as model species due to their current market relevance, relative ease of rearing, and established production protocols. *Lactobacillus plantarum* was chosen for fermentation as it is one of the most extensively studied lactic acid bacteria, particularly in the context of food fermentation.

The moisture content was measured following AOAC 950.46 [45], and dry matter was calculated as 1—the moisture content. Ash, total fat, and protein contents were determined using AOAC 920.153, 991.36, and 928.08 [45], respectively. Protein content was calculated from nitrogen (Kjeldahl method) using a factor of 6.25, after subtracting non-protein nitrogen (NPN).

The NPN extracts were prepared using a modified DeVries et al. (2017) method [46,47]. A known amount (2.00 g) of sample was mixed with 10 mL distilled water and 10 mL 24% trichloroacetic acid, homogenized (Ultra-Turrax, 13,500 rpm, 30 s), sonicated (10 min), centrifuged (4000 rpm, 10 min), and filtered with filter paper discs (grade 389, 125 mm diameter, 8–12 μm pore size, 100% cotton linters, Sartorius, Germany). NPN was determined from the clear filtrate (5 g) by the Kjeldahl method.

Dietary fiber content was determined using the enzymatic–gravimetric AOAC 991.43 method [45], employing heat-stable α-amylase, protease, and amyloglucosidase (Merck KGaA, Germany). Insoluble dietary fiber was recovered by filtration, while soluble dietary fiber was precipitated from filtrates with 96% ethanol. Ash and protein contents of residues were corrected, and total dietary fiber was the sum of insoluble and soluble fractions. Available carbohydrate content (g/100 g) was calculated using Equation (1) [48]:

$$\text{Available carbohydrate} = 100\ \text{g}/100\ \text{g} - (\text{water} + \text{ash} + \text{protein} + \text{fat} + \text{fiber}) \quad (1)$$

where protein is calculated as total nitrogen × 6.25, thereby accounting for both protein and non-protein nitrogen fractions.

The energy value (EV, kJ/100 g) was calculated as the sum of the energy values for each macronutrient using the following equation [48]:

$$EV = EV_{\text{protein}} + EV_{\text{fat}} + EV_{\text{available carbohydrates}} + EV_{\text{dietary fibre}} \quad (2)$$

The energy value of each nutrient was calculated by multiplying its content (g/100 g) by the corresponding energy conversion factor: 17 kJ/g for protein and available carbohydrates, 37 kJ/g for fat, and 8 kJ/g for dietary fiber [48].

2.5. Fatty Acid Analysis

Freeze-dried insect samples (200 mg) were weighed into screw-cap vials. To extract total lipids, 500 µL of dichloromethane and 3 mL of 0.5 M sodium hydroxide in methanol were added. The samples were purged with nitrogen gas, heated for 10 min at 90 °C, and then rapidly cooled in an ice bath. After cooling, 3 mL of boron trifluoride in methanol was introduced to convert the lipids into fatty acid methyl esters (FAMEs). The extract was then purged with nitrogen and heated for 10 min at 90 °C. Once cooled, the FAMEs were extracted using 3 mL Milli-Q water and 1.5 mL hexane. The organic phase was transferred to a GC vial and stored at −20 °C. All samples were prepared in duplicate.

All samples were analyzed using an Agilent 6890N gas chromatograph (Network GC System, Agilent Technology, Santa Clara, CA, USA) with a flame ionization detection (FID). Separation was achieved on a CP-Sil 88 capillary column (100 m × 0.25 mm × 0.20 µm, Agilent J&W). The temperature program was as follows: 100 °C (5 min), ramped to 180 °C at 8 °C/min (9 min), and then increased to 230 °C at 1 °C/min (1 min). Helium was used as the carrier gas in constant flow mode (1.3 mL/min), with an injection volume of 1 µL and a split ratio of 10:1. The inlet and detector were set to 260 °C and 250 °C, respectively.

Fatty acids were identified and quantified by comparing their retention times with the Supelco 37-component FAME mix standard. The results were expressed as relative weight percentages of total fatty acids. Procedural blanks were analyzed alongside each batch of samples, and the FAME mix standard was run after every ten samples to verify system stability.

Fatty acids were categorized as saturated fatty acids (SFA), monounsaturated fatty acids (MUFA), and polyunsaturated fatty acids (PUFA), with a specific focus on $n-3$ and $n-6$ fatty acids. The PUFA/SFA and $n-6/n-3$ ratios were also calculated. Additionally, the atherogenic index (IA), thrombogenic index (IT), and hypocholesterolemic/hypercholesterolemic ratio (h/H) for the total lipid content were calculated according to the following equations [49]:

$$IA = \frac{C12:0 + (4 \times C14:0) + C16:0}{\Sigma MUFA + \Sigma n-6 + \Sigma n-3} \tag{3}$$

$$IT = \frac{C14:0 + C16:0 + C18:0}{(0.5 \times \Sigma MUFA) + (0.5 \times \Sigma n-6) + (3 \times \Sigma n-3) + (n-3/n-6)} \tag{4}$$

$$h/H = \frac{C18:1 + \Sigma PUFA}{C12:0 + C14:0 + C16:0} \tag{5}$$

2.6. Amino Acid Analysis

A known weight of lyophilized insect sample (50 mg) was placed into hydrolysis micro-reaction vessels, and 2.5 mL of 4% thioglycolic acid in 6M HCl and 25 µL of liquid phenol (heated to 80 °C) were added. The mixture was then vortexed. The samples were then purged with nitrogen for 5 min, sealed, and heated at 110 °C for 24 h. During hydrolysis, glutamine was converted to glutamic acid, and asparagine to aspartic acid; therefore, reported values for glutamic and aspartic acids represent the combined total of these amino acids. After hydrolysis, the samples were filtered through CHROMAFIL Xtra PTFE 0.45 µm syringe filters (Macherey-Nagel, Düren, Germany).

Amino acid analysis was performed using the EZ:faastTM kit. First, 50 µL of the filtered hydrolysate was transferred into a sample preparation vial, followed by the addition of 100 μL of 1 M Na_2CO_3 solution and vortexing. A 30 μL aliquot of the pH-adjusted mixture (pH 2) was then transferred into a new vial, and 50 μL of internal standard (200 nmol/mL) was added, followed by vortexing. The sample was then passed through a sorbent tip using a 1.5 mL syringe. After adding 200 μL of Milli-Q water into the vial, the sample was passed through the sorbent tip again, and the liquid was discarded. The sorbent tip was then detached.

The compounds of interest were eluted using 200 μL of a NaOH/N-propanol/3-picoline mixture (3/1.6/0.4, *v*/*v*/*v*). This elution was performed by attaching the sorbent tip to a 0.6 mL syringe, pulling back the plunger halfway, and wetting the sorbent with the eluting medium before expelling the liquid into the sample vial. A 50 μL aliquot of propyl chloroformate/chloroform/iso-octane (1:3:1, *v*/*v*/*v*) was added, and the contents were emulsified by vortexing for 5 s. The reaction was allowed to proceed for 1 min and repeated once more. After this, 100 μL of a 9:1 (*v*/*v*) iso-octane/chloroform mixture was added and vortexed for 1 min. One hundred μL of the organic layer was transferred into a GC autosampler vial, evaporated under nitrogen for 10 min, and reconstituted in 100 μL iso-octane/chloroform (4:1, *v*/*v*). The reconstituted sample was transferred into a GC-MS vial with a glass insert and analyzed. All samples were prepared in duplicate.

Samples were analyzed on a 7890B series GC coupled to a 5977A single quadrupole mass selective detector (Agilent, USA). Separation was achieved using a ZB-AAA GC column (10 m × 0.25 mm, Phenomenex, USA), with helium as the carrier gas (1.1 mL/min). A volume of 1.5 μL of the sample was injected in split mode (1:15). The GC oven temperature program was 110 °C (1 min), then ramped at 30 °C/min to 320 °C (total runtime: 7 min). The inlet and transfer line were set to 250 °C and 260 °C, respectively. The mass spectrometer was operated in electron impact (EI) mode at 70 eV, with a source temperature of 240 °C, a quadrupole temperature of 180 °C, and an auxiliary temperature of 310 °C. The scan range was 45–440 m/z, and the compounds were identified and quantified using selected ion monitoring (SIM) mode (Table S1). The data were processed using MassHunter software (version 10.0., Agilent Technologies, USA).

The ratio of essential to total amino acids (E/T, %) and the essential amino acid index (EAAI) were determined according to FAO (2013) [50] based on valine, leucine, isoleucine, threonine, phenylalanine + tyrosine, methionine, histidine, and lysine. Cystine and tryptophan were excluded from the calculations due to method limitations, since acid hydrolysis degrades tryptophan and cystine. The formula for EAAI is as follows:

$$\mathrm{EAAI} = \sqrt[8]{\frac{\mathrm{goflysinein100\ gofanalysedprotein}}{\mathrm{goflysinein100\ gofreferenceprotein}} \times (\mathrm{etc.forother\ 7EAA}) \times 100} \quad (6)$$

The index is based on the sum of all amino acids compared to those of the reference protein for older children (≥3 years) and adults [50]. The amino acid with the lowest amino acid score (AAS) is considered the limiting amino acid. The AAS was calculated according to the following equation.

$$\mathrm{AAS} = \frac{\mathrm{mgaminoacidper\ g\ ofprotein}}{\mathrm{mgaminoacidper\ g\ ofreferenceprotein}} \times 100 \quad (7)$$

2.7. Elemental Analysis

Sample digestion for elemental analysis was performed following Vehar et al. (2025) [51]. Briefly, a known quantity of lyophilized insect sample (0.10–0.15 g) was placed into a pre-cleaned Teflon digestion vial. Subsequently, 1 mL of HNO_3 and 250 μL of H_2O_2 were added to each vial. All samples were microwave-digested using an Ultrawave Single Reaction Chamber Microwave Digestion System (Milestone, Italy). The digestion program involved heating the samples to 240 °C over 20 min, followed by a 15-min hold at 100 bars with a maximum power of 1500 W. After digestion, the digestate was quantitatively transferred to a polyethylene-graduated tube and diluted to a final volume of 10 mL with Milli-Q water. For measuring macro elements, samples were further diluted with 5% HNO_3.

The determination of selected elements (Ag, Al, As, B, Ba, Ca, Cd, Co, Cs, Cu, Fe, Hg, K, Mg, Mn, Mo, Na, Ni, P, Pb, Rb, S, Sb, Se, Sn, Sr, U, V, and Zn) was conducted using an Agilent 8800 triple quadrupole inductively coupled plasma mass spectrometer, QQQ-ICP-MS (Agilent, Japan, Tokyo). Different reaction gases, H_2, He and O_2 were used. Internal standards (Y, Rh, Sc, and Gd) were added online in all modes, and external calibration was used to quantify the results. All samples were prepared in duplicate. Additionally, blanks and certified reference materials (SRM NIST 1566b Oyster tissue, Typical Diet NIST 1548b) were prepared in each run in the same manner to ensure the quality and accuracy of the results. The limit of detection (LOD) was calculated as three times the standard deviation of the blank samples.

A risk assessment for toxic elements was performed based on an adult (70 kg) ingesting 100 g of fresh insects. The intake of each element was first calculated (Table S6) and then compared with reference values (Table S4) obtained from OpenFoodTox: EFSA's chemical hazards database [52]. Based on the concentrations of essential elements and their respective reference daily intakes (RDI; Table S5), the percentage contribution to daily intake for adults was reported, based on a consumption scenario of 100 g of fresh insects per day.

2.8. Oxidation Stability Analysis

To assess the oxidation stability, a known amount (3.5 g) of lyophilized sample was placed in the test chamber of a RapidOxy 100 Oxidation Stability Tester (Anton Paar, Austria). The induction time was measured at 120 °C and 700 kPa, with continuous monitoring of oxygen consumption. The induction time was recorded when the pressure drop reached 10%, indicating oxidation stability. The measurement error is 2–3%.

2.9. Statistical Analysis

Data processing and visualization were performed using R-based software (version 4.2.3). All data sets were checked for normal distribution using the Shapiro-Wilk test and for homoscedasticity using Levene's test. For proximate composition, where only yellow mealworm and house cricket before and after fermentation with *L. plantarum* were analyzed, significant nutritional parameters were assessed using pairwise comparisons. Depending on the assumptions for parametric testing (normality and homoscedasticity of variance), either a *t*-test or the Mann-Whitney *U* test was applied. If the assumptions were not met, the Mann-Whitney *U* test was used; otherwise, the *t*-test was applied. For all other analyses, where three insect species were evaluated before and after fermentation with *L. plantarum* and a commercial starter culture, significant nutritional and compositional parameters were assessed using one-way ANOVA or the Kruskal-Wallis test, depending on whether the assumptions for parametric testing were satisfied. When assumptions were violated, the Kruskal-Wallis test was applied; otherwise, one-way ANOVA was used. The statistical significance level was 0.05 with $p \leq 0.05$, indicating rejection of the null hypothesis. Fermented and unfermented samples from the same species were grouped to ensure reliable statistical comparisons between species. However, due to the insufficient number of samples, a statistical comparison between fermented and unfermented samples was not possible.

3. Results and Discussion

3.1. Proximate Composition

Due to limited sample availability, the proximate composition was assessed only for yellow mealworm and house cricket before and after 48 h fermentation with *L. plantarum*. Proximate composition in terms of ash, total fat and proteins (Table 1) was comparable to those obtained by Cho et al. (2018), Khatun et al. (2021), Kulma et al. (2019), Messina et al. (2019) and Yi et al. (2013), while Kittibunchakul, Whanmek, and Santivarangkna (2023) and Vasilica et al. (2022) report higher values for total protein and lower values for total fat [35,42,43,53–56]. Statistical analyses (Table S2) showed that yellow mealworm had statistically significantly higher ($p < 0.05$) ash concentrations (4.35 to 4.50 g/100 g) and lower total nitrogen content (8.95 to 9.15 g/100 g) compared to house cricket (ash: 4.23 to 4.30 g/100 g, total nitrogen: 9.24 to 9.30 g/100 g).

Table 1. Proximate composition of unfermented and fermented (48 h) yellow mealworm and house cricket (lyophilized).

	YM	YMP	HC	HCP
Dry matter [g/100 g]	97.8 ± 0.1	96.3 ± 0.1	97.9 ± 0.1	97.3 ± 0.0
Water [g/100 g]	2.20 ± 0.06	3.70 ± 0.08	2.06 ± 0.08	2.68 ± 0.02
Ash [g/100 g]	4.35 ± 0.02	4.50 ± 0.01	4.30 ± 0.04	4.23 ± 0.02
Total fat [g/100 g]	28.2 ± 1.9	24.4 ± 2.3	22.9 ± 4.6	23.6 ± 2.7
Insoluble dietary fiber [g/100 g]	7.05 ± 0.35	4.38 ± 0.10	6.19 ± 0.53	5.22 ± 0.30
Soluble dietary fiber [g/100 g]	0.025 ± 0.029	0.366 ± 0.259	0.178 ± 0.198	0.366 ± 0.290
Total dietary fiber [g/100 g]	7.08 ± 0.33	4.75 ± 0.35	6.37 ± 0.35	5.59 ± 0.05
Total nitrogen [g/100 g]	9.15 ± 0.03	8.95 ± 0.01	9.24 ± 0.01	9.30 ± 0.03
Non-protein nitrogen [g/100 g]	0.061 ± 0.002	0.196 ± 0.005	0.069 ± 0.001	0.179 ± 0.002
Protein nitrogen [g/100 g]	9.09 ± 0.02	8.75 ± 0.02	9.17 ± 0.01	9.12 ± 0.03
Proteins (6.25) [g/100 g]	56.8 ± 0.1	54.7 ± 0.1	57.3 ± 0.1	57.0 ± 0.2
Available carbohydrates [g/100 g]	0.98 ± 1.90	6.71 ± 2.37	6.64 ± 4.62	5.85 ± 2.66
Energy value [kJ/100 g]	2082 ± 3	1985 ± 3	1985 ± 7	1984 ± 4

YM—yellow mealworm, YMP—yellow mealworm fermented using *L. plantarum*, HC—house cricket, HCP—house cricket fermented using *L. plantarum*.

The results show a decrease in total dietary fiber content following fermentation in both insect species (33% for yellow mealworm and 12% for house cricket), primarily due to a reduction in insoluble dietary fiber (38% and 16%, respectively). Although the values for soluble dietary fiber are very low and therefore imprecise, the measurement showed a slight increase in the yellow mealworm (from 0.025 g/100 g to 0.366 g/100 g). The increase in soluble fiber after fermentation with *L. plantarum* has been previously reported in other commodities, as the enzymatic breakdown of insoluble fiber during fermentation leads to its conversion into soluble forms [57–59].

Regarding nitrogen composition, total nitrogen and protein nitrogen remained unchanged in yellow mealworm and house cricket, while non-protein nitrogen increased (221% and 159%, respectively). Despite the rise in non-protein nitrogen, the total nitrogen and protein nitrogen contents stayed constant, suggesting that microbial activity was breaking down proteins, but not enough to reduce the protein content in the insects significantly. Instead, the nitrogen is likely redistributed into various forms, such as amino acids and peptides, maintaining the overall nitrogen balance. This redistribution has already been shown, for example, in *Broussonetia papyrifera* fermented using *L. plantarum* during ensiling [59].

Following fermentation, only yellow mealworm exhibited an increase in available carbohydrates (from 0.98 g/100 g to 6.71 g/100 g), while no significant differences were observed in protein content, energy value, or total fat for either species. The significant increase in available carbohydrates may be due to the fermentation process breaking down complex polysaccharides and fibers into simpler sugars, which are then released as available carbohydrates [60]. The lack of significant changes in protein content, energy value, or total fat suggests the limited impact of *L. plantarum* on protein and fat molecules, as these macronutrients are more resistant to microbial degradation compared to carbohydrates and fibers. Additionally, the fermentation conditions might not have been optimized for significant lipid breakdown or protein denaturation, which can require more specialized conditions or longer fermentation periods.

Comparison with existing literature on edible insects is complicated by the variability in experimental design, insect species, fermentation culture, and analytical methodology in the limited number of available studies. For instance, Cho et al. (2018) fermented raw and defatted yellow mealworm using a soy sauce fermentation process involving *Aspergillus oryzae* and *Bacillus licheniformis*, reporting increased total and protein nitrogen levels after six days, which remained stable over the 13- and 20-day fermentation period [35]. Liu et al. (2022) fermented black soldier fly larvae for four days with *Bacillus subtilis* and *Aspergillus niger*, observing a reduction in actual protein (40%) and fat (30%) content [39]. In the study by Kittibunchakul et al. (2023), house cricket was fermented using a traditional Thai shrimp paste method, employing Kapi as a starter culture, which predominantly contains lactic acid bacteria [42]. After 4 weeks of fermentation, they noted significant reductions in energy (24%), protein (23%), fat (21%), and carbohydrates (43%). Meng et al. (2023), who fermented black soldier fly larvae for 24 h using *Lactobacillus crispatus* and *Pichia kudriavzevii* cultures, also observed a decrease in crude fat content, while crude protein remained unchanged [40]. Also, Pérez-Rodríguez, Ibarra-Herrera, and Pérez-Carrillo (2023) fermented yellow mealworm and Mexican grasshopper with *Aspergillus oryzae*, finding that after four days, protein content decreased (49%, 16%, respectively) in both species, while fat content increased (14%, 119%, respectively), and crude fiber remained unaffected [37].

3.2. Fatty Acid Composition

A total of seven FAs were detected in amounts >1% in the studied insects (Table 2), including three SFAs (myristic, palmitic, stearic), two MUFAs (palmitoleic and oleic acid) and two PUFAs (linoleic—LA and α-linoleic acid—ALA). Their relative levels (>1%) for yellow mealworm were as follows: oleic > LA > palmitic > myristic > stearic > palmitoleic > ALA. The relative levels for house cricket were: oleic > LA > palmitic > stearic > ALA > palmitoleic > myristic acid, while for migratory locust: LA > oleic > palmitic > stearic > ALA > myristic > palmitoleic. All insects contained both essential fatty acids, LA (26.4% to 31.4%) and ALA (0.959% to 9.10%). Overall, the relative concentrations of essential fatty acids are consistent with previously published data, where reported concentrations range from 13.8% to 27.7% for LA and 6.7% to 13.9% for ALA in migratory locust; 16.6% to 33.8% for LA and 0.23% to 1.43% for ALA in yellow mealworm; and 26.26% to 36.67% for LA and 0.7% to 5.13% for ALA in house cricket [51,54,61–64]. Statistical analyses showed that insect species significantly differ ($p < 0.05$) in most fatty acids, except for arachidic, 11-eicosenoic acid and lignoceric acid (Table S2).

All investigated insect species demonstrated favorable PUFA/SFA ratios, ranging from 1.13 to 1.16 for yellow mealworm, 0.935 to 0.993 for house cricket, and 1.14 to 1.23 for migratory locust. These values are consistent with previous reports for edible insects [43,51], where PUFA/SFA close to 1 is considered beneficial for cardiovascular health due to their role in improving plasma lipid profiles and reducing low-density lipoprotein (LDL) cholesterol levels. At the same time, a ratio < 0.33 can have an atherogenic effect [65–67]. Among the species examined, migratory locust exhibited the most optimal n-6/n-3 fatty acid ratio (3.14–3.29), which falls within the range (<4) for lowering chronic disease risk [68]. In contrast, yellow mealworm and house cricket displayed considerably higher ratios (26.6–

27.9 and 17.1–17.5, respectively), reflecting a predominance of n-6 fatty acids that has also been reported for these species in earlier studies [51,64,69].

Table 2. Contents of the main fatty acids (>1%) saturated fatty acids (SFA), monounsaturated fatty acids (MUFA), polyunsaturated fatty acids (PUFA), n-6 fatty acids, n-3 fatty acids, n-6/n-3 ratio, index of atherogenicity (IA), index of thrombogenicity (IT), hypocholesterolemic/hypercholesterolemic ratio (h/H) in unfermented and fermented insects.

	YM	YMP	YMM	HC	HCP	HCM	ML	MLP	MLM
Myristic acid [%]	3.79 ± 0.02	3.76 ± 0.05	3.76 ± 0.01	0.504 ± 0.005	0.533 ± 0.006	0.525 ± 0.003	1.29 ± 0.01	1.41 ± 0.00	1.41 ± 0.01
Palmitic acid [%]	16.0 ± 0.0	16.2 ± 0.2	16.2 ± 0.1	23.8 ± 0.2	25.0 ± 0.1	24.7 ± 0.2	19.8 ± 0.1	20.9 ± 0.0	20.9 ± 0.1
Palmitoleic acid [%]	1.68 ± 0.00	1.67 ± 0.00	1.67 ± 0.00	0.763 ± 0.013	0.795 ± 0.003	0.794 ± 0.028	0.805 ± 0.013	0.801 ± 0.002	0.813 ± 0.019
Stearic acid [%]	3.20 ± 0.00	3.35 ± 0.03	3.34 ± 0.01	8.61 ± 0.01	8.81 ± 0.02	8.81 ± 0.09	9.59 ± 0.03	9.83 ± 0.04	9.87 ± 0.04
Oleic acid [%]	46.5 ± 0.0	46.0 ± 0.3	46.0 ± 0.1	31.9 ± 0.5	31.4 ± 0.0	31.6 ± 0.1	28.2 ± 0.3	28.0 ± 0.0	27.9 ± 0.0
Linoleic acid [%]	26.4 ± 0.0	26.5 ± 0.0	26.5 ± 0.0	31.4 ± 0.1	30.8 ± 0.0	30.9 ± 0.2	29.8 ± 0.2	28.4 ± 0.0	28.5 ± 0.0
α-linolenic acid [%]	1.00 ± 0.01	0.959 ± 0.002	0.968 ± 0.000	1.85 ± 0.01	1.77 ± 0.01	1.77 ± 0.01	9.06 ± 0.06	9.06 ± 0.00	9.10 ± 0.00
SFA [%]	23.9 ± 0.0	24.4 ± 0.3	24.4 ± 0.1	33.6 ± 0.4	34.9 ± 0.0	34.6 ± 0.3	31.6 ± 0.1	33.1 ± 0.1	33.1 ± 0.1
MUFA [%]	48.5 ± 0.0	47.9 ± 0.3	48.0 ± 0.1	32.9 ± 0.5	32.4 ± 0.0	32.6 ± 0.1	29.5 ± 0.2	29.4 ± 0.0	29.2 ± 0.1
PUFA [%]	27.6 ± 0.0	27.7 ± 0.0	27.7 ± 0.0	33.4 ± 0.1	32.6 ± 0.0	32.8 ± 0.2	38.9 ± 0.2	37.5 ± 0.0	37.7 ± 0.1
PUFA/SFA	1.16 ± 0.00	1.13 ± 0.01	1.14 ± 0.00	0.993 ± 0.009	0.935 ± 0.002	0.947 ± 0.015	1.23 ± 0.00	1.14 ± 0.00	1.14 ± 0.01
n-6 [%]	26.6 ± 0.0	26.7 ± 0.0	26.7 ± 0.0	31.6 ± 0.1	30.9 ± 0.0	31.0 ± 0.2	29.9 ± 0.2	28.5 ± 0.0	28.6 ± 0.1
n-3 [%]	1.00 ± 0.01	0.959 ± 0.002	0.968 ± 0.000	1.85 ± 0.01	1.77 ± 0.01	1.77 ± 0.01	9.06 ± 0.06	9.06 ± 0.00	9.10 ± 0.00
n-6/n-3	26.6 ± 0.2	27.9 ± 0.1	27.6 ± 0.0	17.1 ± 0.0	17.4 ± 0.1	17.5 ± 0.1	3.29 ± 0.05	3.14 ± 0.00	3.14 ± 0.01
IA	0.414 ± 0.001	0.417 ± 0.007	0.417 ± 0.002	0.390 ± 0.005	0.417 ± 0.002	0.411 ± 0.006	0.367 ± 0.003	0.399 ± 0.000	0.399 ± 0.003
IT	0.342 ± 0.000	0.342 ± 0.004	0.343 ± 0.001	0.601 ± 0.006	0.631 ± 0.003	0.624 ± 0.009	0.509 ± 0.003	0.542 ± 0.001	0.542 ± 0.003
h/H	3.68 ± 0.01	3.63 ± 0.05	3.64 ± 0.02	2.68 ± 0.04	2.51 ± 0.01	2.55 ± 0.04	3.16 ± 0.02	2.92 ± 0.00	2.92 ± 0.02

YM: yellow mealworm, YMP: yellow mealworm fermented using *L. plantarum*, YMM: yellow mealworm fermented using commercial starter culture, HC: house cricket, HCP: house cricket fermented using *L. plantarum*, HCM: house cricket fermented using commercial starter culture, ML: migratory locust, MLP: migratory locust fermented using *L. plantarum*, MLM: migratory locust fermented using commercial starter culture. SFA, MUFA, PUFA, n-6, n-3, IA, IT, h/H are calculated based on all detected fatty acids (Table S3).

All insects presented a favorable low index of atherogenicity (IA < 0.5), with values of 0.414–0.417 for yellow mealworm, 0.390–0.417 for house cricket, and 0.367–0.399 for migratory locust. However, only the yellow mealworm demonstrated a thrombogenicity index (IT) below 0.5 (0.342–0.343), compared to house cricket (0.601–0.631) and migratory locust (0.509–0.542). Lower values of IA (<0.5) and IT (<0.5) are associated with a reduced risk of developing cardiovascular disease [49,70]. Finally, all species exhibited a desirable hypocholesterolemic/hypercholesterolemic ratio (h/H > 1), with yellow mealworm showing the highest values (3.63–3.68), followed by migratory locust (2.92–3.16) and house cricket (2.51–2.68). A higher h/H ratio is associated with a more favorable lipid profile, as hypocholesterolemic fatty acids (PUFA and MUFA) promote the reduction of serum cholesterol, whereas hypercholesterolemic fatty acids (mainly C12:0, C14:0, C16:0) elevate it [49,71].

Interestingly, despite successful fermentation, the fatty acid profile of the insect species remained unaffected compared to unfermented insects, regardless of the insect species or fermentation culture used. This finding contrasts with those of other previous studies, although those studies involved different insect species and fermentation cultures. For instance, Liu et al. (2022) fermented black soldier fly larvae for four days using *Bacillus subtilis* and *Aspergillus niger*, resulting in a 25-fold increase in free short-chain fatty acids post-fermentation [39]. Similarly, Hadj Saadoun et al. (2020) fermented prepupae, puparia, and dead adults of black soldier flies with lactic acid bacteria (*Lacticaseibacillus rhamnosus*, *Lactiplantibacillus plantarum*) for 72 h, noting a reduction in typical fatty acids like lauric acid (C12:0) and an increase in minor fatty acids such as short-chain fatty acids, odd-chain fatty acids, and branched-chain fatty acids, which are characteristic of bacterial cell walls [38]. Vasilica et al. (2022) fermented insect flour for 48 h using *L. plantarum* and spontaneous fermentation, observing that controlled fermentation enriched SFA and MUFA, while PUFA decreased during fermentation [43].

Several possible explanations exist for the lack of change in the fatty acid profiles observed in this study. First, the relatively short fermentation duration (48 h) may not have been sufficient for microbial activity to alter lipid components significantly. Second, the microbial strains employed—*L. plantarum* and a commercial starter culture composed of *Staphylococcus xylosus*, *Staphylococcus carnosus*, *Lactobacillus sakei*, *Pediococcus pentosaceus*, and *Debaryomyces hansenii*—are primarily adapted to metabolize carbohydrates and amino acids, rather than lipids [72–74]. Consequently, their enzymatic potential for lipid transformation is limited. Third, the processing of insect substrates prior to fermentation could have influenced microbial access and activity. For example, Aleknavičius et al. (2022) reported higher microbial abundance and metabolic activity in non-processed fermented insects compared to processed forms, which may enhance interactions with lipid components [75].

3.3. Amino Acid Composition

Taking into account that arginine, cystine and tryptophan were not determined in this study, the total amino acid content (Table 3) ranged from 38.3 g/100 g DW (MLM) to 50.6 g/100 g DW (MLP), while total essential amino acid content ranged from 19.6 g/100 g DW (MLM) to 24.4 g/100 g DW (YM). Essential to total amino acids (E/T) ranged from 44.1% (MLP) to 54.6% (YMP), while essential amino acid index (EAAI, Table 4) ranged from 150 (MLP) to 184 (YMP). Analyzed insects were the highest in valine (3.81–4.58 g/100 g DW), alanine (3.38–4.71 g/100 g DW), glutamic acid (2.74–8.09 g/100 g DW), leucine (3.61–4.23 g/100 g DW), aspartic acid (2.95–4.66 g/100 g DW), but lower in methionine (0.478–0.774 g/100 g DW), ornithine (0.250–3.20 g/100 g DW) and histidine (0.973–1.64 g/100 g DW). Results align with previously reported values for yellow mealworm, house cricket and migratory locust [51,54,63,64,76].

Table 3. Concentrations of amino acids (g/100 g DW), essential and nonessential amino acid content (g/100 g DW) and essential total amino acids ratio (%) in unfermented and fermented insects.

	YM	YMP	YMM	HC	HCP	HCM	ML	MLP	MLM
Alanine	3.38 ± 0.12	3.90 ± 0.06	3.84 ± 0.02	4.04 ± 0.07	3.98 ± 0.16	4.07 ± 0.15	4.71 ± 0.23	4.64 ± 0.17	4.54 ± 0.56
Aspartic acid	4.66 ± 0.30	3.52 ± 0.10	3.84 ± 0.09	4.34 ± 0.15	3.89 ± 0.08	4.01 ± 0.15	4.01 ± 0.21	3.60 ± 0.14	2.95 ± 0.08
Glutamic acid	4.67 ± 0.48	2.84 ± 0.10	2.74 ± 0.09	4.09 ± 0.54	3.76 ± 0.33	3.88 ± 0.23	3.63 ± 0.39	8.09 ± 0.31	3.30 ± 0.18
Glycine	2.65 ± 0.09	2.82 ± 0.03	2.77 ± 0.04	2.78 ± 0.14	2.66 ± 0.07	2.69 ± 0.01	2.83 ± 0.08	3.08 ± 0.15	2.69 ± 0.09
Histidine	1.57 ± 0.06	1.64 ± 0.02	1.60 ± 0.02	1.14 ± 0.01	0.973 ± 0.018	1.14 ± 0.04	1.20 ± 0.05	1.46 ± 0.05	1.05 ± 0.06
Isoleucine	2.10 ± 0.05	2.25 ± 0.05	2.19 ± 0.02	1.94 ± 0.01	1.92 ± 0.04	1.96 ± 0.04	2.03 ± 0.08	2.25 ± 0.05	1.91 ± 0.07
Leucine	4.12 ± 0.16	4.23 ± 0.09	4.16 ± 0.06	4.03 ± 0.05	3.88 ± 0.08	3.95 ± 0.10	4.05 ± 0.17	4.14 ± 0.08	3.61 ± 0.16
Lysine	3.58 ± 0.21	2.56 ± 0.09	2.51 ± 0.06	3.87 ± 0.06	3.05 ± 0.04	3.49 ± 0.04	3.83 ± 0.17	3.25 ± 0.38	3.01 ± 0.10
Methionine	0.670 ± 0.036	0.677 ± 0.039	0.652 ± 0.003	0.774 ± 0.027	0.742 ± 0.016	0.748 ± 0.026	0.672 ± 0.045	0.660 ± 0.024	0.478 ± 0.020
Ornithine	<0.250	0.537 ± 0.037	0.542 ± 0.112	<0.250	0.699 ± 0.227	0.774 ± 0.234	<0.250	3.20 ± 1.10	1.15 ± 0.24
Phenylalanine	1.97 ± 0.10	1.99 ± 0.02	2.03 ± 0.06	1.93 ± 0.07	1.78 ± 0.03	1.81 ± 0.02	2.04 ± 0.08	1.92 ± 0.20	1.66 ± 0.05
Proline	3.00 ± 0.12	3.20 ± 0.06	3.11 ± 0.04	2.92 ± 0.06	2.76 ± 0.09	2.86 ± 0.10	3.14 ± 0.09	4.03 ± 0.07	2.80 ± 0.16
Serine	2.73 ± 0.08	2.05 ± 0.06	2.09 ± 0.07	2.14 ± 0.33	1.88 ± 0.07	1.93 ± 0.09	2.32 ± 0.10	1.66 ± 0.32	1.31 ± 0.41
Threonine	2.43 ± 0.11	1.88 ± 0.02	1.92 ± 0.03	2.09 ± 0.17	1.90 ± 0.04	1.88 ± 0.08	2.35 ± 0.14	2.02 ± 0.39	1.40 ± 0.45
Tyrosine	3.84 ± 0.13	3.07 ± 0.07	3.12 ± 0.05	3.16 ± 0.06	2.40 ± 0.08	2.56 ± 0.11	3.53 ± 0.02	2.01 ± 0.28	2.42 ± 0.48
Valine	4.11 ± 0.18	4.40 ± 0.07	4.36 ± 0.01	3.90 ± 0.04	3.81 ± 0.09	3.91 ± 0.16	4.38 ± 0.15	4.58 ± 0.06	4.05 ± 0.22
Total EAA	24.4 ± 1.0	22.7 ± 0.3	22.5 ± 0.1	22.8 ± 0.3	20.5 ± 0.4	21.4 ± 0.5	24.1 ± 0.8	22.3 ± 0.9	19.6 ± 0.7
Total AA	45.5 ± 1.5	41.6 ± 0.3	41.5 ± 0.3	43.1 ± 0.9	40.1 ± 0.3	41.7 ± 1.0	44.7 ± 1.2	50.6 ± 1.5	38.3 ± 1.7
E/T [%]	53.6 ± 0.7	54.6 ± 0.4	54.4 ± 0.3	52.9 ± 0.7	51.0 ± 0.7	51.5 ± 0.3	53.8 ± 0.5	44.1 ± 1.6	51.1 ± 0.5

YM: yellow mealworm, YMP: yellow mealworm fermented using *L. plantarum*, YMM: yellow mealworm fermented using commercial starter culture, HC: house cricket, HCP: house cricket fermented using *L. plantarum*, HCM: house cricket fermented using commercial starter culture, ML: migratory locust, MLP: migratory locust fermented using *L. plantarum*, MLM: migratory locust fermented using commercial starter culture.

Table 4. Amino acid scores and essential amino acid index [%].

	YM	YMP	YMM	HC	HCP	HCM	ML	MLP	MLM
Histidine	2.15 ± 0.07	2.47 ± 0.03	2.41 ± 0.03	1.65 ± 0.02	1.52 ± 0.02	1.70 ± 0.02	1.67 ± 0.07	1.80 ± 0.09	1.71 ± 0.03
Isoleucine	1.54 ± 0.02	1.80 ± 0.02	1.76 ± 0.02	1.50 ± 0.04	1.60 ± 0.02	1.57 ± 0.01	1.51 ± 0.02	1.49 ± 0.05	1.66 ± 0.02
Leucine	1.54 ± 0.01	1.73 ± 0.03	1.70 ± 0.01	1.58 ± 0.02	1.64 ± 0.03	1.61 ± 0.01	1.53 ± 0.02	1.39 ± 0.04	1.59 ± 0.01
Lysine	1.75 ± 0.07	1.37 ± 0.04	1.34 ± 0.04	1.99 ± 0.03	1.69 ± 0.01	1.86 ± 0.05	1.90 ± 0.04	1.43 ± 0.15	1.75 ± 0.02
Methionine	0.669 ± 0.014	0.74 ± 0.04	0.714 ± 0.006	0.816 ± 0.013	0.842 ± 0.018	0.816 ± 0.010	0.682 ± 0.027	0.593 ± 0.015	0.568 ± 0.010
Phenylalanine + Tyrosine	3.36 ± 0.02	3.20 ± 0.06	3.27 ± 0.01	3.11 ± 0.03	2.75 ± 0.04	2.76 ± 0.04	3.28 ± 0.05	2.05 ± 0.13	2.80 ± 0.21
Threonine	2.33 ± 0.08	1.97 ± 0.02	2.01 ± 0.02	2.11 ± 0.16	2.06 ± 0.04	1.96 ± 0.04	2.28 ± 0.07	1.73 ± 0.31	1.60 ± 0.57
Valine	2.32 ± 0.05	2.71 ± 0.04	2.70 ± 0.02	2.32 ± 0.02	2.44 ± 0.04	2.40 ± 0.05	2.51 ± 0.09	2.32 ± 0.05	2.71 ± 0.04
EAAI [%]	179 ± 2	184 ± 2	182 ± 1	177 ± 2	172 ± 2	174 ± 1	177 ± 2	150 ± 6	164 ± 6

YM: yellow mealworm, YMP: yellow mealworm fermented using *L. plantarum*, YMM: yellow mealworm fermented using commercial starter culture, HC: house cricket, HCP: house cricket fermented using *L. plantarum*, HCM: house cricket fermented using commercial starter culture, ML: migratory locust, MLP: migratory locust fermented using *L. plantarum*, MLM: migratory locust fermented using commercial starter culture.

According to amino acid scores calculated based on reference protein composition for older children (≥3 years) and adults [50], methionine is limiting amino acid for all unfermented and fermented insects, which is in line with previous studies [77,78], while some studies recognized cystine together with methionine [53,62] and tryptophan [76] as limiting amino acids in insect species analyzed in this study. Also, nitrogen to protein conversion factor (Kp) was determined for yellow mealworm (5.41), yellow mealworm fermented with *L. plantarum* (5.06), house cricket (5.05) and house cricket fermented with *L. plantarum* (4.67), which was comparable to Kp factors for yellow mealworm (4.64–7.35) and house cricket (4.17–5.71) reported in previous studies [51,56,62–64,79–86]. When calculating Kp, it was taken into account that unmeasured amino acids have an 8% contribution to total amino acids [51], and ornithine was excluded from the total amino acids.

Species differed significantly ($p < 0.05$, Table S2) in their alanine, aspartic acid, histidine, isoleucine, leucine, lysine, methionine, ornithine, phenylalanine, proline, serine, tyrosine and valine, while total essential and total amino acids were similar. The reasons for these differences could be attributed to various factors such as species-specific metabolism, dietary differences, environmental factors (temperature, humidity, habitat) and adaptations to specific ecological roles or functions (flight, reproduction, or survival mechanisms) as suggested in previous studies [87–89].

The observed changes in amino acid profiles following fermentation varied significantly across insect species and microbial cultures. In the case of the yellow mealworm, fermentation with both *L. plantarum* and the commercial starter culture resulted in comparable trends, i.e., an increase in alanine (14–15%) and ornithine (115–117%), but reductions in several key amino acids, including aspartic acid (18–25%), glutamic acid (39–41%), lysine (28–30%), serine (24–25%), threonine (21–23%) and tyrosine (19–20%). These results contrast with those of An et al. (2019), who detected an increase in aspartic and glutamic acid when yellow mealworm was fermented with *L. plantarum* [34].

In the case of the house cricket, both fermentation with *L. plantarum* and the commercial starter culture produced similar changes in the amino acid profile. An increase was observed only in ornithine content (115–117%), while several essential and non-essential amino acids decreased. Specifically, reductions were noted for aspartic acid (8–10%), histidine (0–15%), lysine (10–21%), serine (10–12%), threonine (9–10%), and tyrosine (19–24%). Additionally, the total essential amino acid content decreased by 6–10%. These results suggest a potentially adverse impact on protein quality and overall nutritional value, particularly due to the loss of essential amino acids. These results contradict the findings of Kittibunchakul et al. (2023) and Vasilica et al. (2022), who reported exclusively an increase in amino acids after fermentation with lactic acid bacteria and *L. plantarum* [42,43]. Kittibunchakul et al. (2023) mimicked the process of making shrimp paste Kapi, which also involves salting the substrate and a longer fermentation time, which could be the reason for the differences compared to this study, while Vasilica et al. (2022) used the same fermentation time (48 h) and the same *L. plantarum* culture as in this study.

In the case of the migratory locust, the effects of fermentation with *L. plantarum* and the commercial starter culture differed considerably. Fermentation with *L. plantarum* resulted in increased levels of glutamic acid (123%), histidine (22%), isoleucine (11%), ornithine (1181%), and proline (28%), as well as an overall rise in total amino acid content (13%). However, this treatment also led to reductions in aspartic acid (10%), lysine (15%), serine (28%), threonine (14%), tyrosine (43%), and the essential-to-total amino acid (E/T) ratio (18%). In contrast, fermentation with the commercial starter culture resulted in a substantial increase only in ornithine (359%), while most other amino acids declined in concentration. Notable reductions were observed for aspartic acid (27%), histidine (13%), leucine (11%), lysine (21%), methionine (29%), phenylalanine (19%), proline (11%), serine (44%), threonine (40%), tyrosine (31%), as well as total essential amino acids (19%) and total amino acid content (14%). Overall, the results from both fermentation treatments suggest a potentially adverse impact on protein quality and nutritional value, largely due to the depletion of multiple essential amino acids.

As observed in all three insect species, ornithine concentrations significantly increased following fermentation with both microbial cultures. A potential explanation for this could be the involvement of the arginine deiminase (ADI) pathway, which is strongly associated with lactic acid bacteria [90]. In this pathway, arginine is initially converted to citrulline through the action of the enzyme arginine deiminase. Subsequently, citrulline is converted to ornithine by the enzyme ornithine carbamoyl transferase. This pathway plays a crucial role in bacterial survival, particularly under acidic or nitrogen-limited conditions, and also contributes to adenosine triphosphate (ATP) production through the conversion of arginine, with ammonia and carbon dioxide as by-products [91].

Overall, the nutritional consequences of these shifts are mixed. Although increases in amino acids like ornithine are beneficial, the reduction of essential amino acids, such as lysine, methionine, and threonine, is more concerning. Lysine, methionine + cystine, threonine and tryptophan dominate as limiting amino acids in plant-based diets [92], so their reduction could impair protein quality if insect meals are used in food or feed applications. The results also underscore the importance of strain selection and species compatibility in fermentation design to avoid the degradation of nutritionally important amino acids.

3.4. Elemental Composition

Elemental concentrations of fermented and unfermented insects are presented in Tables 5–7 and daily intake values for 100 g of fresh insects for adults in Table 8. The elemental concentrations of unfermented insects are in agreement with data obtained by Finke (2002, 2015), Machado et al. (2024), and Vehar et al. (2025), indicating that species differ significantly ($p < 0.05$, Table S2) in most elements, except for Na, Ni, and Zn [51,62,63,93]. Species-specific patterns in elemental composition have been reported before and are attributed to growing environments (geographic and ecological factors) and processing mechanisms [51,88,94]. However, elements are mainly obtained from dietary sources.

Due to the small number of samples, differences in elemental concentrations were not statistically evaluated; instead, a 10% threshold value was used to confirm the presence of differences, which depended on the insect species and the fermentation culture used. In the case of the yellow mealworm, fermentations with *L. plantarum* and commercial starter culture resulted in increased elemental concentrations. Fermentation with *L. plantarum* increased Al (61%), V (29%), Ni (27%), Ag (34%), Cd (16%), Sn (26%), Sb (14%), Cs (29%), Hg (15%) and U (19%), while commercial starter culture increased Mn (96%), Ag (85%), Cd

(13%), Cs (28%), Hg (14%) and U (11%). In case of house cricket, *L. plantarum* increased B (37%) and Ag (14%) and decreased Al (31%), V (16%), Co (14%), Zn (14%), Cd (12%), Hg (20%) and Pb (21%), while commercial starter culture increased B (18%), Mg (12%), Mn (42%), Ni (11%), Sb (19%) and decreased Al (29%), Zn (12%), Cd (15%) and Pb (16%). In the case of the migratory locust, both fermentation cultures mainly increased the concentrations of elements. *L. plantarum* increased Na (20%), Mn (36%), Fe (23%) and decreased Se (15%), while commercial starter culture increased Na (49%), Mg (23%), P (22%), K (25%), Mn (378%), Fe (19%), Co (20%), Zn (12%), Rb (12%) and Sn (25%) and decreased Cu (16%) and Ag (19%). In general, fermentation resulted in increased concentrations of essential elements Fe (19–23%), K (25%), Mg (12–23%), Mn (36–378%), Na (20–49%), P (22%), decreased Se (24%) and Cu (16%), while Zn increased in migratory locust fermented with commercial starter culture (12%) and decreased in house cricket fermented with both cultures (12–14%).

Table 5. Makro elements (g/kg FW) in unfermented and fermented insects.

Elements	YM	YMP	YMM	HC	HCP	HCM	ML	MLP	MLM
Ca	0.155 ± 0.016	0.148 ± 0.005	0.154 ± 0.004	0.379 ± 0.030	0.368 ± 0.012	0.414 ± 0.002	0.271 ± 0.025	0.277 ± 0.006	0.305 ± 0.033
K	2.71 ± 0.15	2.66 ± 0.00	2.62 ± 0.14	2.90 ± 0.02	2.79 ± 0.03	3.00 ± 0.15	3.13 ± 0.00	3.11 ± 0.23	3.90 ± 0.02
Mg	1.04 ± 0.05	0.960 ± 0.043	1.00 ± 0.05	0.216 ± 0.007	0.213 ± 0.006	0.242 ± 0.015	0.268 ± 0.016	0.272 ± 0.023	0.329 ± 0.015
Na	0.388 ± 0.023	0.383 ± 0.007	0.396 ± 0.025	1.07 ± 0.01	1.04 ± 0.01	1.13 ± 0.02	0.232 ± 0.004	0.279 ± 0.009	0.346 ± 0.030
P	2.79 ± 0.10	2.67 ± 0.06	2.69 ± 0.15	2.14 ± 0.02	2.01 ± 0.00	2.19 ± 0.04	2.07 ± 0.05	2.12 ± 0.21	2.51 ± 0.02
S	1.04 ± 0.07	1.03 ± 0.02	1.02 ± 0.03	1.38 ± 0.01	1.34 ± 0.04	1.39 ± 0.00	1.09 ± 0.04	1.08 ± 0.10	1.14

YM: yellow mealworm, YMP: yellow mealworm fermented using L. plantarum, YMM: yellow mealworm fermented using commercial starter culture, HC: house cricket, HCP: house cricket fermented using L. plantarum, HCM: house cricket fermented using commercial starter culture, ML: migratory locust, MLP: migratory locust fermented using L. plantarum, MLM: migratory locust fermented using commercial starter culture.

Table 6. Mikro elements (mg/kg FW) in unfermented and fermented insects.

Elements	YM	YMP	YMM	HC	HCP	HCM	ML	MLP	MLM
Al	0.607 ± 0.133	0.803 ± 0.070	0.646 ± 0.052	24.9 ± 4.1	17.2 ± 3.5	17.8 ± 2.9	23.3 ± 8.9	37.6 ± 26.7	27.1 ± 6.6
B	1.11 ± 0.17	1.05 ± 0.06	1.03 ± 0.01	0.257 ± 0.036	0.351 ± 0.024	0.303 ± 0.024	0.337 ± 0.029	0.356 ± 0.001	0.329
Cu	5.07 ± 0.33	4.95 ± 0.03	4.95 ± 0.26	5.96 ± 0.12	5.67 ± 0.11	6.27 ± 0.38	10.0 ± 0.4	10.5 ± 0.4	8.44 ± 0.02
Fe	13.8 ± 1.3	12.9 ± 0.5	12.9 ± 0.5	17.4 ± 0.3	15.7 ± 0.8	18.4 ± 0.3	15.7 ± 0.1	19.3 ± 1.7	18.7 ± 0.0
Mn	2.93 ± 0.25	2.88 ± 0.06	5.75 ± 0.22	11.4 ± 0.2	10.8 ± 0.1	16.2 ± 0.7	0.775 ± 0.048	1.05 ± 0.20	3.70 ± 1.23
Mo	0.332 ± 0.003	0.353 ± 0.010	0.336 ± 0.006	0.249 ± 0.001	0.255 ± 0.000	0.251 ± 0.004	0.311 ± 0.063	0.267 ± 0.001	0.277 ± 0.025
Ni	0.147 ± 0.021	0.187 ± 0.030	0.16 ± 0.00	0.0645 ± 0.0063	0.0743 ± 0.0187	0.0716 ± 0.0012	0.256 ± 0.033	0.235 ± 0.004	0.254 ± 0.014
Rb	0.239 ± 0.014	0.226 ± 0.004	0.237 ± 0.018	0.967 ± 0.019	0.883 ± 0.021	0.975 ± 0.012	0.798 ± 0.001	0.796 ± 0.038	0.986 ± 0.013
Sr	0.542 ± 0.075	0.525 ± 0.010	0.551 ± 0.052	0.590 ± 0.046	0.556 ± 0.058	0.614 ± 0.027	0.697 ± 0.057	0.662 ± 0.023	0.759 ± 0.036
Zn	40.0 ± 3.4	39.3 ± 0.2	40.4 ± 3.3	70.2 ± 5.4	60.2 ± 6.8	61.7 ± 1.2	35.5 ± 0.5	35.4 ± 2.7	39.8 ± 0.3

YM: yellow mealworm, YMP: yellow mealworm fermented using *L. plantarum*, YMM: yellow mealworm fermented using commercial starter culture, HC: house cricket, HCP: house cricket fermented using *L. plantarum*, HCM: house cricket fermented using commercial starter culture, ML: migratory locust, MLP: migratory locust fermented using *L. plantarum*, MLM: migratory locust fermented using commercial starter culture.

Table 7. Trace elements (μg/kg FW) in unfermented and fermented insects.

Elements	YM	YMP	YMM	HC	HCP	HCM	ML	MLP	MLM
Ag	0.507 ± 0.029	0.681 ± 0.044	0.937 ± 0.116	0.866 ± 0.028	0.989 ± 0.012	0.850 ± 0.202	2.62 ± 0.54	2.71 ± 0.37	2.11 ± 0.04
As	15.4 ± 2.3	16.1 ± 0.6	16.1 ± 0.4	4.81 ± 2.00	5.09 ± 1.32	4.01 ± 0.75	6.97 ± 0.40	7.57 ± 1.22	19.4 ± 18.5
Cd	25.6 ± 0.5	29.8 ± 0.8	29.0 ± 0.4	11.2 ± 1.3	9.82 ± 0.40	9.52 ± 0.37	12.7 ± 8.0	15.6 ± 3.1	11.2 ± 4.4
Co	17.5 ± 1.6	18.0 ± 0.8	17.3 ± 0.1	23.8 ± 0.7	20.4 ± 0.7	21.6 ± 0.8	8.60 ± 1.00	10.6 ± 3.4	10.3 ± 0.4
Cs	0.382 ± 0.016	0.494 ± 0.052	0.490 ± 0.076	4.46 ± 0.04	4.29 ± 0.39	4.53 ± 0.10	3.68 ± 2.87	1.86	4.35 ± 3.25
Hg	0.282 ± 0.033	0.325 ± 0.002	0.321 ± 0.004	0.641 ± 0.042	0.516 ± 0.025	0.617 ± 0.068	1.06 ± 0.23	1.24 ± 0.18	0.970 ± 0.230
Pb	2.16 ± 0.16	2.34 ± 0.10	2.06 ± 0.03	33.2 ± 0.2	26.3 ± 3.2	28.0 ± 1.1	31.4 ± 5.4	43.2 ± 18.0	34.0 ± 2.67
Sb	0.532 ± 0.018	0.607 ± 0.034	0.493 ± 0.034	1.37 ± 0.20	1.41 ± 0.12	1.63 ± 0.16	2.55 ± 0.41	3.37 ± 0.81	2.78 ± 0.58
Se	42.4 ± 0.6	45.4 ± 0.1	43.6 ± 1.7	98.0 ± 2.3	99.7 ± 4.6	95.1 ± 6.8	83.7 ± 10.1	71.0 ± 5.70	88.9 ± 1.1
Sn	6.13 ± 1.29	7.72 ± 0.09	7.70 ± 1.42	15.7 ± 2.1	14.9 ± 2.9	13.6 ± 1.2	10.1 ± 0.4	12.9 ± 4.0	12.6 ± 2.9
U	23.6 ± 0.4	28.2 ± 1.4	26.2 ± 0.5	31.6 ± 8.4	28.4 ± 2.7	27.8 ± 0.6	1.29 ± 0.21	1.55 ± 0.45	1.21 ± 0.23
V	52.9 ± 7.5	61.6 ± 0.0	58.9 ± 1.1	63.8 ± 9.1	53.9 ± 1.6	56.7 ± 0.7	18.2 ± 4.8	23.5 ± 6.0	17.7 ± 0.7

YM: yellow mealworm, YMP: yellow mealworm fermented using *L. plantarum*, YMM: yellow mealworm fermented using commercial starter culture, HC: house cricket, HCP: house cricket fermented using *L. plantarum*, HCM: house cricket fermented using commercial starter culture, ML: migratory locust, MLP: migratory locust fermented using *L. plantarum*, MLM: migratory locust fermented using commercial starter culture.

Table 8. Daily intake values (%) for 100 g of fresh insects, calculated for adults.

	RDI (mg/day)	YM	YMP	YMM	HC	HCP	HCM	ML	MLP	MLM
Ca	1000	1.5 ± 0.2	1.5 ± 0.1	1.5 ± 0.0	3.8 ± 0.3	3.7 ± 0.1	4.1 ± 0.0	2.7 ± 0.3	2.8 ± 0.1	3.1 ± 0.3
Cu	1.3	39.0 ± 2.6	38.1 ± 0.2	38.1 ± 2.0	45.9 ± 0.9	43.6 ± 0.8	48.3 ± 2.9	76.9 ± 3.0	81.1 ± 2.9	64.9 ± 0.1
Fe	16	8.6 ± 0.8	8.1 ± 0.3	8.1 ± 0.5	10.8 ± 0.2	9.8 ± 0.5	11.5 ± 0.2	9.8 ± 0.0	12.1 ± 1.1	11.7 ± 0.0
K	3500	7.7 ± 0.4	7.6 ± 0.0	7.5 ± 0.4	8.3 ± 0.0	8.0 ± 0.1	8.6 ± 0.4	8.9 ± 0.0	8.9 ± 0.7	11.2 ± 0.1
Mg	300	34.7 ± 1.7	32.0 ± 1.4	33.5 ± 1.6	7.2 ± 0.2	7.1 ± 0.2	8.1 ± 0.5	8.9 ± 0.6	9.1 ± 0.8	11.0 ± 0.5
Mn	3	9.8 ± 0.8	9.6 ± 0.2	19.2 ± 0.7	38.0 ± 0.6	36.1 ± 0.3	53.9 ± 2.3	2.6 ± 0.2	3.5 ± 0.7	12.3 ± 4.1
Na	2000	1.9 ± 0.1	1.9 ± 0.0	2.0 ± 0.1	5.4 ± 0.0	5.2 ± 0.0	5.7 ± 0.1	1.2 ± 0.0	1.4 ± 0.0	1.7 ± 0.1
P	500	55.8 ± 2.0	53.4 ± 1.2	53.8 ± 3.0	42.8 ± 0.4	40.2 ± 0.1	43.8 ± 0.8	41.4 ± 1.1	42.3 ± 4.1	50.3 ± 0.5
Se	0.07	6.1 ± 0.1	6.5 ± 0.0	6.2 ± 0.2	14.0 ± 0.3	14.2 ± 0.7	13.6 ± 1.0	12.0 ± 1.4	10.1 ± 0.8	12.7 ± 0.2
Zn	11	36.3 ± 3.1	35.7 ± 0.2	36.7 ± 3	63.8 ± 4.9	54.7 ± 6.2	56.1 ± 1.1	32.3 ± 0.4	32.1 ± 2.5	36.2 ± 0.3

YM: yellow mealworm, YMP: yellow mealworm fermented using *L. plantarum*, YMM: yellow mealworm fermented using commercial starter culture, HC: house cricket, HCP: house cricket fermented using *L. plantarum*, HCM: house cricket fermented using commercial starter culture, ML: migratory locust, MLP: migratory locust fermented using *L. plantarum*, MLM: migratory locust fermented using commercial starter culture.

Vasilica et al. (2022) examined essential elements, Cr and Ni, in fermented house cricket using atomic absorption spectrophotometry after ash recovery in 20% HCl [43]. They found that *L. plantarum* increased all elemental concentrations after 24 h, while spontaneous sourdough fermentation increased all except P and Zn, which is only partially consistent with our findings. The authors attributed the increase to a reduction in pH, which may activate phytase and reduce mineral-binding compounds, such as phytates and tannins [95]. In contrast, the present study measured total elemental content after acid digestion with ICP-MS. Therefore, the observed increases likely reflect relative concentration effects, such as a reduction in dry matter during fermentation [96], which concentrates remaining minerals, as well as microbial accumulation or redistribution within the biomass. These effects appear to be species- and culture-dependent.

The results show that consuming 100 g of fermented and unfermented insects in fresh weight (FW) would contribute a substantial portion of the recommended daily intake for adults of Cu (38.1–81.1%), P (40.2–55.8%), and Zn (32.1–63.8%). Specifically, 100 g of fermented and unfermented yellow mealworm would provide 32.0–34.7% of the recommended daily intake of Mg, while house cricket would supply 36.1–53.9% of the requirement for Mn. Daily intake values for all essential elements, except Na and Zn, differ statistically significantly between species ($p < 0.05$, Table S2). A risk assessment indicates that consuming 100 g of fresh fermented or unfermented insects poses no health risk to adults. Furthermore, based on the maximum levels established by Directive 2002/32/EC of the European Parliament and of the Council of 7 May 2002 on undesirable substances in animal feed for Cu (15–35 mg/kg FW), Fe (30 mg/kg FW), Cd (2 mg/kg FW), Pb (10 mg/kg FW), As (2 mg/kg FW) and Hg (0.1 mg/kg FW) in the feedstuff, all tested insects comply with feed safety standards [97].

3.5. Oxidation Stability

Numerous studies have explored the impact of fermentation on oxidation stability in various food products, highlighting its potential to enhance the shelf life and stability of lipids and bioactive compounds [98–101]. However, limited research has been conducted specifically on the oxidation stability of edible insects, particularly under the fermentation conditions used in this study. Given the constraints on sample material, oxidation stability measurements were conducted with a single measurement per condition (Figure 1). Among the three insect species analyzed, yellow mealworm exhibited the highest oxidation stability (induction time = 209 min), followed by migratory locust (78 min), then house cricket (66 min). The effect of fermentation on oxidation stability depended on the species. In yellow mealworm, fermentation notably reduced the induction time by 42% with *L. plantarum* and by 41% with the commercial starter culture. A similar, though less pronounced decrease, was observed for migratory locust, with induction times reduced by 21% and 29%, respectively. In contrast, house cricket exhibited improved oxidation stability after fermentation, with an increase in induction time of 153% with *L. plantarum* and 167% with the commercial starter culture.

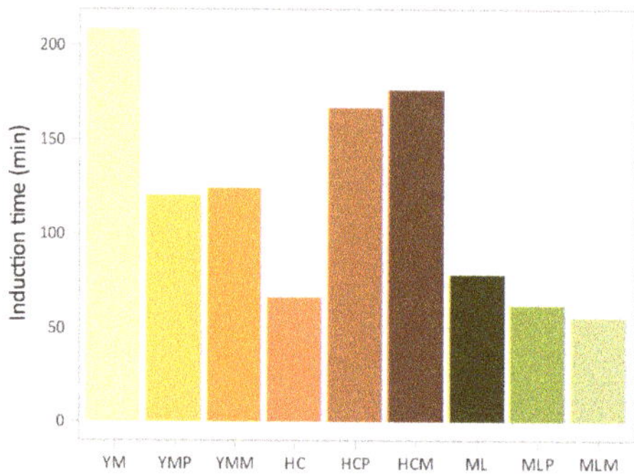

Figure 1. Induction times for insects before and after fermentation (YM: yellow mealworm, YMP: YM + *L. plantarum*, YMM: YM + commercial starter culture, HC: house cricket, HCP: HC + *L. plantarum*, HCM: HC + commercial starter culture, ML: migratory locust, MLP: ML + *L. plantarum*, MLM: ML + commercial starter culture).

These findings suggest that the impact of fermentation on lipid oxidation stability is primarily dependent on the insect species, while the type of microbial culture used plays a comparatively minor role. Similar observations were reported by Borremans, Smets, and Van Campenhout (2020), who noted the onset of secondary lipid oxidation in yellow mealworm during lactic acid fermentation, with further progression over time [102]. This finding aligns with the results of the current study, which found that fermentation reduced the oxidation stability of yellow mealworm. Bernardo and Conte-Junior (2024) also emphasized that the oxidation stability of edible insects is strongly influenced by species-specific lipid composition and the processing methods used [103]. Insects vary in PUFA content, which contributes to differences in susceptibility to oxidation. In this study, yellow mealworm contained the lowest amount of PUFA (Table 2), potentially explaining its initially higher oxidation stability compared to the other two species, except for fermented house cricket, which showed an unexpected increase in induction time.

Processing steps, such as freezing, freeze-drying, cooking and grinding, can also affect oxidation stability. These treatments may alter the nutritional and physicochemical properties of food products through physical, thermal or chemical mechanisms, consequently decreasing their oxidation stability [104]. In the present study, all insect samples were killed by freezing, freeze-dried, ground, and stored frozen prior to fermentation. These methods are known to increase sample porosity and the specific surface area, making the lipids more vulnerable to oxidation [53,103,105].

Overall, the observed changes in oxidation stability were influenced not only by fermentation but also by prior processing methods. Treatments such as freezing, freeze-drying, and grinding likely contributed to increased lipid exposure and susceptibility to oxidation. Thus, oxidation stability in fermented insect products reflects the combined effects of species-specific lipid composition, fermentation conditions, and processing techniques, all of which should be considered in product development.

4. Conclusions

This study demonstrated that the effects of fermenting three edible insect species (yellow mealworm, house cricket, and migratory locust), with *L. plantarum* and a commercial starter culture, are highly dependent on both the insect species and the specific fermentation conditions. Fermentation resulted in a decrease in total dietary fiber and an increase in non-protein nitrogen in both yellow mealworm and house cricket, while other parameters, such as total nitrogen, protein nitrogen, and total fat, remained largely unaffected. No notable improvements were observed in fatty acid profiles, likely due to the relatively short fermentation time. Although increases in specific amino acids, such as ornithine, were beneficial, the reduction of essential amino acids, including lysine, methionine, and threonine, which are often limiting in plant-based diets, was less desirable. Fermentation had mixed effects on mineral composition, enhancing the levels of essential elements like Fe, K, Mg, Mn, Na, P, and Zn in some cases, while reducing others, including Zn, Se, and Cu, with changes reflecting relative concentration effects. Oxidation stability also varied, decreasing in yellow mealworm and migratory locust but improving significantly in house cricket.

These findings highlight the complex interplay between insect species, microbial cultures, and processing conditions in determining nutritional and functional outcomes. Given the limited sample size in this study, future research should include larger sample sets and longer fermentation periods to understand better the mechanisms involved and optimize fermentation strategies for edible insect products. Additionally, it would be valuable to assess bioaccessibility and elemental speciation in future studies to explore how these factors influence the nutritional bioavailability of key micronutrients in fermented insect products.

Supplementary Materials: The following supporting information can be downloaded at: https://www.mdpi.com/article/10.3390/foods14172929/s1, Table S1: Monitored ions for amino acid analysis; Table S2: Significant differences in all measured parameters between all species and yellow mealworm vs. house cricket for nutritional composition (* = $p < 0.05$, x = not statistically analyzed); Table S3: Fatty acid profile (%); Table S4: Reference values for risk assessment; Table S5: Population reference intake and adequate intake values; Table S6: Intake values calculated for the consumption of 100 g of fresh insects for risk assessment.

Author Contributions: A.V.: writing—original draft, investigation, formal analysis, visualization; D.P.: methodology, investigation; M.M.: investigation; M.K. (Mojca Korošec): resources; B.F.: investigation; M.J.H.: investigation; P.J.: methodology; A.O.: investigation; L.K.: funding acquisition, M.K. (Martin Kulma): investigation, D.J.H.: writing—review & editing, funding acquisition, N.O.: writing—review & editing, supervision, funding acquisition. All authors have read and agreed to the published version of the manuscript.

Funding: The authors kindly acknowledge the financial support from the Slovene Research and Innovation Agency (ARIS, program groups P1-0143, P3-0395, P4-0234 and project number J7-3155) and the Czech Science Foundation (GAČR, project number 21-47159L). Anja Vehar also acknowledges ARIS for the Young Researcher Grant.

Data Availability Statement: Data will be made available on request.

Acknowledgments: The authors would like to express their deep gratitude to the late Polona Jamnik for her significant contribution to this work.

Conflicts of Interest: Author Ajda Ota is employed by the company Jata Emona. The remaining authors declare that the research was conducted in the absence of any commercial or financial relationships that could be construed as a potential conflict of interest.

Abbreviations

The following abbreviations are used in this manuscript:

AAS	amino acid score
ADI	arginine deiminase
ALA	α-linolenic acid
AOAC	Association of Official Agricultural Chemists
ATP	adenosine triphosphate
CFU	colony-forming units
DW	dry weight
E/T	essential amino acid to total amino acid ratio
EAA	essential amino acid
EAAI	essential amino acid index
EFSA	European Food Safety Authority
EI	electron impact
EU	European Union
EV	energy value
FAME	fatty acid methyl ester
FAO	Food and Agriculture Organization
FID	flame ionization detection
FW	fresh weight
GC	gas chromatography
GC-MS	gas chromatography coupled to mass spectrometry
h/H	hypocholesterolemic/hypercholesterolemic ratio
HC	house cricket
HCM	house cricket fermented using commercial starter culture
HCP	house cricket fermented using *L. plantarum*
IA	atherogenic index
IT	thrombogenic index
LA	linoleic acid
LOD	limit of detection
ML	migratory locust
MLM	migratory locust fermented using a commercial starter culture
MLP	migratory locust fermented using *L. plantarum*
MRS	De Man, Rogosa and Sharpe broth
MUFA	monounsaturated fatty acid
n-3	omega-3 fatty acids
n-6	omega-6 fatty acids
NPN	non-protein nitrogen
NIST	National Institute of Standards and Technology
PUFA	polyunsaturated fatty acid
QQQ-ICP-MS	triple quadrupole inductively coupled plasma mass spectrometer
RDI	reference daily intake
SFA	saturated fatty acid
SIM	selected ion monitoring
spp.	species
SRM	standard reference material
YM	yellow mealworm
YMM	yellow mealworm fermented using commercial starter culture
YMP	yellow mealworm fermented using *L. plantarum*

References

1. Ordoñez-Araque, R.; Egas-Montenegro, E. Edible Insects: A Food Alternative for the Sustainable Development of the Planet. *Int. J. Gastron. Food Sci.* **2021**, *23*, 100304. [CrossRef]

2. Lange, K.W.; Nakamura, Y. Edible Insects as Future Food: Chances and Challenges. *J. Future Foods* **2021**, *1*, 38–46. [CrossRef]
3. Antoniadis, V.; Molla, A.; Grammenou, A.; Apostolidis, V.; Athanassiou, C.G.; Rumbos, C.I.; Levizou, E. Insect Frass as a Novel Organic Soil Fertilizer for the Cultivation of Spinach (*Spinacia oleracea*): Effects on Soil Properties, Plant Physiological Parameters, and Nutrient Status. *J. Soil Sci. Plant Nutr.* **2023**, *23*, 5935–5944. [CrossRef]
4. Chavez, M.; Uchanski, M. Insect Left-over Substrate as Plant Fertiliser. *JIFF* **2021**, *7*, 683–694. [CrossRef]
5. Dobermann, D.; Swift, J.A.; Field, L.M. Opportunities and Hurdles of Edible Insects for Food and Feed. *Nutr. Bull.* **2017**, *42*, 293–308. [CrossRef]
6. Kuo, C.; Fisher, B.L. A Literature Review of the Use of Weeds and Agricultural and Food Industry By-Products to Feed Farmed Crickets (Insecta; Orthoptera; Gryllidae). *Front. Sustain. Food Syst.* **2022**, *5*, 810421. [CrossRef]
7. Mannaa, M.; Mansour, A.; Park, I.; Lee, D.-W.; Seo, Y.-S. Insect-Based Agri-Food Waste Valorization: Agricultural Applications and Roles of Insect Gut Microbiota. *Environ. Sci. Ecotechnology* **2024**, *17*, 100287. [CrossRef]
8. Ojha, S.; Bußler, S.; Schlüter, O.K. Food Waste Valorisation and Circular Economy Concepts in Insect Production and Processing. *Waste Manag.* **2020**, *118*, 600–609. [CrossRef] [PubMed]
9. Varelas, V. Food Wastes as a Potential New Source for Edible Insect Mass Production for Food and Feed: A Review. *Fermentation* **2019**, *5*, 81. [CrossRef]
10. Čičková, H.; Newton, G.L.; Lacy, R.C.; Kozánek, M. The Use of Fly Larvae for Organic Waste Treatment. *Waste Manag.* **2015**, *35*, 68–80. [CrossRef] [PubMed]
11. Imathiu, S. Benefits and Food Safety Concerns Associated with Consumption of Edible Insects. *NFS J.* **2020**, *18*, 1–11. [CrossRef]
12. Oonincx, D.G.A.B.; De Boer, I.J.M. Environmental Impact of the Production of Mealworms as a Protein Source for Humans—A Life Cycle Assessment. *PLoS ONE* **2012**, *7*, e51145. [CrossRef]
13. Smetana, S.; Schmitt, E.; Mathys, A. Sustainable Use of *Hermetia illucens* Insect Biomass for Feed and Food: Attributional and Consequential Life Cycle Assessment. *Resour. Conserv. Recycl.* **2019**, *144*, 285–296. [CrossRef]
14. Van Broekhoven, S.; Oonincx, D.G.A.B.; Van Huis, A.; Van Loon, J.J.A. Growth Performance and Feed Conversion Efficiency of Three Edible Mealworm Species (Coleoptera: Tenebrionidae) on Diets Composed of Organic by-Products. *J. Insect Physiol.* **2015**, *73*, 1–10. [CrossRef] [PubMed]
15. *Commission Implementing Regulation (EU) 2021/882 of 1 June 2021 Authorising the Placing on the Market of Dried Tenebrio molitor Larva as a Novel Food under Regulation (EU) 2015/2283 of the European Parliament and of the Council, and Amending Commission Implementing Regulation (EU) 2017/2470*; Commission Regulation 2021/882; European Union: Brussels, Belgium, 2021.
16. *Commission Implementing Regulation (EU) 2022/169 of 8 February 2022 Authorising the Placing on the Market of Frozen, Dried and Powder Forms of Yellow Mealworm (Tenebrio molitor Larva) as a Novel Food under Regulation (EU) 2015/2283 of the European Parliament and of the Council, and Amending Commission Implementing Regulation (EU) 2017/2470*; Commission Regulation 2022/169; European Union: Brussels, Belgium, 2022.
17. *Commission Implementing Regulation (EU) 2025/89 of 20 January 2025 Authorising the Placing on the Market of UV-Treated Powder of Whole Tenebrio molitor Larvae (Yellow mealworm) as a Novel Food and Amending Implementing Regulation (EU) 2017/2470*; Commission Regulation 2025/89; European Union: Brussels, Belgium, 2025.
18. *Commission Implementing Regulation (EU) 2021/1975 of 12 November 2021 Authorising the Placing on the Market of Frozen, Dried and Powder Forms of Locusta migratoria as a Novel Food under Regulation (EU) 2015/2283 of the European Parliament and of the Council and Amending Commission Implementing Regulation (EU) 2017/2470*; Commission Regulation 2021/1975; European Union: Brussels, Belgium, 2021.
19. *Commission Implementing Regulation (EU) 2022/188 of 10 February 2022 Authorising the Placing on the Market of Frozen, Dried and Powder Forms of Acheta domesticus as a Novel Food under Regulation (EU) 2015/2283 of the European Parliament and of the Council, and Amending Commission Implementing Regulation (EU) 2017/2470*; Commission Regulation 2022/188; European Union: Brussels, Belgium, 2022.
20. *Commission Implementing Regulation (EU) 2023/5 of 3 January 2023 Authorising the Placing on the Market of Acheta domesticus (House Cricket) Partially Defatted Powder as a Novel Food and Amending Implementing Regulation (EU) 2017/2470*; Commission Regulation 2023/5; European Union: Brussels, Belgium, 2023.
21. *Commission Implementing Regulation (EU) 2023/58 of 5 January 2023 Authorising the Placing on the Market of the Frozen, Paste, Dried and Powder Forms of Alphitobius diaperinus Larvae (Lesser Mealworm) as a Novel Food and Amending Implementing Regulation (EU) 2017/2470*; Commission Regulation 2023/58; European Union: Brussels, Belgium, 2023.
22. *Commission Regulation (EU) 2021/1372 of 17 August 2021 Amending Annex IV to Regulation (EC) No 999/2001 of the European Parliament and of the Council as Regards the Prohibition to Feed Non-Ruminant Farmed Animals, Other than Fur Animals, with Protein Derived from Animals*; Commission Regulation 2021/1372; European Union: Brussels, Belgium, 2021.
23. Olivadese, M.; Dindo, M.L. Edible Insects: A Historical and Cultural Perspective on Entomophagy with a Focus on Western Societies. *Insects* **2023**, *14*, 690. [CrossRef]

24. Kröger, T.; Dupont, J.; Büsing, L.; Fiebelkorn, F. Acceptance of Insect-Based Food Products in Western Societies: A Systematic Review. *Front. Nutr.* **2022**, *8*, 759885. [CrossRef] [PubMed]
25. Rehman, N.; Ogrinc, N. Consumer Perceptions and Acceptance of Edible Insects in Slovenia. *Foods* **2024**, *13*, 2629. [CrossRef]
26. Borremans, A.; Lenaerts, S.; Crauwels, S.; Lievens, B.; Van Campenhout, L. Marination and Fermentation of Yellow Mealworm Larvae (*Tenebrio molitor*). *Food Control* **2018**, *92*, 47–52. [CrossRef]
27. Hernández-Álvarez, A.-J.; Mondor, M.; Piña-Domínguez, I.-A.; Sánchez-Velázquez, O.-A.; Melgar Lalanne, G. Drying Technologies for Edible Insects and Their Derived Ingredients. *Dry. Technol.* **2021**, *39*, 1991–2009. [CrossRef]
28. Melgar-Lalanne, G.; Hernández-Álvarez, A.-J.; Salinas-Castro, A. Edible Insects Processing: Traditional and Innovative Technologies. *Compr. Rev. Food Sci. Food Saf.* **2019**, *18*, 1166–1191. [CrossRef]
29. Nyangena, D.N.; Mutungi, C.; Imathiu, S.; Kinyuru, J. Effects of Traditional Processing Techniques on the Nutritional and Microbiological Quality of Four Edible Insect Species Used for Food and Feed in East Africa. *Foods* **2020**, *9*, 574. [CrossRef]
30. Castro-López, C.; Santiago-López, L.; Vallejo-Cordoba, B.; González-Córdova, A.F.; Liceaga, A.M.; García, H.S.; Hernández-Mendoza, A. An Insight to Fermented Edible Insects: A Global Perspective and Prospective. *Food Res. Int.* **2020**, *137*, 109750. [CrossRef] [PubMed]
31. Di Cagno, R.; Coda, R.; De Angelis, M.; Gobbetti, M. Exploitation of Vegetables and Fruits through Lactic Acid Fermentation. *Food Microbiol.* **2013**, *33*, 1–10. [CrossRef] [PubMed]
32. Gaggia, F.; Di Gioia, D.; Baffoni, L.; Biavati, B. The Role of Protective and Probiotic Cultures in Food and Feed and Their Impact in Food Safety. *Trends Food Sci. Technol.* **2011**, *22*, S58–S66. [CrossRef]
33. Kewuyemi, Y.O.; Kesa, H.; Chinma, C.E.; Adebo, O.A. Fermented Edible Insects for Promoting Food Security in Africa. *Insects* **2020**, *11*, 283. [CrossRef]
34. An, B.; Sam, C.; Dries, V.; Ruben, S.; Christel, V.; Mik, V.; Bart, L.; Leen, V. Comparison of Six Commercial Meat Starter Cultures for the Fermentation of Yellow Mealworm (*Tenebrio molitor*) Paste. *Microorganisms* **2019**, *7*, 540. [CrossRef]
35. Cho, J.-H.; Zhao, H.-L.; Kim, J.-S.; Kim, S.-H.; Chung, C.-H. Characteristics of Fermented Seasoning Sauces Using *Tenebrio molitor* Larvae. *Innov. Food Sci. Emerg. Technol.* **2018**, *45*, 186–195. [CrossRef]
36. Mendoza-Salazar, A.; Santiago-López, L.; Torres-Llanez, M.J.; Hernández-Mendoza, A.; Vallejo-Cordoba, B.; Liceaga, A.M.; González-Córdova, A.F. In Vitro Antioxidant and Antihypertensive Activity of Edible Insects Flours (Mealworm and Grasshopper) Fermented with *Lactococcus lactis* Strains. *Fermentation* **2021**, *7*, 153. [CrossRef]
37. Pérez-Rodríguez, E.; Ibarra-Herrera, C.C.; Pérez-Carrillo, E. Effect of Incorporation of Solid-State Fermented Edible Insects *Tenebrio molitor* and *Sphenarium purpurascens* with *Aspergillus oryzae* in the Elaboration of Bread. *LWT* **2023**, *184*, 115003. [CrossRef]
38. Hadj Saadoun, J.; Luparelli, A.V.; Caligiani, A.; Macavei, L.I.; Maistrello, L.; Neviani, E.; Galaverna, G.; Sforza, S.; Lazzi, C. Antimicrobial Biomasses from Lactic Acid Fermentation of Black Soldier Fly Prepupae and Related By-Products. *Microorganisms* **2020**, *8*, 1785. [CrossRef]
39. Liu, H.; Yang, X.; Mai, L.; Lin, J.; Zhang, L.; Wang, D.; Li, Q. Comparative Proteomic Analysis of *Bacillus subtilis* and *Aspergillus niger* in Black Soldier Fly Co-Fermentation. *Fermentation* **2022**, *8*, 593. [CrossRef]
40. Meng, L.; Ma, L.; Xu, J.; Rong, K.; Peng, N.; Zhao, S. Effect of Enzyme-Assisted Fermentation on Quality, Safety, and Microbial Community of Black Soldier Fly Larvae (*Hermetia illucens* L.) as a Novel Protein Source. *Food Res. Int.* **2023**, *174*, 113624. [CrossRef]
41. Liu, H.; Yang, X.; Yu, X.; Lin, J.; Peng, S.; Li, Q.; Yang, Q.; Wang, D.; Li, Q. Untargeted Metabolomics and PacBio Analysis on Bioactive Components and Microbial Community in Co-Fermentation of Black Soldier Fly Larva. *Food Res. Int.* **2024**, *197*, 115304. [CrossRef] [PubMed]
42. Kittibunchakul, S.; Whanmek, K.; Santivarangkna, C. Physicochemical, Microbiological and Nutritional Quality of Fermented Cricket (*Acheta domesticus*) Paste. *LWT* **2023**, *189*, 115444. [CrossRef]
43. Vasilica, B.B.; Chiș, M.S.; Alexa, E.; Pop, C.; Păucean, A.; Man, S.; Igual, M.; Haydee, K.M.; Dalma, K.E.; Stănilă, S.; et al. The Impact of Insect Flour on Sourdough Fermentation-Fatty Acids, Amino-Acids, Minerals and Volatile Profile. *Insects* **2022**, *13*, 576. [CrossRef]
44. Jamnik, P.; Mahnič, N.; Ekselenski, S.; Pogačnik da Silva, L.; Čadež, N.; Membrino, V.; Poklar Ulrih, N.; Plateis, Z.; Toplak, N.; Koren, S.; et al. Microbial and Biochemical Characterisation of Fermented House Crickets (*Acheta domesticus*) and Mealworm Larvae (*Tenebrio molitor*). *J. Insects Food Feed* **2025**, *1*, 1–21. [CrossRef]
45. AOAC. *Official Methods of Analysis of AOAC International*, 16th ed.; AOAC International: Washington, DC, USA, 1997.
46. DeVries, J.W.; Greene, G.W.; Payne, A.; Zbylut, S.; Scholl, P.F.; Wehling, P.; Evers, J.M.; Moore, J.C. Non-Protein Nitrogen Determination: A Screening Tool for Nitrogenous Compound Adulteration of Milk Powder. *Int. Dairy J.* **2017**, *68*, 46–51. [CrossRef]
47. Mihaly Cozmuta, A.; Nicula, C.; Peter, A.; Mihaly Cozmuta, L.; Nartea, A.; Kuhalskaya, A.; Pacetti, D.; Silvi, S.; Fiorini, D.; Pruteanu, L. Cricket and Yellow Mealworm Powders Promote Higher Bioaccessible Fractions of Mineral Elements in Functional Bread. *J. Funct. Foods* **2022**, *99*, 105310. [CrossRef]

48. FAO. *Food Energy: Methods of Analysis and Conversion Factors*; FAO, Ed.; Report of a Technical Workshop, Rome, Italy, 3–6 December 2002; FAO Food and Nutrition Paper; Food and Agriculture Organization of the United Nations: Rome, Italy, 2003; ISBN 978-92-5-105014-9.
49. Chen, J.; Liu, H. Nutritional Indices for Assessing Fatty Acids: A Mini-Review. *Int. J. Mol. Sci.* **2020**, *21*, 5695. [CrossRef]
50. FAO. *Dietary Protein Quality Evaluation in Human Nutrition*; Report of an FAO Expert Consultation [on Protein Quality Evaluation in Human Nutrition], 31 March–2 April, 2011, Auckland, New Zealand; FAO Food and Nutrition Paper; Food and Agriculture Organization of the United Nations: Rome, Italy, 2013; ISBN 978-92-5-107417-6.
51. Vehar, A.; Potočnik, D.; Strojnik, L.; Zuliani, T.; Heath, D.; Mencin, M.; Vrhovšek, U.; Škvorová, P.; Kouřimská, L.; Kulma, M.; et al. Nutritional Composition of Farmed Insects: Impact of Species, Developmental Stage, and Sex. *J. Insects Food Feed* **2025**, 1–23. [CrossRef]
52. Carnesecchi, E.; Mostrag, A.; Ciacci, A.; Roncaglioni, A.; Tarkhov, A.; Gibin, D.; Sartori, L.; Benfenati, E.; Yang, C.; Dorne, J.L.C.M. *OpenFoodTox: EFSA's Chemical Hazards Database*; European Food Safety Authority: Parma, Italy, 2023.
53. Khatun, H.; Claes, J.; Smets, R.; De Winne, A.; Akhtaruzzaman, M.; Van Der Borght, M. Characterization of Freeze-Dried, Oven-Dried and Blanched House Crickets (*Acheta domesticus*) and Jamaican Field Crickets (*Gryllus assimilis*) by Means of Their Physicochemical Properties and Volatile Compounds. *Eur. Food Res. Technol.* **2021**, *247*, 1291–1305. [CrossRef]
54. Kulma, M.; Kouřimská, L.; Plachý, V.; Božik, M.; Adámková, A.; Vrabec, V. Effect of Sex on the Nutritional Value of House Cricket, *Acheta domestica* L. *Food Chem.* **2019**, *272*, 267–272. [CrossRef]
55. Messina, C.M.; Gaglio, R.; Morghese, M.; Tolone, M.; Arena, R.; Moschetti, G.; Santulli, A.; Francesca, N.; Settanni, L. Microbiological Profile and Bioactive Properties of Insect Powders Used in Food and Feed Formulations. *Foods* **2019**, *8*, 400. [CrossRef] [PubMed]
56. Yi, L.; Lakemond, C.M.M.; Sagis, L.M.C.; Eisner-Schadler, V.; Van Huis, A.; Van Boekel, M.A.J.S. Extraction and Characterisation of Protein Fractions from Five Insect Species. *Food Chem.* **2013**, *141*, 3341–3348. [CrossRef]
57. Christensen, P.; Glitsø, V.; Pettersson, D.; Wischmann, B. Fibre Degrading Enzymes and *Lactobacillus plantarum* Influence Liquid Feed Characteristics and the Solubility of Fibre Components and Dry Matter in Vitro. *Livest. Sci.* **2007**, *109*, 100–103. [CrossRef]
58. Yin, H.; Zhong, Y.; Xia, S.; Hu, J.; Nie, S.; Xiong, T.; Xie, M. Effects of Fermentation with *Lactobacillus plantarum* NCU137 on Nutritional, Sensory and Stability Properties of Coix (*Coix Lachryma-jobi* L.) Seed. *Food Chem.* **2020**, *314*, 126037. [CrossRef]
59. Yu, Q.; Xu, J.; Li, M.; Xi, Y.; Sun, H.; Xie, Y.; Cheng, Q.; Li, P.; Chen, C.; Yang, F.; et al. Synergistic Effects of Ferulic Acid Esterase-producing Lactic Acid Bacteria, Cellulase and Xylanase on the Fermentation Characteristics, Fibre and Nitrogen Components and Microbial Community Structure of *Broussonetia papyrifera* during Ensiling. *J. Sci. Food Agric.* **2024**, *104*, 3543–3558. [CrossRef]
60. Huang, G.; Su, D.; Lee, Y.; Zou, X.; Dong, L.; Deng, M.; Zhang, R.; Huang, F.; Zhang, M. Accumulation of Water-Soluble Polysaccharides during Lychee Pulp Fermentation with *Lactiplantibacillus plantarum* Involves Endoglucanase Expression. *J. Agric. Food Chem.* **2025**, *73*, 3669–3679. [CrossRef]
61. Dreassi, E.; Cito, A.; Zanfini, A.; Materozzi, L.; Botta, M.; Francardi, V. Dietary Fatty Acids Influence the Growth and Fatty Acid Composition of the Yellow Mealworm *Tenebrio molitor* (Coleoptera: Tenebrionidae). *Lipids* **2017**, *52*, 285–294. [CrossRef] [PubMed]
62. Finke, M.D. Complete Nutrient Composition of Commercially Raised Invertebrates Used as Food for Insectivores. *Zoo Biol.* **2002**, *21*, 269–285. [CrossRef]
63. Finke, M.D. Complete Nutrient Content of Four Species of Commercially Available Feeder Insects Fed Enhanced Diets during Growth. *Zoo Biol.* **2015**, *34*, 554–564. [CrossRef]
64. Perez-Santaescolastica, C.; De Pril, I.; Van De Voorde, I.; Fraeye, I. Fatty Acid and Amino Acid Profiles of Seven Edible Insects: Focus on Lipid Class Composition and Protein Conversion Factors. *Foods* **2023**, *12*, 4090. [CrossRef]
65. Simopoulos, A.P. The Importance of the Ratio of Omega-6/Omega-3 Essential Fatty Acids. *Biomed. Pharmacother.* **2002**, *56*, 365–379. [CrossRef]
66. Hamułka, J.; Głąbska, D.; Guzek, D.; Białkowska, A.; Sulich, A. Intake of Saturated Fatty Acids Affects Atherogenic Blood Properties in Young, Caucasian, Overweight Women Even without Influencing Blood Cholesterol. *Int. J. Environ. Res. Public Health* **2018**, *15*, 2530. [CrossRef]
67. Czernichow, S.; Thomas, D.; Bruckert, E. *N*-6 Fatty Acids and Cardiovascular Health: A Review of the Evidence for Dietary Intake Recommendations. *Br. J. Nutr.* **2010**, *104*, 788–796. [CrossRef]
68. Milićević, D.; Vranić, D.; Mašić, Z.; Parunović, N.; Trbović, D.; Nedeljković-Trailović, J.; Petrović, Z. The Role of Total Fats, Saturated/Unsaturated Fatty Acids and Cholesterol Content in Chicken Meat as Cardiovascular Risk Factors. *Lipids Health Dis.* **2014**, *13*, 42. [CrossRef]
69. Lawal, K.G.; Kavle, R.R.; Akanbi, T.O.; Mirosa, M.; Agyei, D. Lipid Nutritional Indices, Regioisomeric Distribution, and Thermal Properties of *Tenebrio molitor* and *Hermetia illucens* Larvae Fat. *J. Asia-Pac. Entomol.* **2022**, *25*, 101951. [CrossRef]
70. Ulbricht, T.L.V.; Southgate, D.A.T. Coronary Heart Disease: Seven Dietary Factors. *The Lancet* **1991**, *338*, 985–992. [CrossRef] [PubMed]

71. Santos-Silva, J.; Bessa, R.J.B.; Santos-Silva, F. Effect of Genotype, Feeding System and Slaughter Weight on the Quality of Light Lambs II. Fatty Acid Composition of Meat. *Livest. Prod. Sci.* **2002**, *77*, 187–194. [CrossRef]
72. Breuer, U.; Harms, H. *Debaryomyces hansenii*—An Extremophilic Yeast with Biotechnological Potential. *Yeast* **2006**, *23*, 415–437. [CrossRef]
73. Chaillou, S.; Champomier-Vergès, M.-C.; Cornet, M.; Crutz-Le Coq, A.-M.; Dudez, A.-M.; Martin, V.; Beaufils, S.; Darbon-Rongère, E.; Bossy, R.; Loux, V.; et al. The Complete Genome Sequence of the Meat-Borne Lactic Acid Bacterium *Lactobacillus sakei* 23K. *Nat. Biotechnol.* **2005**, *23*, 1527–1533. [CrossRef] [PubMed]
74. Søndergaard, A.K.; Stahnke, L.H. Growth and Aroma Production by *Staphylococcus xylosus*, *S. carnosus* and *S. equorum*—A Comparative Study in Model Systems. *Int. J. Food Microbiol.* **2002**, *75*, 99–109. [CrossRef] [PubMed]
75. Aleknavičius, D.; Lukša, J.; Strazdaitė-Žielienė, Ž.; Servienė, E. The Bacterial Microbiota of Edible Insects *Acheta domesticus* and *Gryllus assimilis* Revealed by High Content Analysis. *Foods* **2022**, *11*, 1073. [CrossRef]
76. Köhler, R.; Kariuki, L.; Lambert, C.; Biesalski, H.K. Protein, Amino Acid and Mineral Composition of Some Edible Insects from Thailand. *J. Asia-Pac. Entomol.* **2019**, *22*, 372–378. [CrossRef]
77. Škvorová, P.; Kulma, M.; Božik, M.; Kurečka, M.; Plachý, V.; Slavíková, D.; Šebelová, K.; Kouřimská, L. Evaluation of Rapeseed Cake as a Protein Substitute in the Feed of Edible Crickets: A Case Study Using *Gryllus assimilis*. *Food Chem.* **2024**, *441*, 138254. [CrossRef] [PubMed]
78. Araújo, R.R.S.; Fagundes, M.M.A.; Viana, A.M.F.; Paulino, A.H.S.; Silva, M.E.; Santos, E.M. Protein Quality Evaluation in Vivo of Cricket Flour (*Gryllus assimilis*) Reared in Brazil. *J. Insects Food Feed.* **2024**, *8*, 409–416. [CrossRef]
79. Belghit, I.; Lock, E.-J.; Fumière, O.; Lecrenier, M.-C.; Renard, P.; Dieu, M.; Berntssen, M.H.G.; Palmblad, M.; Rasinger, J.D. Species-Specific Discrimination of Insect Meals for Aquafeeds by Direct Comparison of Tandem Mass Spectra. *Animals* **2019**, *9*, 222. [CrossRef]
80. Boulos, S.; Tännler, A.; Nyström, L. Nitrogen-to-Protein Conversion Factors for Edible Insects on the Swiss Market: *T. molitor*, *A. domesticus*, and *L. migratoria*. *Front. Nutr.* **2020**, *7*, 89. [CrossRef]
81. Finke, M.D. Estimate of Chitin in Raw Whole Insects. *Zoo Biol.* **2007**, *26*, 105–115. [CrossRef]
82. Hussain, I.; Khan, S.; Sultan, A.; Chand, N.; Khan, R.; Alam, W.; Ahmad, N. Mealworm (*Tenebrio molitor*) as Potential Alternative Source of Protein Supplementation in Broiler. *Int. J. Biosci.* **2017**, *10*, 255–262. [CrossRef]
83. Janssen, R.H.; Vincken, J.-P.; Van Den Broek, L.A.M.; Fogliano, V.; Lakemond, C.M.M. Nitrogen-to-Protein Conversion Factors for Three Edible Insects: *Tenebrio molitor*, *Alphitobius diaperinus*, and *Hermetia illucens*. *J. Agric. Food Chem.* **2017**, *65*, 2275–2278. [CrossRef] [PubMed]
84. Lampová, B.; Kopecká, A.; Šmíd, P.; Kulma, M.; Kurečka, M.; Ogrinc, N.; Heath, D.; Kouřimská, L.; Doskočil, I. Evaluating Protein Quality in Edible Insects: A Comparative Analysis of House Cricket, Yellow Mealworm, and Migratory Locust Using DIAAS Methodologies. *LWT* **2024**, *213*, 117062. [CrossRef]
85. Nakagaki, B.J.; Sunde, M.L.; Defoliart, G.R. Protein Quality of the House Cricket, *Acheta domesticus*, When Fed to Broiler Chicks. *Poult. Sci.* **1987**, *66*, 1367–1371. [CrossRef]
86. Ritvanen, T.; Pastell, H.; Welling, A.; Raatikainen, M. The Nitrogen-to-Protein Conversion Factor of Two Cricket Species—*Acheta domesticus* and *Gryllus bimaculatus*. *AFSci* **2020**, *29*, 1–5. [CrossRef]
87. Hughes, A.L. Evolutionary Conservation of Amino Acid Composition in Paralogous Insect Vitellogenins. *Gene* **2010**, *467*, 35–40. [CrossRef]
88. Meyer-Rochow, V.B.; Gahukar, R.T.; Ghosh, S.; Jung, C. Chemical Composition, Nutrient Quality and Acceptability of Edible Insects Are Affected by Species, Developmental Stage, Gender, Diet, and Processing Method. *Foods* **2021**, *10*, 1036. [CrossRef] [PubMed]
89. Oonincx, D.G.A.B.; Finke, M.D. Nutritional Value of Insects and Ways to Manipulate Their Composition. *JIFF* **2021**, *7*, 639–659. [CrossRef]
90. Hwang, H.; Lee, J.-H. Characterization of Arginine Catabolism by Lactic Acid Bacteria Isolated from Kimchi. *Molecules* **2018**, *23*, 3049. [CrossRef]
91. Novák, L.; Zubáčová, Z.; Karnkowska, A.; Kolisko, M.; Hroudová, M.; Stairs, C.W.; Simpson, A.G.B.; Keeling, P.J.; Roger, A.J.; Čepička, I.; et al. Arginine Deiminase Pathway Enzymes: Evolutionary History in Metamonads and Other Eukaryotes. *BMC Evol. Biol.* **2016**, *16*, 197. [CrossRef]
92. Hambræus, L. Protein and Amino Acids in Human Nutrition. In *Reference Module in Biomedical Sciences*; Elsevier: Amsterdam, The Netherlands, 2014; p. B9780128012383000283. ISBN 978-0-12-801238-3.
93. Machado, I.; Priede, A.S.; Rodríguez, M.C.; Heath, D.; Heath, E.; Kouřimská, L.; Kulma, M.; Bettmer, J.; Montes-Bayón, M. Bioaccessibility of Trace Elements and Fe and Al Endogenic Nanoparticles in Farmed Insects: Pursuing Quality Sustainable Food. *Food Chem.* **2024**, *458*, 140229. [CrossRef]
94. Haque, M.M.; Hossain, N.; Jolly, Y.N.; Tareq, S.M. Probabilistic Health Risk Assessment of Toxic Metals in Chickens from the Largest Production Areas of Dhaka, Bangladesh. *Environ. Sci. Pollut. Res.* **2021**, *28*, 51329–51341. [CrossRef]

95. Chiş, M.S.; Păucean, A.; Man, S.M.; Vodnar, D.C.; Teleky, B.-E.; Pop, C.R.; Stan, L.; Borsai, O.; Kadar, C.B.; Urcan, A.C.; et al. Quinoa Sourdough Fermented with *Lactobacillus plantarum* ATCC 8014 Designed for Gluten-Free Muffins—A Powerful Tool to Enhance Bioactive Compounds. *Appl. Sci.* **2020**, *10*, 7140. [CrossRef]
96. Lau, N.; Hummel, J.; Kramer, E.; Hünerberg, M. Fermentation of Liquid Feed with Lactic Acid Bacteria Reduces Dry Matter Losses, Lysine Breakdown, Formation of Biogenic Amines, and Phytate-Phosphorus. *Transl. Anim. Sci.* **2022**, *6*, txac007. [CrossRef]
97. The European Parliament and the Council. *Directive 2002/32/EC of the European Parliament and of the Council of 7 May 2002 on Undesirable Substances in Animal Feed*; The European Parliament and the Council: Strasbourg, France, 2002.
98. Hur, S.J.; Lee, S.Y.; Kim, Y.-C.; Choi, I.; Kim, G.-B. Effect of Fermentation on the Antioxidant Activity in Plant-Based Foods. *Food Chem.* **2014**, *160*, 346–356. [CrossRef]
99. Feng, L.; Tang, N.; Liu, R.; Gong, M.; Wang, Z.; Guo, Y.; Wang, Y.; Zhang, Y.; Chang, M. The Relationship between Flavor Formation, Lipid Metabolism, and Microorganisms in Fermented Fish Products. *Food Funct.* **2021**, *12*, 5685–5702. [CrossRef]
100. Slaný, O.; Klempová, T.; Cibulková, Z.; Marcinčák, S.; Shapaval, V.; Čertík, M. Evaluation of Stability and Quality of Bioproducts Derived from Solid-state Fermentation of Wheat Bran Using *Mortierella alpina*. *J. Food Sci.* **2025**, *90*, e70188. [CrossRef] [PubMed]
101. Nassu, R.T.; Gonçalves, L.A.G.; Pereira Da Silva, M.A.A.; Beserra, F.J. Oxidative Stability of Fermented Goat Meat Sausage with Different Levels of Natural Antioxidant. *Meat Sci.* **2003**, *63*, 43–49. [CrossRef]
102. Borremans, A.; Smets, R.; Van Campenhout, L. Fermentation Versus Meat Preservatives to Extend the Shelf Life of Mealworm (*Tenebrio molitor*) Paste for Feed and Food Applications. *Front. Microbiol.* **2020**, *11*, 1510. [CrossRef]
103. Bernardo, Y.A.A.; Conte-Junior, C.A. Oxidative Stability in Edible Insects: Where Is the Knowledge Frontier? *Trends Food Sci. Technol.* **2024**, *148*, 104518. [CrossRef]
104. Ojha, S.; Bußler, S.; Psarianos, M.; Rossi, G.; Schlüter, O.K. Edible Insect Processing Pathways and Implementation of Emerging Technologies. *JIFF* **2021**, *7*, 877–900. [CrossRef]
105. Keil, C.; Grebenteuch, S.; Kröncke, N.; Kulow, F.; Pfeif, S.; Kanzler, C.; Rohn, S.; Boeck, G.; Benning, R.; Haase, H. Systematic Studies on the Antioxidant Capacity and Volatile Compound Profile of Yellow Mealworm Larvae (*T. molitor* L.) under Different Drying Regimes. *Insects* **2022**, *13*, 166. [CrossRef] [PubMed]

foods

MDPI

Article

Novel Insight into Metabolism Mechanism of Biogenic Amines During Fermentation of Chinese Traditional Fermented Mandarin Fish (*Chouguiyu*) Based on Metabolism Pathway and Correlation Network

Jun Li [1], Daqiao Yang [2,3,*], Yongqiang Zhao [2], Di Wang [2], Hui Huang [2] and Chunsheng Li [2,*]

1 Guangdong Provincial Key Laboratory of Lingnan Specialty Food Science and Technology, Key Laboratory of Green Processing and Intelligent Manufacturing of Lingnan Specialty Food, Ministry of Agriculture and Rural, College of Light Industry and Food, Zhongkai University of Agriculture and Engineering, Guangzhou 510225, China; lijun@zhku.edu.cn

2 Key Laboratory of Aquatic Product Processing, Ministry of Agriculture and Rural Affairs, National R&D Center for Aquatic Product Processing, South China Sea Fisheries Research Institute, Chinese Academy of Fishery Sciences, Guangzhou 510300, China; zhaoyq@scsfri.ac.cn (Y.Z.); wangdi@scsfri.ac.cn (D.W.); huanghuigd@aliyun.com (H.H.)

3 College of Life Sciences, Linyi University, Linyi 276005, China

* Correspondence: yangdaqiao233@163.com (D.Y.); lichunsheng@scsfri.ac.cn (C.L.); Tel.: +86-20-89020911 (C.L.)

Abstract: A complex microbial community gives the possibility to produce biogenic amines in traditional fermented foods. In this study, the metabolism mechanisms of biogenic amines during fermentation of fermented mandarin fish *Chouguiyu* were revealed based on the metabolic pathways and correlation analysis. Functional genes based on KEGG orthology related to biogenic amine metabolism were selected from the metagenome and were used to construct the biogenic amine metabolic pathways in *Chouguiyu*. A total of 91 and 75 genera were related to the synthesis and degradation of biogenic amines, respectively. High concentrations of cadaverine and putrescine were observed, while the other biogenic amines were detected in relatively low concentrations. The metabolic mechanisms of various biogenic amines were illuminated by correlation network maps between biogenic amines and microbial synthesis/degradation enzymes. *Lactococcus*, *Flavobacterium*, *Tessaracoccus*, and *Yoonia* could only degrade and not produce biogenic amines. *Acinetobacter* and *Enterococcus* possessed more abundant enzymes for degradation than synthesis. Amine oxidase (K00276), diamine N-acetyltransferase (K00657), and gamma-glutamylputrescine synthase (K09470) were the main biogenic amine degradation enzymes in the microbial community. *Lactococcus garvieae*, *Flavobacterium gelidilacus*, *Tessaracoccus antarcticus*, *Yoonia vestfoldensis*, *Acinetobacter haemolyticus*, and *Enterococcus ureasiticus* were the main microbial species for biogenic amine degradation and could be isolated as the potential strains for biogenic amine degradation in fermented foods.

Keywords: fermented mandarin fish; biogenic amine; synthesis; degradation; metagenome; correlation network; food safety

1. Introduction

Biogenic amines (BAs) in foods are mainly produced by the decarboxylation of amino acids under the action of microbial decarboxylases [1]. Excessive BAs can cause a variety of harmful reactions, such as headaches, rashes, nausea, diarrhea, flushing, and neurotoxicity [2]. BAs are easily produced in fermented foods because of the high concentrations of free amino acids and complex microbial community [3]. Many researchers have studied the BA synthesis and degradation ability of microorganisms in fermented foods after isolation and identification [4,5]. However, it is difficult to give a comprehensive understanding of the microbial metabolism mechanism of BAs in fermented foods.

With the development of microbial analysis, high-throughput sequencing technology can comprehensively analyze the complex microbial community in fermented foods [6,7]. Many studies have discovered BA-producing and BA-degrading microorganisms through correlation analysis between microbial communities and BAs [8,9]. However, correlation analysis only considers whether the abundance of the microbial community is consistent with the change in BA content and ignores the relevant enzymes for BA metabolism in these microorganisms. It cannot determine the core microorganisms that contribute to the synthesis and degradation of BAs in fermented foods. Metagenomics can predict microbial function by analysis of the whole genetic material of microbial community [10,11]. It provides an effective method to select the enzymes related to BA metabolism from microorganisms in fermented foods [12].

Chouguiyu is a traditional fermented food made from mandarin fish in China, and it has a unique flavor and taste. Most studies have focused on the study of changes in free amino acids [13], volatile flavor compounds [14], and microbial community [11,15,16] during fermentation of *Chouguiyu*. There are plenty of microorganisms in *Chouguiyu* due to the low-salt fermentation process, which can not only generate unique flavor and taste substances but also produce BAs. However, there is a lack of research on the changes in BAs during fermentation of *Chouguiyu*. Moreover, the synthesis and degradation mechanisms of BAs based on microbial metabolism have rarely been reported.

Therefore, in this study, metagenome sequencing was performed using an Illumina Hiseq Xten to analyze the whole genetic material of the microbial community during *Chouguiyu* fermentation. The key genes that were associated with the synthesis and degradation of various BAs in the microbial community were selected after KEGG orthology (KO) annotation. Combined with the results of BA concentrations, the metabolism mechanisms of different BAs during fermentation of *Chouguiyu* were revealed based on the BA metabolic pathways and correlation networks between microorganisms and BAs. The core microorganisms for BA degradation in *Chouguiyu* were also identified. This study is expected to provide a comprehensive understanding of the synthesis and degradation mechanisms of BAs in *Chouguiyu*, which is helpful in the isolation of functional strains for the degradation of BAs.

2. Materials and Methods

2.1. Fermentation of Chouguiyu

Chouguiyu was made by fermenting mandarin fish coming from Guangdong, China, (*Siniperca chuatsi*, av. 500 g, frozen and thawed before fermentation) at 15–20 °C for 8 days in Anhui, China, according to previous studies [11,15,16]. Briefly, frozen mandarin fish was thawed, gutted, and cleaned. Then, the prepared fish was soaked in brine (6% salt and 0.02% spices). Fermentation used wild microbiota (no starter cultures). Three samples of *Chouguiyu* were collected on Day 1 (D1), Day 2 (D2), Day 4 (D4), Day 6 (D6), and Day 8 (D8) for further study.

2.2. Metagenome Sequencing of Microbial Community in Chouguiyu

The microorganisms in *Chouguiyu* were obtained after centrifugation of fermented liquid in each sample. Shotgun metagenomics was adopted to analyze the microbial community, as per previous studies [11,15,16]. Briefly, total genomic DNA was extracted from microorganisms using the E.Z.N.A.® Soil DNA Kit (Omega Bio-tek, Norcross, GA,

USA). After the verification of DNA purity, the genomic DNA from microorganisms was fragmented to construct the DNA library using NEXTFLEX® Rapid DNA-Seq (Bioo Scientific, Austin, TX, USA). The DNA library was then used for paired-end sequencing through the Illumina Hiseq Xten (Illumina Inc., San Diego, CA, USA).

2.3. Gene Assembly and Functional Annotation

Sequencing data were assembled using MEGAHIT (version 1.1.2) to obtain the contigs. Open reading frames (ORFs) from each assembled contig were predicted using MetaGene. CD-HIT (version 4.6.1) was used to cluster the gene sequences (90% identity and 90% coverage). The longest gene in each class was selected as the representative sequence to construct the non-redundant gene set. The obtained genes were aligned to the KEGG and COG databases to obtain their annotation. The BA metabolic pathways in the microbial community of *Chouguiyu* were constructed based on the KO in the KEGG pathways. The abundance of enzymes from microbial genera was calculated based on the sum of KO abundance related to BA metabolism. Representative sequences of non-redundant gene catalogs were aligned to the NCBI NR database with an e-value cutoff of 1e-5 using Diamond (version 0.8.35).

2.4. Determination of Biogenic Amines

The quantification of tryptamine, β-phenylethylamine, putrescine, cadaverine, histamine, tyramine, spermidine, and spermine was carried out as previously described [17] with some modifications. Briefly, a total of 5.00 g of samples was added to centrifuge tubes with 10 mL perchloric acid (0.4 mol/L) and was ground for 20 min at 40 Hz using a Multi-Tube Vortexer (EFAA-HM-01, Yalin, Shanghai, China). After centrifugation at 4000 r/min for 20 min using a high-speed tabletop refrigerated centrifuge (TDZ5-WS, Xiangyi, Changsha, Hunan, China), the supernatant (1 mL) was mixed with 100 μL sodium hydroxide (2 mol/L) and 300 μL saturated sodium bicarbonate. The derivative reaction was performed by adding dansyl chloride (1 mL) into the mixture, followed by incubation in the dark at 40 °C for 45 min. The BAs were detected using a high-performance liquid chromatograph (LC-20AD, Shimadzu, Nakagyo-ku, Kyoto, Japan) equipped with an ultraviolet detector (SPD-M20A, Shimadzu, Nakagyo-ku, Kyoto, Japan) on the WondaCract ODS-2 column (Shimadzu-GL, Nakagyo-ku, Kyoto, Japan). Solvent A and solvent B were acetonitrile and ammonium acetate, respectively. Elution program was performed as follows: 0.1 min, 55% B; 7 min, 65% B; 14 min, 70% B; 20 min, 70% B; 27 min, 90% B; 30 min, 100% B; 31 min, 100% B, 32 min, 55% B, and 37 min, 55% B with a flow rate of 0.80 mL/min. The BA contents of each sample were calculated according to the standard curves (Table S1).

2.5. Statistical Analysis

The statistical analysis was conducted using SPSS 20.0 software (SPSS Inc., Chicago, IL, USA) with one-way analysis of variance and multiple comparison Tukey tests. The correlation network maps between microbial BA metabolism enzymes and BA contents were built according to Pearson's correlation by Cytoscape v3.9.1. Heatmaps were produced using Excel (Microsoft Office Home and Student 2019, Redmond, WA, USA).

3. Results and Discussion

3.1. Metagenomic Analysis of Microbial Community in *Chouguiyu*

BAs are potential risks in *Chouguiyu* due to high concentrations of free amino acids and the complex microbial community. To investigate this, we analyzed the metagenome of the microbial community during *Chouguiyu* fermentation (Figure 1). According to the taxonomic analysis of obtained genes of the metagenome, Firmicutes and Proteobacteria were the main phyla and their total abundance accounted for over 85% during fermentation (Figure 1A). With increasing fermentation time, Firmicutes obviously increased while Proteobacteria obviously decreased, in accordance with the results in other fermented aquatic products [7,18], reflecting a successional pattern common to fermented fish. As shown in Figure 1B, there were a total of 66 microbial genera (relative abundance > 0.1%). *Vagococcus*, *Acinetobacter*, *Psychrobacter*, *Peptostreptococcus*, *Enterococcus*, *Myroides*, and *Streptococcus* were the dominant genera during the fermentation of *Chouguiyu*. Among these genera, *Acinetobacter* was the most abundant genus in the initial fermentation but was significantly inhibited thereafter. A similar trend was also found in *Myroides*. With an increase in fermentation time, the abundance of *Vagococcus*, *Peptostreptococcus*, and *Enterococcus* obviously increased, while *Psychrobacter* first increased and then decreased. *Vagococcus*, *Peptostreptococcus*, *Psychrobacter*, and *Enterococcus* became dominant at the end of fermentation.

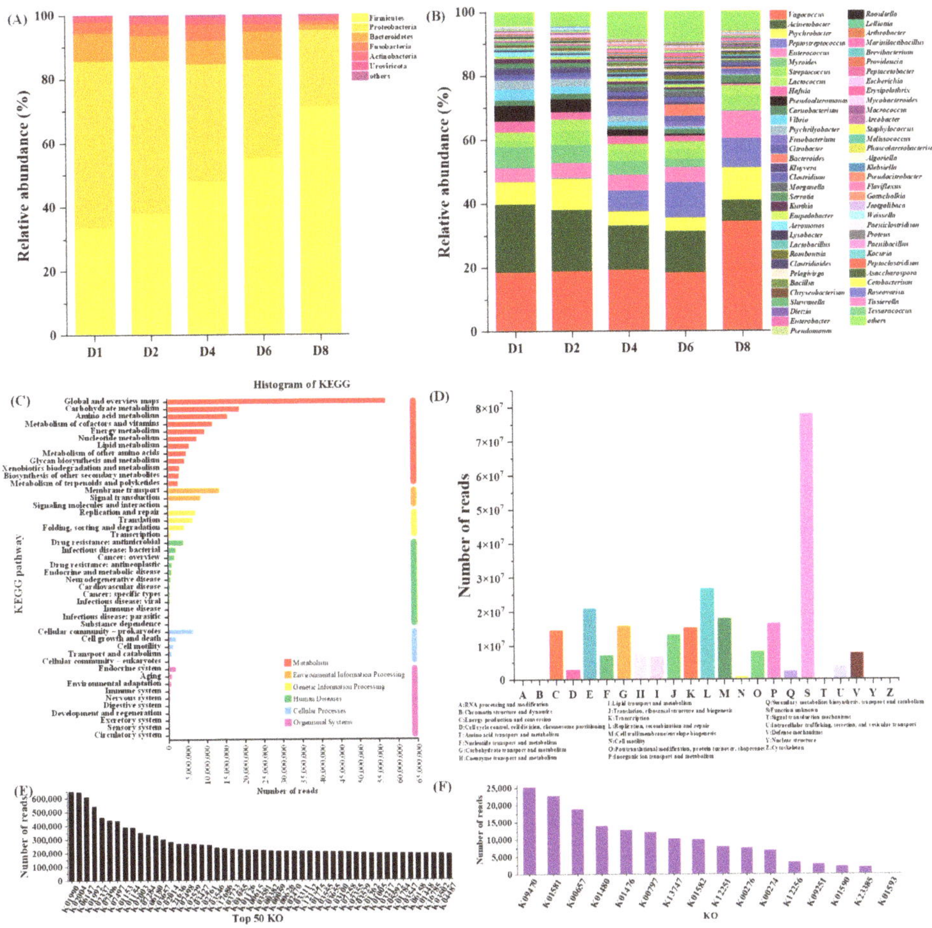

Figure 1. Metagenomic analysis of microbial community during fermentation of *Chouguiyu*. (**A**) Microbial taxonomic composition at the (**A**) phylum and (**B**) genus levels. Gene annotations of metagenome with the (**C**) KEGG and (**D**) COG databases. The (**E**) top 50 KO functions of the gene sequences and the (**F**) KO functions related to BA metabolism.

In order to obtain gene annotations of the metagenome of the microbial community, the gene sequences were further annotated using the KEGG and COG databases. As shown in Figure 1C, the KEGG pathways mainly focused on metabolism, among which carbohydrate metabolism and amino acid metabolism had the highest abundance. As shown in Figure 1D, except for function unknown (S), most gene sequences are annotated as replication, recombination, and repair (L) and amino acid transport and metabolism (E). The high abundance of genes related to amino acid metabolism provided the basis for analyzing the metabolism process of BAs. KEGG orthology (KO) is a straight-homologous classification system that can group genes with similar sequences and functions and use the functions of known genes as the KO functions for cross-species annotation. All the gene sequences of the microbial community in *Chouguiyu* were totally annotated with 13,968 KO functions. The top 50 KO functions are shown in Figure 1E. K01990, K02004, K06147, and K01992 had the highest gene abundance during the fermentation of *Chouguiyu*. The KO functions related to BA metabolism were further identified (Figure 1F). A total of 16 KO functions were found in the metagenome of the microbial community. Among these, K00797, K01476, K01480, K01581, K01582, K01590, K01593, K10536, K12251, K13746, K13747, and K23385 were related to BA synthesis, while K00274, K00276, K00657, K00797, K09251, K09470, K12256, K13746, and K13747 contributed to BA degradation. K00797, K13746, and K13747 function in both synthesis and degradation pathways (Figure 2).

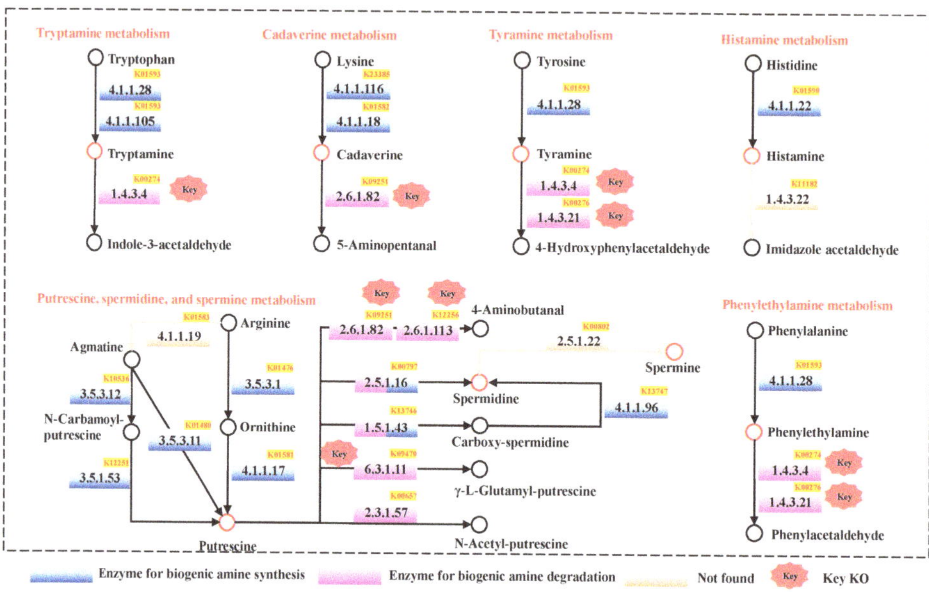

Figure 2. Biogenic amine metabolic pathways of microbial community in *Chouguiyu*.

3.2. Enzymes Related to Biogenic Amine Metabolism in Chouguiyu Based on KO Functions

In order to fully understand the microbial metabolism mechanism and related enzymes of BAs in *Chouguiyu*, the BA metabolic pathways in *Chouguiyu* were constructed according to KO functions (Figure 2, Tables S2 and S3). For tryptamine metabolism, tryptamine was produced from tryptophan by L-tryptophan decarboxylase (EC 4.1.1.28 and EC 4.1.1.105) in K01593, while monoamine oxidase (EC 1.4.3.4) in K00274 could degrade tryptamine. Cadaverine was mainly produced from lysine by D-lysine decarboxylase (EC 4.1.1.116) in K23385 and lysine decarboxylase (EC 4.1.1.18) in K01582, while putrescine aminotransferase (EC 2.6.1.82) in K09251 could degrade cadaverine into 5-aminopentanal. Tyramine was obtained from tyrosine using L-tryptophan decarboxylase (EC 4.1.1.28) in K01593 and could be degraded by monoamine oxidase (EC 1.4.3.4) in K00274 and primary-amine oxidase (EC 1.4.3.21) in K00276. Histamine was generated from histidine by histidine decarboxylase (EC 4.1.1.22) in K01590, while no histamine-degrading enzymes were identified in *Chouguiyu*.

As one of the most abundant BAs, putrescine could be synthesized in a variety of ways. Arginine was first catalyzed by arginase (EC 3.5.3.1, K01476) into ornithine, which was further transferred into putrescine by ornithine decarboxylase (EC 4.1.1.17, K01581). This pathway was also the main putrescine-producing pathway in fermented foods [19]. Putrescine could also be synthesized through agmatine deiminase (EC 3.5.3.12) in K10536 and N-carbamoylputrescine amidase (EC 3.5.1.53) in K12251 or agmatinase (EC 3.5.3.11) in K01480 using agmatine as the substrate. Meanwhile, many enzymes contributed to the degradation of putrescine in *Chouguiyu*, including putrescine aminotransferase (EC 2.6.1.82) in K09251 and pyruvate transaminase (EC 2.6.1.113) in K12256 for 4-aminobutanal production, gamma-glutamylputrescine synthase (EC 6.3.1.11) in K09470 for γ-L-glutamyl-putrescine production, and diamine N-acetyltransferase (EC 2.3.1.57) in K00657 for N-acetyl-putrescine production.

In addition, putrescine was also catalyzed into spermidine by spermidine synthase (EC 2.5.1.16) in K00797 or was first degraded into carboxy-spermidine by carboxynorspermidine synthase (EC 1.5.1.43) in K13746 and transferred into spermidine by carboxynorspermidine decarboxylase (EC 4.1.1.96) in K13747. No spermidine-degrading enzyme, i.e., spermine-synthesizing enzyme, was observed in *Chouguiyu*.

Phenylethylamine was produced from phenylalanine by L-tryptophan decarboxylase (EC 4.1.1.28) in K01593. The enzymes for phenylethylamine degradation were the same as those for tyramine degradation, resulting in the production of phenylacetaldehyde.

3.3. Microbial Genus Related to Biogenic Amine Metabolism in Chouguiyu Based on KO

3.3.1. Microbial Genera for BA Synthesis

After analyzing the KO abundance, there were a total of 111 microbial genera related to BA metabolism (Figure 3). Among these, 91 microbial genera possessed BA synthesis ability (Figure 3A). *Pseudomonas* was the only microbial genus that possessed L-tryptophan decarboxylase (K01593), which was the most important enzyme for producing tryptamine, tyramine, and phenylethylamine. The gene abundance of *Pseudomonas* first increased and then decreased to none. Similarly, previous studies found that *Pseudomonas* was one of the main microorganisms to produce tryptamine [20], tyramine [5], and phenylethylamine [21].

The lysine decarboxylase in *Hafnia* can effectively produce cadaverine [22]. In this study, *Hafnia* was the most important microbial genus for the production of D-lysine decarboxylase (K23385) and lysine decarboxylase (K01582) for cadaverine synthesis. In addition, a high abundance of lysine decarboxylase (K01582) was found in *Obesumbacterium*, *Citrobacter*, *Raoultella*, *Serratia*, and *Enterobacter*, contributing to abundant cadaverine synthesis. Similar results were found showing that the expression of the decarboxylase gene in *Serratia* [23], as well as *Citrobacter* and *Enterobacter* [24–26], was associated with cadaverine production in foods. There were five microbial genera related to histamine production, among which *Morganella* and *Raoultella* contained the most abundant histidine decarboxylase (K01590). The histamine-producing ability of *Morganella* [27] and *Raoultella* [28] has also been reported in other studies.

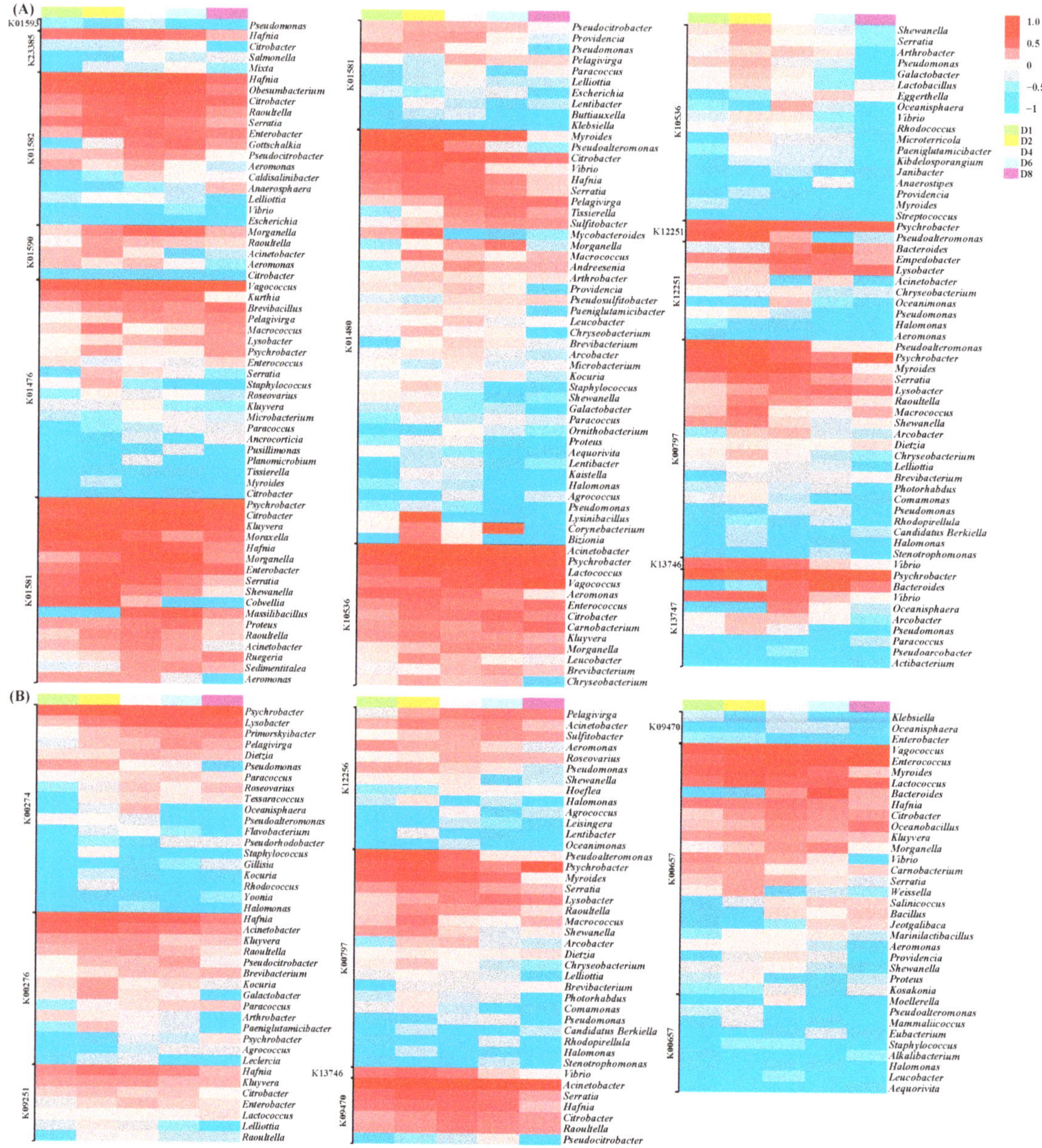

Figure 3. Abundance of enzymes for the (**A**) synthesis and (**B**) degradation of biogenic amines from the microbial community in *Chouguiyu*.

Putrescine was mainly produced by a two-step reaction of agmatinase (K01476) and ornithine decarboxylase (K01581). A total of 20 microbial genera were responsible for the production of agmatinase, and most arginase-producing genera increased along with the fermentation process. Among these, *Vagococcus*, *Kurthia*, *Brevibacillus*, and *Pelagivirga* were the main microbial genera for agmatinase production. More microbial genera (27 kinds) could produce ornithine decarboxylase, and *Psychrobacter*, *Citrobacter*, *Kluyvera*, and *Moraxella* were the main microbial genera. However, there were only six microbial genera that could produce agmatinase and ornithine decarboxylase together, including *Pelagivirga*, *Psychrobacter*, *Serratia*, *Kluyvera*, *Paracoccus*, and *Citrobacter*. For putrescine synthesis from agmatine, there were 38 microbial genera secreting agmatinase (K01480), among which *Myroides*, *Pseudoalteromonas*, *Citrobacter*, and *Vibrio* played an important role in enzyme secretion. Most agmatinase-producing genera with high abundance decreased in the late fermentation.

In addition, putrescine could also be produced by a two-step reaction of agmatine deiminase (K10536) and N-carbamoylputrescine amidase (K12251) from agmatine. A total of 31 and 11 microbial genera were responsible for the production of agmatine deiminase and N-carbamoylputrescine amidase. Most genera with a high abundance of agmatine deiminase production increased with the increasing fermentation time. *Acinetobacter*, *Psychrobacter*, *Lactococcus*, and *Vagococcus* were the main microbial genera for agmatine deiminase production, while *Psychrobacter*, *Pseudoalteromonas*, *Bacteroides*, and *Empedobacter* were the main microbial genera for N-carbamoylputrescine amidase. Only *Acinetobacter*, *Psychrobacter*, *Aeromonas*, *Chryseobacterium*, and *Pseudomonas* produced these two enzymes together. The pathways of putrescine production are more complicated than those of other BAs and many studies have reported that putrescine can be generated from similar microorganisms, such as *Enterobacter* [29], *Citrobacter* [10], *Proteus* [24,30], *Serratia* [31], *Aeromonas* [32], and *Lactococcus* [32].

A total of 20 microbial genera possessed a high abundance of spermidine synthase (K00797) to produce spermidine, among which *Pseudoalteromonas*, *Psychrobacter*, *Myroides*, *Serratia*, *Lysobacter*, *Raoultella*, *Macrococcus*, and *Shewanella* were the main microbial genera. Spermidine was also produced by carboxynorspermidine synthase (K13746) and carboxynorspermidine decarboxylase (K13747) from putrescine. However, only *Vibrio* could produce carboxynorspermidine synthase, while there were nine microbial genera that possessed the ability to produce carboxynorspermidine decarboxylase, such as *Psychrobacter*, *Bacteroides*, and *Vibrio*. Similarly, *Psychrobacter* [33], *Bacteroides* [34], *Serratia* [35], and *Shewanella* [36] showed greater capacity to produce spermidine. Spermidine was further degraded into spermine using spermine synthase. However, there was no microbial metabolic gene for spermine in *Chouguiyu*.

3.3.2. Microbial Genera for BA Degradation

Besides the abundant microorganisms for the production of BAs, many microorganisms (75 microbial genera) possessed various enzymes related to the degradation of BAs in *Chouguiyu* (Figure 3B). In this study, a total of 73 microbial genera possessed BA degradation ability and 19 microbial genera were potentially able to produce monoamine oxidase (K00274) for tryptamine, tyramine, and phenylethylamine degradation, among which *Psychrobacter*, *Lysobacter*, *Primorskyibacter*, *Pelagivirga*, and *Dietzia* possessed the highest abundance and increased along with the fermentation process. Tyramine and phenylethylamine could also be degraded by primary-amine oxidase (K00276). A total of 14 microbial genera generated primary-amine oxidase. *Hafnia*, *Acinetobacter*, *Kluyvera*, and *Raoultella* were the main microbial genera to produce the enzyme. Most genera with a high abundance of primary-amine oxidase production decreased with the prolongation of fermentation time. A total of 7 microbial genera played an important role in secreting putrescine aminotransferase (K09251) for cadaverine and putrescine degradation, and *Hafnia*, *Kluyvera*, *Citrobacter*, *Enterobacter*, and *Lactococcus* possessed the highest abundance. *Lactococcus* has previously been found to possess the enzyme activities responsible for BA degradation in cheese [37]. In addition, putrescine could be degraded by pyruvate transaminase (K12256), spermidine synthase (K00797), carboxynorspermidine synthase (K13746), gamma-glutamylputrescine synthase (K09470), and diamine N-acetyltransferase (K00657). Pyruvate transaminase was produced from 13 microbial genera, such as *Pelagivirga*, *Acinetobacter*, *Sulfitobacter*, and *Aeromonas*. The microbial genera for putrescine degradation using spermidine synthase and carboxynorspermidine synthase were the same as those for spermidine synthesis. There were nine genera that produced gamma-glutamylputrescine synthase, among which *Acinetobacter*, *Serratia*, *Hafnia*, *Citrobacter*, and *Raoultella* were the main genera. More genera (32 kinds) had the ability to produce diamine N-acetyltransferase, and *Vagococcus*, *Enterococcus*, *Myroides*, and *Lactococcus* were the main genera with high abundance. It was reported that *Enterococcus*'s multicopper oxidases could degrade phenylethylamine, putrescine, histamine, and tyramine [25]. Interestingly, some microbial genera possessed various enzymes for the degradation of various BAs in *Chouguiyu*, such as *Acinetobacter*, *Aeromonas*, *Arthrobacter*, *Hafnia*, *Lysobacter*, *Raoultella*, and *Pseudomonas*. Similar results were found showing that *Arthrobacter* was widely used to produce various amine oxidases for phenylethylamine, tyramine, and histamine degradation [38].

Interestingly, huge differences in microbial composition were found using total genes of the metagenome (Figure 1B) and genes related to BA metabolism (Figure 3). It was more scientific to analyze the mechanism of BA production and degradation based on functional gene abundance.

3.4. Change in Biogenic Amines During Fermentation of *Chouguiyu*

In order to clarify the actual influence of microbial genera on BAs of *Chouguiyu*, the change rules of eight BAs in *Chouguiyu* at different fermentation time are shown in Figure 4. As the fermentation time increased, tryptamine and phenylethylamine first increased and then decreased, and all reached maximums at 2 d of 4.30 and 2.86 mg/kg, respectively. Interestingly, the gene abundance of K01593 in *Pseudomonas* was low in *Chouguiyu* (Figure 3), resulting in the relatively low content of these two BAs. Cadaverine, tyramine, histamine, and putrescine markedly increased with the increasing fermentation time, while spermidine and spermine decreased. Histamine is considered the most toxic BA in fermented foods [7]. Worldwide, the legal limit for histamine in aquatic products varies by country and product type [39]. Among these, the minimum limit is 30 mg/kg and the United States suggests that the histamine safe limit in aquatic products is 50 mg/kg [40]. In this study, the histamine content reached a maximum of 15.91 mg/kg after fermentation for 8 d, indicating the safety of *Chouguiyu*. And the abundance of histidine decarboxylase genes decreased in the late fermentation, which might be an important reason for the low production of histamine. Tyramine also shows high toxicity, and its excessive intake can cause headaches and high blood pressure [24]. In this study, the maximal tyramine content was 25.65 mg/kg after 8 d fermentation.

Cadaverine and putrescine, despite being non-toxic in themselves, can inhibit the metabolic enzyme activity of histamine and tyramine (such as monoamine oxidase and diamine oxidase), therefore enhancing the toxicity of histamine and tyramine [31]. In this study, cadaverine and putrescine were the highest BAs in *Chouguiyu* at D8, reaching their maximums of 105.73 mg/kg and 105.30 mg/kg, respectively. The smelly flavor of these two BAs might contribute to the unique flavor of *Chouguiyu*. Interestingly, the abundance of the cadaverine-producing genes in almost all microbial genera first increased and then decreased during fermentation of *Chouguiyu*. There were low contents of spermidine, spermine, tryptamine, and phenylethylamine in the late fermentation of *Chouguiyu*, especially spermidine, which was not even detected.

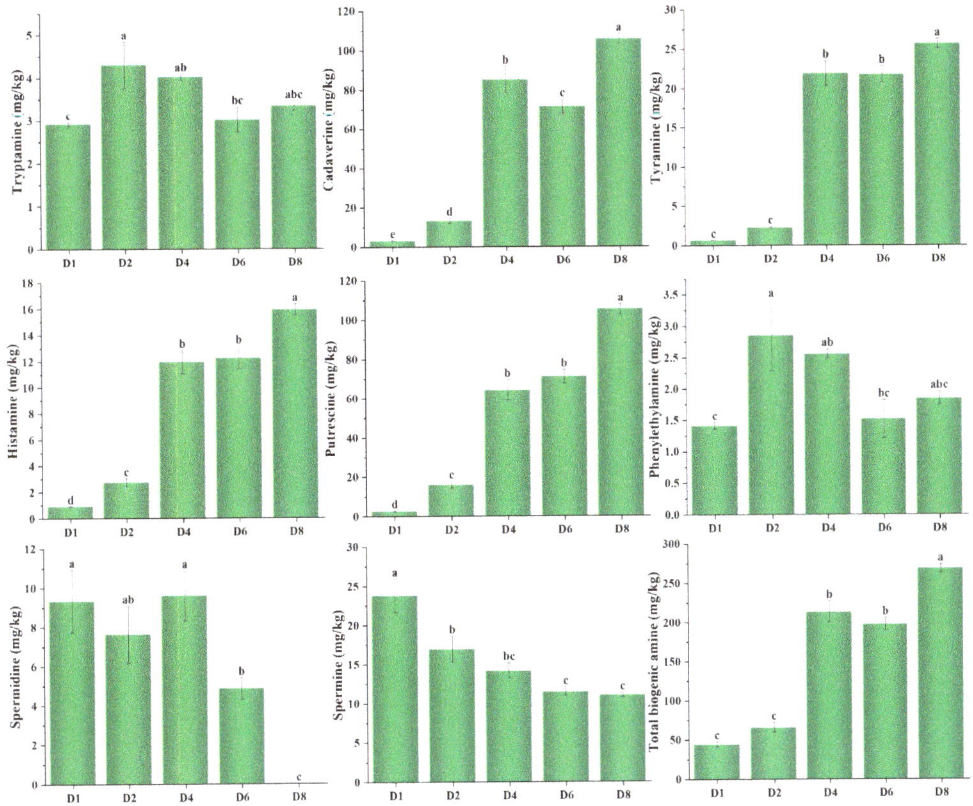

Figure 4. Change in biogenic amines during fermentation of *Chouguiyu*. Bars with different lowercase letters differ at $p < 0.05$ by one-way analysis of variance and Tukey test.

3.5. Relationship of Biogenic Amine and Metabolic Pathway in Chouguiyu

The relationship between microorganisms and their metabolites can be clearly visualized by correlation network maps [41,42]. To elucidate the microbial sources of biogenic amines (BAs), Pearson's correlation between microbial synthesis enzymes and BAs was analyzed within each KO (Figures 5 and S1 and Table S4). L-tryptophan decarboxylase (K01593) from *Pseudomonas* was positively correlated with tryptamine and phenylethylamine, while a negative correlation was observed with tyramine. *Pseudomonas* was similarly identified as an important genus for tryptamine production in correlative analyses [43]. Cadaverine levels were significantly positively correlated with D-lysine decarboxylase (K23385) from *Salmonella* but negatively correlated with the enzyme from *Hafnia*. Cadaverine was significantly positively correlated with lysine decarboxylase (K01582) from *Caldisalinibacter*, *Gottschalkia*, *Anaerosphaera*, and *Citrobacter* but was negatively correlated with that from *Serratia*, *Hafnia*, and *Escherichia*.

Salmonella, *Caldisalinibacter*, *Gottschalkia*, *Anaerosphaera*, and *Citrobacter* might play a major role in cadaverine synthesis. Similarly, cadaverine in *Dongbei Suancai* (fermented cabbage from Dongbei) was highly related to *Citrobacter* [9]. The change in putrescine was significantly positively correlated with arginase (K01476) from *Pelagivirga*, *Lysobacter*, *Vagococcus*, *Paracoccus*, and *Brevibacillus*, ornithine decarboxylase (K01581) from *Ruegeria*, *Sedimentitalea*, *Paracoccus*, and *Pelagivirga*, N-carbamoylputrescine amidase (K12251) from *Lysobacter*, and agmatinase (K01480) from *Pseudosulfitobacter*, *Pelagivirga*, *Sulfitobacter*,

Andreesenia, *Tissierella*, and *Agrococcus*. The enhanced abundance of enzymes from these genera might play an important role in putrescine production.

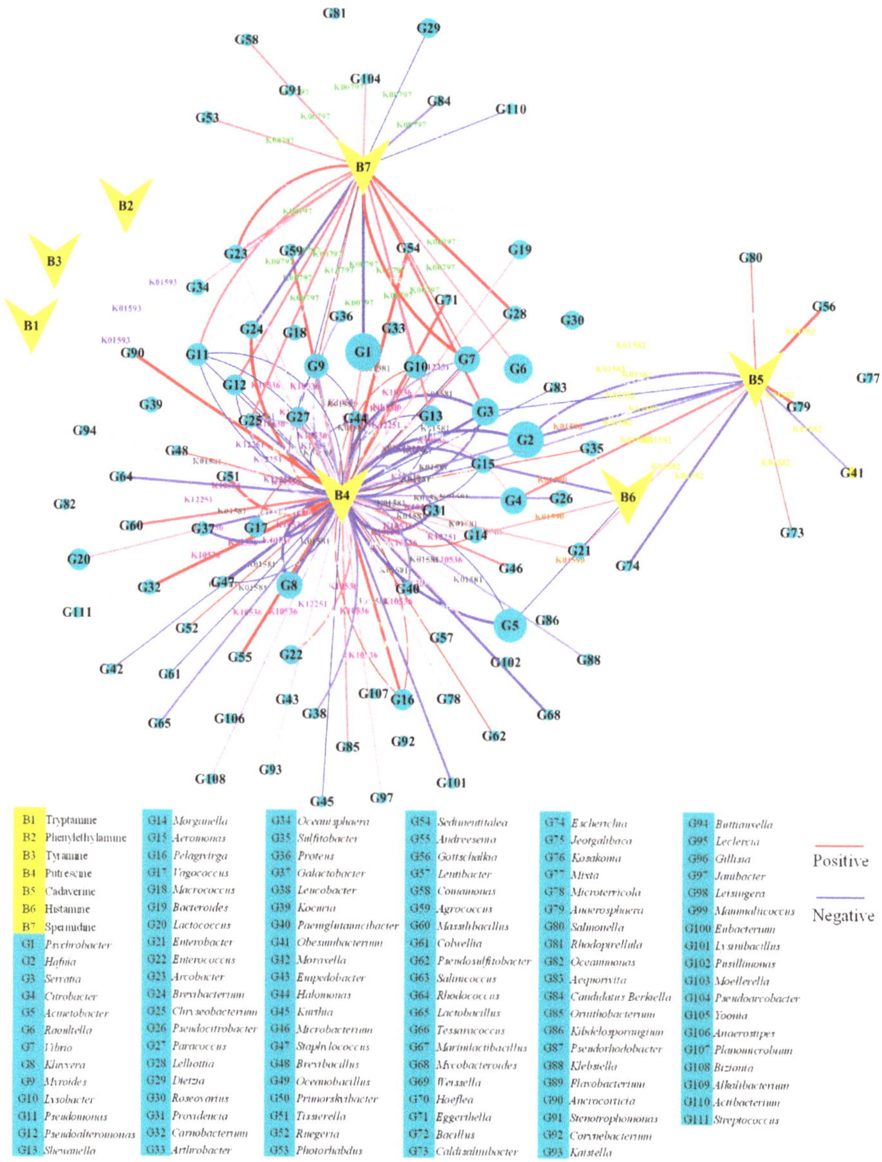

Figure 5. Correlation network map between microbial synthesis enzymes and biogenic amines. The red and blue lines, respectively, represent positive and negative correlation.

Histamine was positively correlated with histidine decarboxylase (K01590) from *Morganella* and *Raoultella*, with *Morganella*—an established histamine-producing bacterium in seafood—being particularly notable [44]. Spermidine levels were significantly positively correlated with carboxynorspermidine decarboxylase (K13747) from *Vibrio* and *Arcobacter* and with spermidine synthase (K00797) from *Myroides*, *Lelliottia*, and *Serratia*. *Vibrio*, *Arcobacter*, *Myroides*, *Lelliottia*, and *Serratia* might be the major spermidine-synthesizing bacteria. Based on correlation analysis, *Serratia* was positively correlated with most of the BAs in soybean paste [26].

For the correlation between BAs and their degradation enzymes from microorganisms in *Chouguiyu*, Pearson's correlation between microbial degradation enzymes and BAs in each KO was analyzed (Figures 6 and S2 and Table S5). Tryptamine levels were significantly positively correlated with monoamine oxidase (K00274) from *Kocuria, Flavobacterium*, and *Tessaracoccus*. Cadaverine was significantly positively correlated with putrescine aminotransferase (K09251) from *Lactococcus* but was negatively correlated with that from *Citrobacter* and *Lelliottia*. Declining *Citrobacter* and *Lelliottia* abundances correlated with cadaverine accumulation, suggesting reduced degradation capacity, and the decrease in *Lactococcus* was beneficial to the control of cadaverine accumulation. Tyramine levels were significantly positively correlated with monoamine oxidase from *Primorskyibacter, Lysobacter, Roseovarius, Dietzia, Paracoccus, Gillisia, Tessaracoccus, Pseudorhodobacter*, and *Pelagivirga* and positively related with primary-amine oxidase from *Brevibacterium, Agrococcus, Paracoccus*, and *Psychrobacter*. The increasing enzymes in these genera might contribute to the inhibition of tyramine accumulation. Conversely, tyramine concentration was significantly negatively correlated with monoamine oxidase from *Pseudomonas* and *Pseudoalteromonas* and with primary-amine oxidase from *Acinetobacter, Arthrobacter, Galactobacter, Kocuria*, and *Hafnia*, indicating their role in tyramine increase.

The change in putrescine was also significantly positively correlated with putrescine aminotransferase (K09251) from *Lactococcus*, pyruvate transaminase (K12256) from *Pelagivirga, Sulfitobacter, Roseovarius*, and *Hoeflea*, spermidine synthase (K00797) from *Lysobacter, Brevibacterium*, and *Dietzia*, and diamine N-acetyltransferase (K00657) from *Oceanobacillus, Bacillus, Enterococcus, Salinicoccus, Jeotgalibaca, Vagococcus*, and *Aequorivita*. These increasing enzymes in the above genera were important for the inhibition of putrescine in *Chouguiyu*. Meanwhile, putrescine was significantly negatively correlated with putrescine aminotransferase from *Citrobacter, Lelliottia, Kluyvera*, and *Hafnia*, pyruvate transaminase from *Pseudomonas, Aeromonas, Halomonas*, and *Shewanella*, spermidine synthase from *Pseudoalteromonas, Shewanella, Lelliottia, Myroides, Chryseobacterium*, and *Comamonas*, carboxynorspermidine synthase (K13746) from *Vibrio*, gamma-glutamylputrescine synthase (K09470) from *Enterobacter, Serratia, Acinetobacter*, and diamine N-acetyltransferase from *Vibrio, Aeromonas, Pseudoalteromonas, Carnobacterium, Kluyvera, Myroides, Hafnia, Serratia*, and *Weissella*. The decreasing enzymes in these genera contributed most to the increase in putrescine during fermentation of *Chouguiyu*.

The change in phenylethylamine was significantly positively correlated with monoamine oxidase from *Kocuria, Flavobacterium*, and *Tessaracoccus* and correlated with primary-amine oxidase from *Leclercia* and *Paeniglutamicibacter*. These microbial genera played a crucial role in the maintenance of phenylethylamine concentration. Meanwhile, the KOs of K09251, K00657, K00274, K12256, K09470, and K00276 were the major biological amine degradation pathways (Figure 2).

Interestingly, microbial degraders of tryptamine, tyramine, and phenylethylamine outnumbered producers, explaining their low concentrations. Conversely, putrescine and cadaverine accumulated due to scarce degraders (Figure 6).

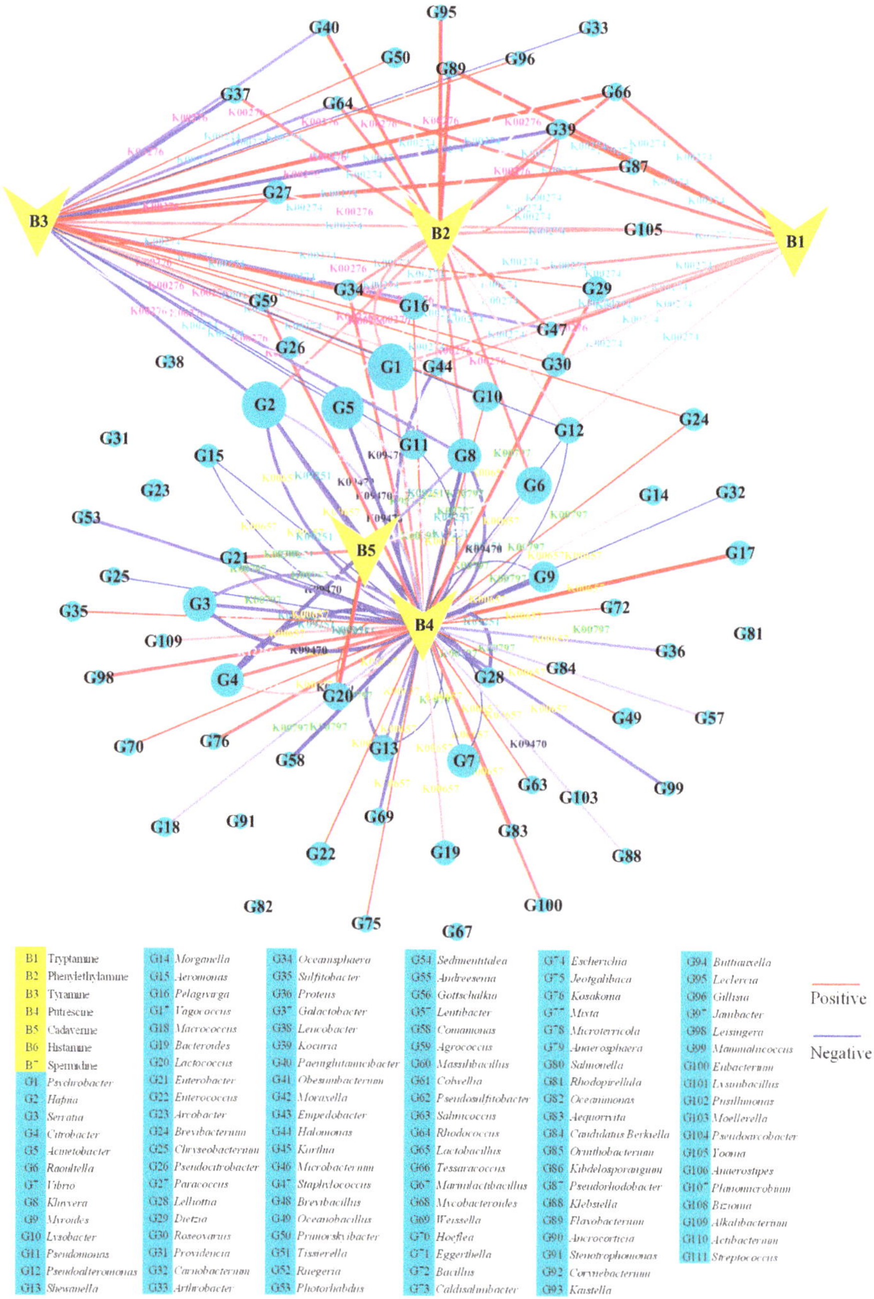

Figure 6. Correlation network map between microbial degradation enzymes and biogenic amines. The red and blue lines, respectively, represent positive and negative correlation.

3.6. Identification of Core Microorganisms for Biogenic Amine Degradation in *Chouguiyu*

In order to more clearly focus on BA-degrading microbial species, the microorganisms associated with key degradation pathways are shown in Figure 7. In this study, the synthesis of cadaverine by microorganisms in *Chouguiyu* was higher than its degradation process, resulting from the increasing cadaverine concentration along with fermentation (Figure 7A). *Salmonella*, *Gottschalkia*, *Caldisalinibacter*, and *Anaerosphaera* contributed most to the production of cadaverine, while *Enterobacter* and *Lactococcus* were the main cadaverine-degrading microbial genera.

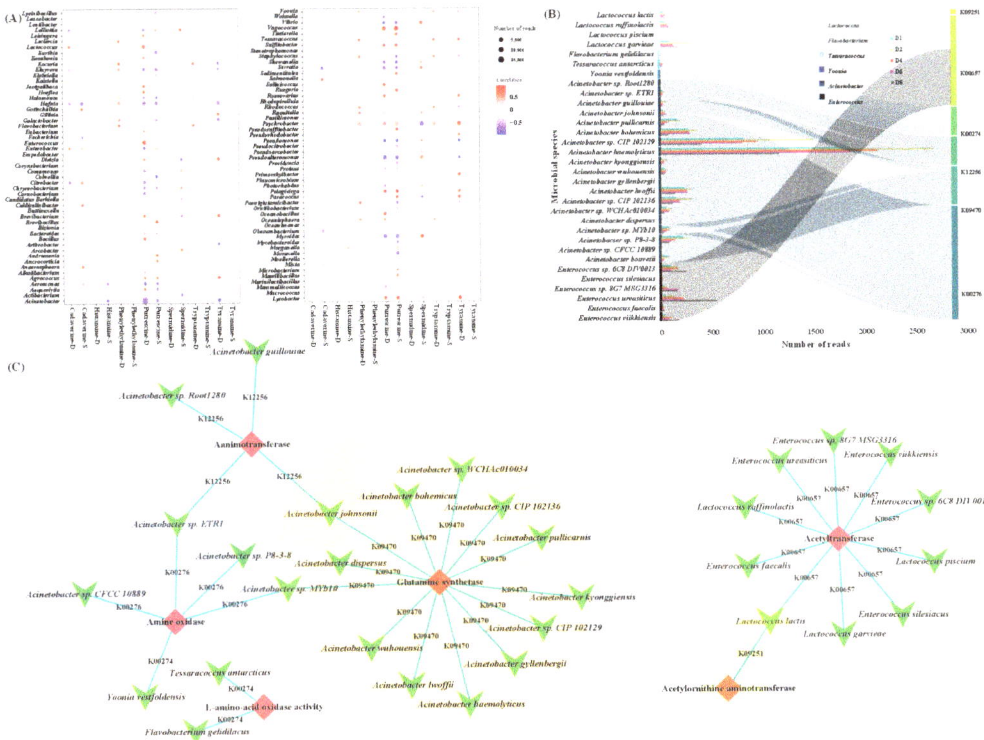

Figure 7. Core microbial genera for the degradation of biogenic amines. (**A**) Total abundance of functional enzymes related to biogenic amine synthesis and degradation in different microbial genera and their correlation with biogenic amines. (**B**) Core microbial species with the ability to degrade biogenic amines during the fermentation of *Chouguiyu*. (**C**) Relationship of core microbial species and enzymes related to biogenic amine degradation.

Among the histamine-producing genera, only *Morganella* showed a close relationship with histamine, and its low abundance resulted in the low content of histamine.

Pseudomonas was the only genus that was responsible for the synthesis of phenylethylamine, tryptamine, and tyramine, leading to the relatively low production of these BAs in *Chouguiyu*. More microbial genera were helpful in the degradation of phenylethylamine, among which *Psychrobacter*, *Hafnia*, *Lysobacter*, and *Acinetobacter* were the main degradation bacteria of phenylethylamine, while *Psychrobacter* and *Lysobacter* contributed most to tryptamine and tyramine degradation. *Vagococcus* and *Psychrobacter*, which possessed high enzyme abundance, were the main putrescine synthesis microorganisms, while *Vagococcus* and *Enterococcus* were beneficial for putrescine degradation. The high abundance of synthesis enzymes in *Psychrobacter*, *Pseudoalteromonas*, *Myroides*, and *Bacteroides* was important for spermidine synthesis, while there was no related enzyme for spermidine degradation. Interestingly, *Lactococcus*, *Flavobacterium*, *Tessaracoccus*, and *Yoonia* possessed only the BA degradation enzymes. Similarly to Korean jeotgal [45], *Chouguiyu*'s low-salt process permits diverse BA-producing microbes, but unlike jeotgal, it harbors unique degraders like *Yoonia*. In addition, *Acinetobacter* and *Enterococcus* had a much higher abundance of degradation enzymes than synthesis enzymes for BA metabolism.

The microbial species in the above genera for BA degradation were further explored (Figure 7B,C). For *Lactococcus*, a total of four microbial species had the BA degradation enzyme genes, among which *Lactococcus garvieae* possessed the most abundant putrescine degradation enzymes. *Lactococcus lactis* possessed both putrescine aminotransferase (K09251) and diamine N-acetyltransferase (K00657) for cadaverine and putrescine degradation, while *Lactococcus raffinolactis*, *Lactococcus piscium*, and *Lactococcus garvieae* only possessed diamine N-acetyltransferase for putrescine degradation. *Flavobacterium gelidilacus*, *Tessaracoccus antarcticus*, and *Yoonia vestfoldensis* were the only species that possessed monoamine oxidase (K00274) for tryptamine, tyrosine, and phenylalanine degradation in *Flavobacterium*, *Tessaracoccus*, and *Yoonia*, respectively. For *Acinetobacter*, a total of 19 microbial species possessed plenty of BA degradation genes, and they were observed to have a variety of putrescine, tyrosine, and phenylalanine degradation enzymes, including pyruvate transaminase (K12256), gamma-glutamylputrescine synthase (K09470), and primary-amine oxidase (K00276). Among these species, *Acinetobacter haemolyticus* possessed the highest enzyme activity for tyrosine and phenylalanine degradation. For *Enterococcus*, there were six microbial species that had putrescine degradation enzymes, i.e., diamine N-acetyltransferase. *Enterococcus ureasiticus* possessed more abundant enzymes than the other microbial species in *Enterococcus*.

It can be concluded that *Lactococcus garvieae*, *Flavobacterium gelidilacus*, *Tessaracoccus antarcticus*, *Yoonia vestfoldensis*, *Acinetobacter haemolyticus*, and *Enterococcus ureasiticus* were the main microbial species for BA degradation during the fermentation of *Chouguiyu*, which could be isolated as the potential functional strains for the degradation of BAs in fermented foods. According to the above results, the degradation of undesirable BAs in *Chouguiyu* could be reduced and isolated strains could be applied in fermented foods like fish sauce or cheese to reduce BA risks.

4. Conclusions

Based on KO analysis, there were 91 BA-synthesizing microbial genera, among which *Psychrobacter*, *Vagococcus*, *Pseudoalteromonas*, *Hafnia*, *Citrobacter*, and *Myroides* were the main genera, especially for putrescine synthesis. Seventy-five genera possessed degradation enzymes, primarily *Acinetobacter*, *Psychrobacter*, *Vagococcus*, *Pseudoalteromonas*, *Enterococcus*, *Myroides*, and *Hafnia*, especially for putrescine. After determination of BAs, high concentrations of cadaverine and putrescine were observed in *Chouguiyu*, while histamine and tyramine exhibited relatively low concentrations, and these four BAs increased with the increasing fermentation time, similar to KO analysis. The correlation networks revealed *Lactococcus*, *Flavobacterium*, *Tessaracoccus*, and *Yoonia* as non-producers but potential BA degraders. *Acinetobacter* and *Enterococcus* possessed more abundant degradation enzymes than synthesis enzymes. Amine oxidase from *Acinetobacter*, as well as diamine N-acetyltransferase from *Enterococcus* and *Lactococcus*, played the main role in biogenic amine degradation in *Chouguiyu*. *Lactococcus garvieae*, *Flavobacterium gelidilacus*, *Tessaracoccus antarcticus*, *Yoonia vestfoldensis*, *Acinetobacter haemolyticus*, and *Enterococcus ureasiticus* were the main microbial species for BA degradation and could be isolated as the potential functional strains for the degradation of BAs in fermented foods.

Supplementary Materials: The following supporting information can be downloaded at: https://www.mdpi.com/article/10.3390/foods14162863/s1, Table S1: Standard curves of different biogenic amine; Figure S1: Correlation heatmap between microbial synthesis enzymes and biogenic amines in each KO; Figure S2: Correlation heatmap between microbial degradation enzymes and biogenic amines in each KO. Table S2: Enzymes and relative abundance of reads involved in biogenic amine production in *Chouguiyu*; Table S3: Enzymes and relative abundance of reads involved in biogenic amine degradation in *Chouguiyu*; Table S4: Correlations and *p*-value between synthesis genus and biogenic amine in *Chouguiyu*. Table S5: Correlations and *p*-value between degradation genus and biogenic amine in *Chouguiyu*.

Author Contributions: J.L., D.Y.: formal analysis, methodology, and writing—original draft. Y.Z.: data curation and funding acquisition. D.W.: visualization. H.H.: funding acquisition. C.L.: writing—review and editing, and conceptualization. All authors have read and agreed to the published version of the manuscript.

Funding: This research was financially supported by the Earmarked fund for CARS (CARS-46), the National Natural Science Foundation of China (32272348), the Key Laboratory of Aquatic Product Processing, Ministry of Agriculture and Rural Affairs, China (NYJG202405), the Guangdong Provincial Special Fund for Modern Agriculture Industry Technology Innovation Teams (2024CXTD26), and the Central Public-interest Scientific Institution Basal Research Fund, CAFS (2023TD78).

Informed Consent Statement: Not applicable.

Data Availability Statement: The original contributions presented in this study are included in the article/Supplementary Material. Further inquiries can be directed to the corresponding authors.

Conflicts of Interest: The authors declare no conflicts of interest.

References

1. Dabade, D.S.; Jacxsens, L.; Miclotte, L.; Abatih, E.; Devlieghere, F.; De Meulenaer, B. Survey of multiple biogenic amines and correlation to microbiological quality and free amino acids in foods. *Food Control* **2021**, *120*, 107497. [CrossRef]
2. Turna, N.S.; Chung, R.; Mcintyre, L. A review of biogenic amines in fermented foods: Occurrence and health effects. *Heliyon* **2024**, *10*, e24501. [CrossRef]
3. Li, Y.; Li, W.; Li, C.; Li, L.; Yang, D.; Wang, Y.; Chen, S.; Wang, D.; Wu, Y. Novel insight into flavor and quality formation in naturally fermented low-salt fish sauce based on microbial metabolism. *Food Res. Int.* **2023**, *166*, 112586. [CrossRef]
4. Jiang, W.; Li, C.; Xu, B.; Dong, X.; Ma, N.; Yu, J.; Wang, D.; Xu, Y. Halomonas shantousis sp. nov., a novel biogenic amines degrading bacterium isolated from Chinese fermented fish sauce. *Antonie Leeuwenhoek* **2014**, *106*, 1073–1080. [CrossRef]
5. Moon, J.S.; Cho, S.K.; Choi, H.Y.; Kim, J.E.; Kim, S.; Cho, K.; Han, N.S. Isolation and characterization of biogenic amine-producing bacteria in fermented soybean pastes. *J. Microbiol.* **2010**, *48*, 257–261. [CrossRef] [PubMed]
6. Li, C.; Cui, Q.; Li, L.; Huang, H.; Chen, S.; Zhao, Y.; Wang, Y. Formation and improvement mechanism of physical property and volatile flavor of fermented tilapia surimi by newly isolated lactic acid bacteria based on two dimensional correlation networks. *Food Chem.* **2024**, *440*, 138260. [CrossRef]
7. Zhao, Y.; Wang, Y.; Li, C.; Li, L.; Yang, X.; Wu, Y.; Chen, S.; Zhao, Y. Novel insight into physicochemical and flavor formation in naturally fermented tilapia sausage based on microbial metabolic network. *Food Res. Int.* **2021**, *141*, 110122. [CrossRef] [PubMed]
8. Sang, X.; Ma, X.; Hao, H.; Bi, J.; Zhang, G.; Hou, H. Evaluation of biogenic amines and microbial composition in the Chinese traditional fermented food grasshopper sub shrimp paste. *LWT-Food Sci. Technol.* **2020**, *134*, 109979. [CrossRef]
9. Ye, H.; Lang, X.; Ji, Y.; Li, S.; Xin, N.; Meng, X.; Zhang, T.; Shen, X.; Zhao, C. The interaction between Lactobacillus plantarum SC-5 and its biogenic amine formation with different salt concentrations in Chinese Dongbei Suancai. *Food Res. Int.* **2021**, *150*, 110813. [CrossRef]
10. Li, R.; Zheng, M.; Zheng, M.; Cai, R.; Cui, X.; Wang, Y.; Jiang, X.; Xu, C. Metagenomic analysis reveals the linkages between bacteria and the functional enzymes responsible for potential ammonia and biogenic amine production in alfalfa silage. *J. Appl. Microbiol.* **2022**, *132*, 2594–2604. [CrossRef]
11. Yang, D.; Li, C.; Li, L.; Wang, Y.; Wu, Y.; Chen, S.; Zhao, Y.; Wei, Y.; Wang, D. Novel insight into the formation mechanism of umami peptides based on microbial metabolism in Chouguiyu, a traditional Chinese fermented fish. *Food Res. Int.* **2022**, *157*, 111211. [CrossRef]
12. Hu, M.; Dong, J.; Tan, G.; Li, X.; Zheng, Z.; Li, M. Metagenomic insights into the bacteria responsible for producing biogenic amines in sufu. *Food Microbiol.* **2021**, *98*, 103762. [CrossRef]
13. Yang, Z.; Liu, S.; Lv, J.; Sun, Z.; Xu, W.; Ji, C.; Liang, H.; Li, S.; Yu, C.; Lin, X. Microbial succession and the changes of flavor and aroma in Chouguiyu, a traditional Chinese fermented fish. *Food Biosci.* **2020**, *37*, 100725. [CrossRef]
14. Shen, Y.; Wu, Y.; Wang, Y.; Li, L.; Li, C.; Zhao, Y.; Yang, S. Contribution of autochthonous microbiota succession to flavor formation during Chinese fermented mandarin fish (*Siniperca chuatsi*). *Food Chem.* **2021**, *348*, 129107. [CrossRef]

15. Yang, D.; Li, C.; Li, L.; Chen, S.; Hu, X.; Xiang, H. Taste mechanism of umami peptides from Chinese traditional fermented fish (*Chouguiyu*) based on molecular docking using umami receptor T1R1/T1R3. *Food Chem.* **2022**, *389*, 133019. [CrossRef] [PubMed]
16. Yang, D.; Li, L.; Li, C.; Chen, S.; Deng, J.; Yang, S. Formation and inhibition mechanism of novel angiotensin I converting enzyme inhibitory peptides from *Chouguiyu*. *Front. Nutr.* **2022**, *9*, 851895. [CrossRef] [PubMed]
17. Li, Y.; Cui, L.; Du, F.; Han, X.; Li, J. Impacts of ε-polylysine hydrochloride with thymol on biogenic amines formation and biochemical changes of squid (*Illexargentinus*). *J. Food Process. Preserv.* **2021**, *45*, e15505. [CrossRef]
18. Wang, Y.; Li, C.; Zhao, Y.; Li, L.; Yang, X.; Wu, Y.; Chen, S.; Cen, J.; Yang, S.; Yang, D. Novel insight into the formation mechanism of volatile flavor in Chinese fish sauce (Yu-lu) based on molecular sensory and metagenomics analyses. *Food Chem.* **2020**, *323*, 126839. [CrossRef]
19. Romano, A.; Ladero, V.; Alvarez, M.A.; Lucas, P.M. Putrescine production via the ornithine decarboxylation pathway improves the acid stress survival of *Lactobacillus brevis* and is part of a horizontally transferred acid resistance locus. *Int. J. Food Microbiol.* **2014**, *175*, 14–19. [CrossRef]
20. Kosma, I.; Badeka, A. Determination of six underivatized biogenic amines by LC-MS/MS and study of biogenic amine production during trout (*Salmo trutta*) storage in ice. *Food Addit. Contam. Part A* **2021**, *38*, 476–487. [CrossRef]
21. Geornaras, I.; Dykes, G.A.; von Holy, A. Biogenic amine formation by poultry-associated spoilage and pathogenic bacteria. *Lett. Appl. Microbiol.* **1995**, *21*, 164–166. [CrossRef] [PubMed]
22. Wang, C.; Zhang, K.; Zhongjun, C.; Cai, H.; Honggui, W.; Ouyang, P. Directed evolution and mutagenesis of lysine decarboxylase from *Hafnia alvei* AS1.1009 to improve its activity toward efficient cadaverine production. *Biotechnol. Bioprocess Eng.* **2015**, *20*, 439–446. [CrossRef]
23. Bubelova, Z.; Bunka, F.; Tatakova, M.; Stajnochova, K.; Purevdorj, K.; Bunkova, L. Effects of temperature, pH and NaCl content on in vitro putrescine and cadaverine production through the growth of Serratia marcescens CCM 303. *J. Environ. Sci. Health Part B* **2015**, *50*, 797–808. [CrossRef]
24. Greif, G.; Greifova, M.; Karovicova, J. Effects of NaCl concentration and initial pH value on biogenic amine formation dynamics by *Enterobacter spp.* bacteria in model conditions. *J. Food Nutr. Res.* **2006**, *45*, 21–29.
25. Li, B.; Wang, Y.; Xue, L.; Lu, S. Heterologous expression and application of multicopper oxidases from *Enterococcus spp.* for degradation of biogenic amines. *Protein Pept. Lett.* **2021**, *28*, 183–194. [CrossRef]
26. Li, S.; Du, X.; Feng, L.; Mu, G.; Tuo, Y. The microbial community, biogenic amines content of soybean paste, and the degradation of biogenic amines by *Lactobacillus plantarum* HM24. *Food Sci. Nutr.* **2021**, *9*, 6458–6470. [CrossRef]
27. Podeur, G.; Dalgaard, P.; Leroi, F.; Prévost, H.; Emborg, J.; Martinussen, J.; Hansen, L.H.; Pilet, M. Development of a real-time PCR method coupled with a selective pre-enrichment step for quantification of Morganella morganii and Morganella psychrotolerans in fish products. *Int. J. Food Microbiol.* **2015**, *203*, 55–62. [CrossRef]
28. Lin, C.; Kung, H.; Lin, C.; Tsai, H.; Tsai, Y. Histamine production by Raoultella ornithinolytica in mahi-mahi meat at various storage temperatures. *J. Food Drug Anal.* **2016**, *24*, 305–310. [CrossRef]
29. Lavizzari, T.; Breccia, M.; Bover-Cid, S.; Vidal-Carou, M.C.; Veciana-Nogues, M.T. Histamine, cadaverine, and putrescine produced in vitro by Enterobacteriaceae and Pseudomonadaceae isolated from spinach. *J. Food Prot.* **2010**, *73*, 385–389. [CrossRef]
30. Helinck, S.; Perello, M.; Deetae, P.; de Revel, G.; Spinnler, H. Debaryomyces hansenii, Proteus vulgaris, Psychrobacter sp. and Microbacterium foliorum are able to produce biogenic amines. *Dairy Sci. Technol.* **2013**, *93*, 191–200. [CrossRef]
31. De Filippis, F.; Pennacchia, C.; Di Pasqua, R.; Fiore, A.; Fogliano, V.; Villani, F.; Ercolini, D. Decarboxylase gene expression and cadaverine and putrescine production by Serratia proteamaculans in vitro and in beef. *Int. J. Food Microbiol.* **2013**, *165*, 332–338. [CrossRef] [PubMed]
32. Zhang, Y.; Li, D.; Lv, J.; Li, Q.; Kong, C.; Luo, Y. Effect of cinnamon essential oil on bacterial diversity and shelf-life in vacuum-packaged common carp (*Cyprinus carpio*) during refrigerated storage. *Int. J. Food Microbiol.* **2017**, *249*, 1–8. [CrossRef] [PubMed]
33. Kämpfer, P.; Jerzak, L.; Wilharm, G.; Golke, J.; Busse, H.; Glaeser, S.P. *Psychrobacter ciconiae* sp. nov., isolated from white storks (*Ciconia ciconia*). *Int. J. Syst. Evol. Microbiol.* **2015**, *65 Pt 3*, 772–777. [CrossRef] [PubMed]
34. Sakanaka, M.; Sugiyama, Y.; Nara, M.; Kitakata, A.; Kurihara, S. Functional analysis of arginine decarboxylase gene *speA* of *Bacteroides dorei* by markerless gene deletion. *FEMS Microbiol. Lett.* **2018**, *365*, fny003. [CrossRef]
35. Chaves López, C.; De Angelis, M.; Martuscelli, M.; Serio, A.; Paparella, A.; Suzzi, G. Characterization of the Enterobacteriaceae isolated from an artisanal Italian ewe's cheese (*Pecorino Abruzzese*). *J. Appl. Microbiol.* **2006**, *101*, 353–360. [CrossRef]
36. Wang, S.; Liang, H.; Liu, L.; Jiang, X.; Wu, S.; Gao, H. Promiscuous enzymes cause biosynthesis of diverse siderophores in Shewanella oneidensis. *Appl. Environ. Microbiol.* **2020**, *86*, e00030-20. [CrossRef]
37. Guarcello, R.; De Angelis, M.; Settanni, L.; Formiglio, S.; Gaglio, R.; Minervini, F.; Moschetti, G.; Gobbetti, M. Selection of amine-oxidizing dairy lactic acid bacteria and identification of the enzyme and gene involved in the decrease of biogenic amines. *Appl. Environ. Microbiol.* **2016**, *82*, 6870–6880. [CrossRef]
38. Lee, J.; Kim, Y. Characterization of amine oxidases from *Arthrobacter aurescens* and application for determination of biogenic amines. *World J. Microbiol. Biotechnol.* **2013**, *29*, 673–682. [CrossRef]

39. DeBeeR, J.; Bell, J.W.; Nolte, F.; Arcieri, J.; Correa, G. Histamine Limits by Country: A Survey and Review. *J. Food Prot.* **2021**, *84*, 1610–1628. [CrossRef]
40. FDA. *Fish and Fishery Products Hazards and Controls Guidance*, 4th ed.; U.S. Food and Drug Administration: Silver Spring, MD, USA, 2021.
41. Li, C.; Li, W.; Li, L.; Chen, S.; Wu, Y.; Qi, B. Microbial community changes induced by a newly isolated salt-tolerant *Tetragenococcus muriaticus* improve the volatile flavor formation in low-salt fish sauce. *Food Res. Int.* **2022**, *156*, 111153. [CrossRef]
42. Cui, Q.; Li, L.; Huang, H.; Yang, Y.; Chen, S.; Li, C. Novel insight into the formation and improvement mechanism of physical property in fermented tilapia sausage by cooperative fermentation of newly isolated lactic acid bacteria based on microbial contribution. *Food Res. Int.* **2024**, *187*, 114456. [CrossRef]
43. Yu, Y.; Li, L.; Xu, Y.; An, K.; Shi, Q.; Yu, Y.; Xu, Z. Evaluation of the relationship among biogenic amines, nitrite and microbial diversity in fermented mustard. *Molecules* **2021**, *26*, 6173. [CrossRef]
44. Wang, D.; Li, C.; Pan, C.; Wang, Y.; Xiang, H.; Feng, Y.; Yang, X.; Chen, S.; Zhao, Y.; Wu, Y.; et al. Antimicrobial activity and mechanism of action of oregano essential oil against *Morganella psychrotolerans* and potential application in tuna. *LWT-Food Sci. Technol.* **2022**, *165*, 113758. [CrossRef]
45. Jung, M.Y.; Kim, T.; Lee, C.; Kim, J.Y.; Song, H.S.; Kim, Y.B.; Ahn, S.W.; Kim, J.S.; Roh, S.W.; Lee, S.H. Role of jeotgal, a Korean traditional fermented fish sauce, in microbial dynamics and metabolite profiles during kimchi fermentation. *Food Chem.* **2018**, *265*, 135–143. [CrossRef]

foods

MDPI

Article

Anti-Obesity and Hepatoprotective Effects of Herring–Saury Oil Fermented by *Lactobacillus brevis* KCCM13538P in High-Fat-Diet-Induced Mice

Hyun-Sol Jo [1,2], Tae-Won Goo [3,*] and Sun-Mee Hong [1,*]

1 Marin Industry Research Institute for East Sea Rim, 22 Haeyanggwahak-gil, Uljin-gun 36315, Gyeongsangbuk-do, Republic of Korea; hyunsoljo@kfri.re.kr
2 Korea Food Research Institute, Iseo-myeon, Wanju-Gun 55365, Jeollabuk-do, Republic of Korea
3 Department of Biochemistry, College of Medicine, Dongguk University, Gyeongju 38766, Gyeongsangbuk-do, Republic of Korea
* Correspondence: gootw@dongguk.ac.kr (T.-W.G.); hongsunmee@mire.re.kr (S.-M.H.); Tel.: +82-54-703-7801 (T.-W.G.); +82-54-780-3461 (S.-M.H.)

Abstract: Background: Obesity-associated liver dysfunction is a key feature of metabolic syndrome. Marine by-products, such as fish oils, offer promising dietary interventions. In this study, we aimed to assess the anti-obesity and hepatoprotective effects of herring–saury by-product-derived fermented fish oil—Gwamegi oil (GmO)—and the same oil fermented with *Lactobacillus brevis* KCCM13538P (GmOLb) in a high-fat-diet (HFD)-induced obese mouse model. **Methods**: GmO was extracted and fermented. Anti-obesity and hepatoprotective effects were assessed using in vitro and in vivo studies. For the in vivo study, female C57BL/6J mice were fed an HFD supplemented with lard (control), GmO, or GmOLb for 60 days. Metabolic and liver function parameters were assessed. **Results**: In 3T3-L1 adipocytes, GmOLb significantly reduced lipid accumulation and intracellular triglyceride (TG) levels compared with GmO. In HFD-fed mice, GmOLb significantly reduced body weight gain, ovarian fat mass, serum TG, low-density lipoprotein cholesterol, leptin concentration, atherogenic indices, and cardiac risk factor ratio. Furthermore, it reduced liver damage indicators, including alanine aminotransferase, aspartate aminotransferase, and total bilirubin levels. **Conclusions**: Fermenting herring–saury oil with *L. brevis* KCCM13538P enhanced its anti-obesity and hepatoprotective effects in HFD-fed mice. GmOLb shows strong potential as a functional dietary lipid for preventing and managing metabolic disorders.

Keywords: fermented fish oil; *Lactobacillus brevis*; high-fat diet; obesity; liver function

1. Introduction

Obesity is a major global health concern that significantly increases the risk of metabolic complications, including type 2 diabetes, non-alcoholic fatty liver disease (NAFLD), cardiovascular disease, and certain cancers. It results from a complex interplay of factors, including a sustained energy imbalance, reduced physical activity, and dysregulated lipid and glucose metabolism. The widespread adoption of Western-style diets, characterized by an excessive intake of saturated fats, refined carbohydrates, and ultra-processed foods, has further exacerbated the global prevalence of obesity and its associated complications [1,2]. High-fat diets (HFDs) are widely used in experimental models to replicate the key features of human obesity, such as visceral adiposity, hepatic steatosis, dyslipidemia, and chronic low-grade inflammation [3–5].

Marine-derived oils, particularly those rich in omega-3 polyunsaturated fatty acids (ω-3 PUFAs), have gained increasing attention for their anti-obesity and hepatoprotective properties [6,7]. Herring (*Clupea pallasii*) and saury (*Cololabis saira*) are two underutilized fish species rich in bioactive lipids, particularly eicosapentaenoic acid (EPA) and docosahexaenoic acid (DHA). These long-chain ω-3 PUFAs support lipid metabolism, reduce systemic inflammation, and enhance hepatic function [8–10]. In addition to ω-3 PUFAs, marine oils from herring and saury contain physiologically relevant levels of fat-soluble vitamins, including vitamins A, D, E, and K, which may aid immune regulation, lipid metabolism, and vascular health [11–13]. Among these, vitamin K2 variants such as menaquinone-4 (MK-4) and menaquinone-7 (MK-7) are notable for their high bioavailability and metabolic functions [14–16]. Gwamegi is a traditional Korean seafood product made by semi-drying herring or saury using cold sea winds during winter months. This artisanal preservation method, primarily practiced in coastal areas such as Pohang, enhances flavor and nutritional value while maintaining a soft, chewy texture. Gwamegi is widely consumed in Korea; however, it is relatively unfamiliar in global markets and scientific literature. In this study, oil extracted from surplus herring and saury raw materials—before their use in traditional Gwamegi production—is referred to as Gwamegi oil (GmO).

The functional efficacy of marine oils can be further enhanced through microbial fermentation using probiotic strains, such as *Lactobacillus brevis* [17,18]. Fermentation enhances the oxidative stability and bioavailability of unsaturated fatty acids (USFAs) and generates secondary metabolites with anti-inflammatory and lipid-lowering activities [19,20]. *L. brevis*, a lactic acid bacterium commonly identified in fermented foods, has demonstrated beneficial properties, including antioxidant, anti-obesity, and hepatoprotective effects [21]. Specifically, *L. brevis* possesses esterase and lipase genes that are enzymatically active under fermentation conditions, enabling the hydrolysis of triglycerides and esterified lipids [22]. In this study, we developed a probiotic fermentation system using a blend of herring and saury by-product oils inoculated with *L. brevis* KCCM13538P, supplemented with 0.1% yeast extract as a minimal nitrogen source to maintain microbial viability. In this study, we aimed to evaluate the anti-obesity and hepatoprotective effects of Gwamegi oil (GmO) and its fermented form (GmOLb) in in vitro and in vivo models of metabolic dysfunction. We hypothesized that fermentation with *L. brevis* KCCM13538P would enhance the functional efficacy of fish oil by increasing ω-3 PUFA bioavailability and generating additional bioactive metabolites [23,24]. To test this hypothesis, we assessed lipid accumulation, serum lipid profiles, liver function markers, and histopathological changes in HFD-induced obese mice.

2. Materials and Methods

2.1. Preparation of GmO and GmOLb

GmO was prepared by mixing the by-products of herring (*C. pallasii*) and saury (*C. saira*) at a 1:9 ratio, reflecting the relative availability of raw materials during production, as saury by-products were more abundant. Future studies will compare various blending ratios to evaluate their effects on the physicochemical properties and biological activities of the resulting oils. The raw materials, sourced from Hyun-Dae Feed Co., Ltd. (Pohang, Gyeongsangbuk-do, Republic of Korea), were thermally treated at 95 °C for 5 min to separate lipids from solid residues. Subsequently, the mixture was centrifuged (3000× *g*, 10 min) to obtain the oil fraction. The collected oil (GmO) was filtered using a separatory funnel to remove any insoluble particles, and stored at 4 °C for short-term use or at −20 °C for long-term storage. Prior to the experiments, GmO was equilibrated at room temperature for 1 h to ensure complete liquefaction. For fermentation, GmO supplemented with 0.1% yeast extract (BD Difco™, Sparks, MD, USA) was sterilized at 121 °C for 15 min to eliminate potential microbial contaminants. *L. brevis* KCCM13538P was used as the fermenting strain. A preculture was prepared by inoculating 1% (*v*/*v*) *L. brevis* KCCM13538P into de Man, Rogosa, and Sharpe broth (BD Difco™, Sparks, MD, USA), and incubating at 30 °C for 24 h. The resulting bacterial suspension was adjusted to an optical density of 1.0 (10^8 CFU/mL) at 600 nm (OD_{600}), and 1.0% (*v*/*v*) of this suspension was inoculated into the sterilized GmO. Fermentation was conducted under stationary conditions at 30 °C for 24 h, achieving a viable cell count of approximately 10^{10} to 10^{11} CFU/mL. The fermented oil, designated as GmOLb, was subsequently stored at −20 °C.

2.2. Fatty Acid and Fat-Soluble Vitamin Analysis

Fatty acid composition was determined using gas chromatography (GC) after derivatization to fatty acid methyl esters (FAMEs; Sigma-Aldrich, St. Louis, MO, USA) following the method described by Ichihara et al. [25], with minor modifications. Lipids were methylated with 14% boron trifluoride (Sigma-Aldrich, St. Louis, MO, USA) in methanol (Sigma-Aldrich, St. Louis, MO, USA), followed by extraction with hexane (Sigma-Aldrich, MO, USA). The resulting FAMEs were analyzed using a GC system (Agilent 7890A; Agilent Technologies, Santa Clara, CA, USA) equipped with a flame ionization detector and capillary column (HP-88; 100 m × 0.25 mm, 0.20 μm film thickness). The injector and detector temperatures were set at 250 °C, and the oven temperature was programmed to increase from 140 °C (held for 5 min) to 240 °C at a rate of 4 °C/min. Helium was used as the carrier gas at a flow rate of 1.0 mL/min. Individual fatty acids were identified by comparing retention times with those of authentic standards. Quantitative data were expressed as the mean ± standard deviation (SD), excluding trace or unidentified peaks.

Fat-soluble vitamins (A, D, E, and K2) were extracted using a liquid–liquid extraction method and analyzed by high-performance liquid chromatography (HPLC), following the previously described methods [26,27], with slight modifications. Oil samples were saponified with ethanolic potassium hydroxide and extracted using n-hexane. The extracted vitamins were analyzed using an HPLC system (Waters Alliance e2695; Waters Corp., Milford, MA, USA) equipped with a photodiode array detector. Separation was performed on a C18 reverse-phase column (250 × 4.6 mm, 5 μm). The mobile phase consisted of methanol and water (95:5, *v*/*v*) at a flow rate of 1.0 mL/min. Detection wavelengths were 325, 265, 290, and 248 nm for vitamins A, D, E, and K2, respectively. Vitamin A was expressed as retinol equivalents (RE), vitamin D and K2 (MK-4 and MK-7) as micrograms (μg), and vitamin E as alpha-tocopherol equivalents (α-TE). The vitamin content was normalized to 100 g of oil.

2.3. T3-L1 Cell Culture and Adipogenic Differentiation

Murine 3T3-L1 preadipocytes (American Type Culture Collection CL-173; Manassas, VA, USA) were cultured in Dulbecco's modified Eagle's medium (DMEM; Gibco, Grand Island, NY, USA) supplemented with 10% (*v*/*v*) bovine calf serum (HyClone, Logan, UT, USA) and 1% (*v*/*v*) penicillin–streptomycin (Gibco, Grand Island, NY, USA) at 37 °C in a humidified atmosphere containing 5% CO_2. Upon reaching confluence (designated as day 0), adipogenic differentiation was induced by treating the cells for 48 h with a standard MDI induction cocktail containing 0.5 mM 3-isobutyl-1-methylxanthine (Sigma-Aldrich, St. Louis, MO, USA), 1 μM dexamethasone (Sigma-Aldrich, St. Louis, MO, USA), and 10 μg/mL insulin (Sigma-Aldrich, St. Louis, MO, USA) in DMEM supplemented with 10% (*v*/*v*) fetal bovine serum (FBS; Gibco, Grand Island, NY, USA), as described by Zebisch et al. [28], with minor modifications. As insulin was already included in the MDI cocktail to induce adipogenesis, it was not used separately as a positive control during the differentiation assay. After 48 h (day 2), the medium was replaced with DMEM containing 10% (*v*/*v*) FBS and 10 μg/mL insulin, and the cells were further incubated for six days with media changes every two days. Test samples, including GmO and GmOLb (fermented with *L. brevis* KCCM13538P), were administered at concentrations ranging from 10 to 300 μg/mL throughout the differentiation period (day 0 to day 8). Control cells were treated with vehicle dimethyl sulfoxide (DMSO) under similar differentiation conditions. Adipocyte differentiation and lipid accumulation were assessed on day 8, as described by Gregoire et al. [29].

2.4. Cell Viability and Lipid Accumulation Assays

Cell viability was assessed using the 3-(4,5-dimethylthiazol-2-yl)-2,5-diphenyltetrazolium bromide (MTT) assay (Abcam, Cambridge, UK). Then, 3T3-L1 preadipocytes were seeded in 96-well plates and treated with various concentrations (10–300 μg/mL) of GmO or GmOLb for 24 h. Subsequently, 20 μL of MTT solution (5 mg/mL in phosphate-buffered saline [PBS]; Sigma-Aldrich) was added to each well and incubated at 37 °C for 4 h. The resulting formazan crystals were solubilized in 150 μL of DMSO, and absorbance was measured at 570 nm using a microplate reader (BioTek Instruments, Winooski, VT, USA), as described previously [30]. Lipid accumulation was assessed using Oil Red O staining (Sigma-Aldrich) on day 8 of differentiation. Cells were washed with PBS, fixed with 10% (*v*/*v*) formalin for 30 min, and stained for 1 h at room temperature using a filtered Oil Red O solution (0.5% in isopropanol, diluted 3:2 [*v*/*v*] with distilled water). After thorough washing with distilled water, the stained intracellular lipids were visualized. For quantification, the retained dye was eluted with 100% isopropanol, and absorbance was measured at 520 nm using a microplate reader, as described by Ramirez-Zacarias et al. [31]. Intracellular triglyceride (TG) content was determined using a commercial Triglyceride Quantification Assay Kit (Abcam), following the manufacturer's protocol. Absorbance was recorded at 570 nm and normalized to the total protein content that was measured using the bicinchoninic acid assay (Thermo Fisher Scientific, Waltham, MA, USA), following the method described by Smith et al. [32].

2.5. Animal Study Design

Thirty-two female C57BL/6 mice (5-weeks-old) were obtained from Daehan Bio Link Co., Ltd. (Eumseong, Republic of Korea) and housed under standard laboratory conditions (22 ± 2 °C, 55 ± 10% humidity, and 12 h light/dark cycle) with ad libitum access to food and water. After one week of acclimatization, the mice were randomly divided into four groups (*n* = 8 per group) as follows: ND: normal diet (AIN-76A), HFD_L: HFD containing 45% lard (AIN-76A modified), HFD_GmO: HFD containing 45% GmO, and HFD_GmOLb: HFD containing45 % GmOLb (fermented with *L. brevis* KCCM13538P). All diets were prepared based on the AIN-76A formulation (DooYeol Biotech, Seoul, Republic of Korea) and provided ad libitum for 60 days, adhering to the American Institute of Nutrition guidelines [33]. The HFD was formulated to provide 45% of total energy from fat. In the GmO and GmOLb groups, 45% (*w*/*w*) of the total dietary fat was replaced with either non-fermented or fermented GmO, replacing an equivalent amount of lard. Body weight and food intake were monitored weekly. At the end of the experiment, mice were fasted overnight, anesthetized with isoflurane (Sigma-Aldrich, St. Louis, MO, USA), and euthanized by cardiac exsanguination. Blood samples were collected via cardiac puncture, and the serum was separated by centrifugation (3000 rpm, 15 min, 4 °C). Liver, adipose, and other relevant tissues were excised, weighed, and either snap-frozen in liquid nitrogen or fixed in 10% (*v*/*v*) neutral-buffered formalin for histological and molecular analyses. No explicit inclusion or exclusion criteria were set a priori. All animals completed the study and were included in the analysis, as no signs of illness unrelated to the intervention were observed. The sample size (n = 8 per group) was determined based on conventional standards in similar preclinical models; no formal power calculation was performed. Animals were randomly assigned to groups; however, no specific measures were taken to control for potential confounders such as cage position or treatment order. Group allocation and outcome assessments were not blinded. We acknowledge that the absence of blinding and a formal power calculation is a limitation of this study, and we will incorporate these elements in future experimental designs to enhance rigor and reproducibility. All animal procedures were approved by the Institutional Animal Care and Use Committee (IACUC) of Dongguk University (approval no. IACUC-2024-04; approval date: 25 April 2024).

2.6. Serum Biochemical Analysis

At the end of the experimental period, blood was collected via cardiac puncture under anesthesia and allowed to clot at room temperature for 30 min. Serum was separated by centrifugation (3000 rpm, 15 min, 4 °C) and stored at −80 °C. Serum levels of TG, total cholesterol (TC), high-density lipoprotein cholesterol (HDL-C), low-density lipoprotein cholesterol (LDL-C), alanine aminotransferase (ALT), and aspartate aminotransferase (AST) were measured using commercial enzymatic colorimetric assay kits (Asan Pharmaceutical Co., Ltd., Seoul, Republic of Korea), following the manufacturer's protocol. Serum samples were mixed with specific enzyme reagents in 96-well microplates, and the resulting colorimetric reactions were quantified using a microplate reader (BioTek Instruments, Winooski, VT, USA) at the appropriate wavelengths (550 nm for TG and 500 nm for cholesterol). Serum leptin concentrations were determined using mouse-specific enzyme-linked immunosorbent assay kits (R&D Systems, Minneapolis, MN, USA). Assays were performed following the manufacturer's protocol. OD was measured at 450 nm, and the concentrations were determined using standard curves prepared with known concentrations of recombinant mouse leptin. All measurements were performed in triplicate and are expressed as the mean ± SD. Inter- and intra-assay variability was maintained within 10%.

2.7. Statistical Analysis

Statistical analyses were performed using SPSS software (version 20.0; SPSS Inc., Chicago, IL, USA). Data are presented as the mean ± standard deviation (SD). Normality was assessed using the Shapiro–Wilk test. A one-way analysis of variance (ANOVA), followed by Tukey's honest significant difference (HSD) test, was used to determine statistical significance, which was set at $p < 0.05$. The sample size (n = 8 per group) was based on precedent in similar preclinical studies; no formal power analysis was conducted.

3. Results

3.1. Fatty Acid Composition of GmO and GmOLb

The fatty acid compositions of GmO and GmOLb are presented in Table 1. Fermentation with *L. brevis* KCCM13538P resulted in a modest reduction in total saturated fatty acids (SFAs) and a corresponding increase in USFAs, particularly ω-3 PUFAs. Notably, the levels of EPA (C20:5 ω-3) and DHA (C22:6 ω-3) were slightly elevated in GmOLb than in GmO. However, independent t-test analysis revealed that these differences were not statistically significant ($p > 0.05$). Fatty acid data were analyzed using independent t-tests to compare the means between the GmO and GmOLb groups. Statistical analyses were conducted using SPSS software (version 20.0; SPSS Inc., Chicago, IL, USA), and results are expressed as mean ± SD (n = 3). These long-chain ω-3 PUFAs are well known for their anti-inflammatory, lipid-lowering, and cardiometabolic benefits [34–37]. Thus, the compositional changes suggest a potential fermentation-induced enhancement; however, further studies with larger sample sizes and targeted enzymatic assays are needed to confirm the functional relevance.

Table 1. Fatty acid composition of GmO and GmOLb.

Fatty Acid (g/100 g)	Lipid Number	GmO	GmOLb	Type
SFA (Saturated Fatty Acid)				
Myristic acid	C14:0	4.28 ± 0.21	4.33 ± 0.22	
Pentadecanoic acid	C15:0	0.37 ± 0.02	0.37 ± 0.03	
Palmitic acid	C16:0	13.03 ± 0.65	12.32 ± 0.62	
Heptadecanoic acid	C17:0	0.31 ± 0.02	0.35 ± 0.02	
Stearic acid	C18:0	2.74 ± 0.14	2.6 ± 0.13	
Arachidic acid	C20:0	0.23 ± 0.01	0.21 ± 0.01	
Eicosanoic acid	C20:0	3.33 ± 0.17	3.37 ± 0.17	
Heneicosylic acid	C21:0	0.07 ± 0.01	0.07 ± 0.01	
Behenic acid	C22:0	0.08 ± 0.03	0.08 ± 0.03	
Total SFAs (g/100 g)		24.44 ± 0.72	23.7 ± 0.69	
USFA (Unsaturated Fatty Acid)				
Myristoleic acid	C14:2	0.07 ± 0.01	0.0 7± 0.02	MUFA (ω-5)
Palmitoleic acid	C16:1	6.71 ± 0.34	6.92 ± 0.35	MUFA (ω-7)
Oleic acid	C18:1	19.2 ± 0.96	20.03 ± 1.02	MUFA (ω-9)
Linoleic acid	C18:2	2.74 ± 0.14	2.79 ± 0.14	PUFA (ω-6)
α-Linolenic acid	C18:3	1.33 ± 0.07	1.39 ± 0.07	PUFA (ω-3)
γ-Linolenic acid	C18:3	0.01 ± 0.00	0.15 ± 0.01	PUFA (ω-6)
Elaidic acid	C18:1	0.16 ± 0.01	0.74 ± 0.04	Trans-MUFA (ω-9)
Linolelaidic acid	C18:2	0	0.15 ± 0.02	Trans-PUFA (ω-6)
Eicosadienoic acid	C20:2n-6	0.28 ± 0.01	0.16 ± 0.21	PUFA (ω-6)
Eicosatetraenoic acid	C20:4	0.12 ± 0.01	0.09 ± 0.02	PUFA (ω-3)
Eicosatrienoic acid	C20:3n-3	0.15 ± 0.01	0.13 ± 0.01	PUFA (ω-3)
Arachidonic acid	C20:4	0.61 ± 0.03	0.64 ± 0.02	PUFA (ω-6)
Eicosapentaenoic acid (EPA)	C20:5	7.56 ± 0.38	7.66 ± 0.38	PUFA (ω-3)
Erucic acid	C22:1	0.54 ± 0.03	0.48 ± 0.02	MUFA (ω-9)
Docosapentaenoic acid	C22:5	0.59 ± 0.03	0.62 ± 0.03	PUFA (ω-3)
Docosahexaenoic acid (DHA)	C22:6	8.92 ± 0.45	9.13 ± 0.46	PUFA (ω-3)
Nervonic acid	C24:1	0.44 ± 0.02	0.51 ± 0.03	MUFA (ω-9)
Total USFAs (g/100 g)		49.43 ± 1.19	51.66 ± 1.26	

GmO: Gwamegi oil, GmOLb: fermented Gwamegi oil, and MUFA: monounsaturated fatty acid. Values are expressed as the mean ± standard deviation (estimated). Statistical comparisons were not performed. Data represent the sum of quantified fatty acids detected using gas chromatography; trace-level and unidentified fatty acids are not included.

While previous research has reported lipase and esterase activities in *L. brevis* [22], its substrate specificity toward esterified ω-3 PUFAs remains unclear. Since enzymatic activity was not directly quantified in this study, the observed increase in ω-3 content may instead reflect enhanced oxidative stability or bioavailability post-fermentation. Future investigations using both targeted enzymatic assays and untargeted metabolomics are needed to clarify the underlying biochemical mechanisms.

3.2. Contents of Fat-Soluble Vitamins

Fat-soluble vitamins (A, D, E, and K2) play crucial roles in antioxidant defense, bone metabolism, and overall metabolic regulation [38–40]. Because of the lipophilic nature of marine oils, fermentation may affect the stability, transformation, or biosynthesis of these vitamins through microbial enzymatic pathways, such as esterification, dehydrogenation, or decarboxylation [41,42]. In this study, we assessed the effect of *L. brevis* KCCM13538P fermentation on the fat-soluble vitamin content of GmO to assess the potential for nutritional enhancement through microbial bioprocessing. The concentrations of fat-soluble vitamins in non-fermented GmO and GmOLb are summarized in Table 2. Fermentation significantly altered the vitamin profile, specifically enhancing the bioactive compound levels. The vitamin A content, expressed as RE, increased from 6999.41 ± 34.99 μg/100 g in GmO to 7579.27 ± 37.89 μg/100 g in GmOLb. This may be attributed to the microbial-mediated conversion of provitamin A carotenoids (β-carotene) to retinoid forms or the increased stability of pre-existing retinoids during fermentation [43]. Furthermore, the vitamin D content increased significantly, from 56.91 ± 0.28 μg/100 g in GmO to 67.95 ± 0.34 μg/100 g in GmOLb. This result is nutritionally significant because of the crucial role of vitamin D in calcium homeostasis, immune regulation, and metabolic health [44]. Lactic acid bacteria affect vitamin D metabolism and protect against its degradation in lipid-rich matrices [45]. A significant increase in vitamin E (α-TE) was observed, from 1070 ± 6.85 μg to 4780 ± 23.91 μg/100 g, after fermentation. Due to its role as a major lipid-phase antioxidant, this increased tocopherol level may contribute to enhanced oxidative stability and bioactivity of the fermented oil [46], possibly enhancing its shelf-life and physiological functions. For vitamin K2, both MK-4 (from 42.19 ± 3.26 to 64.13 ± 3.07 μg) and MK-7 (from 19.28 ± 1.48 to 46.56 ± 1.82 μg/100 g) contents were significantly increased in GmOLb. These forms of vitamin K2 are associated with bone health, vascular calcification prevention, and anti-inflammatory effects [47,48]. The observed enrichment indicates that *L. brevis* may harbor metabolic pathways for MK biosynthesis or facilitate conversion from dietary phylloquinone during fermentation [49]. In summary, these results indicate that microbial fermentation of GmO preserves and significantly enhances its nutritional profile, specifically in terms of bioactive fat-soluble vitamins. The increased levels of vitamins A, D, E, and K2 may synergistically contribute to the observed anti-obesity and hepatoprotective effects of GmOLb, potentially through antioxidant activity, lipid metabolism regulation, and systemic metabolic support.

Table 2. Contents of fat-soluble vitamins in GmO and GmOLb.

Vitamin	Abbreviation	Unit (/100 g)	GmO	GmOLb
Vitamin A	Vit A	μg RE	6999.41 ± 34.99	7579.27 ± 37.89
Vitamin D	Vit D	μg	56.91 ± 0.28	67.95 ± 0.34
Vitamin E	Vit E	μg α-TE	1070 ± 6.85	4780 ± 23.91
Vitamin K2	MK-4	μg	42.19 ± 3.26	64.13 ± 3.07
	MK-7	μg	19.28 ± 1.48	46.56 ± 1.82

GmO: Gwamegi oil and GmOLb: fermented Gwamegi oil. Values are expressed as the mean ± standard deviation per 100 g of sample. The experiments were conducted in triplicate. Vitamin A is expressed as retinol equivalents (RE); vitamin D as micrograms (μg); vitamin E as alpha-tocopherol equivalents (α-TE); and vitamin K2 as menaquinone-4 (MK-4) and menaquinone-7 (MK-7), both in μg.

3.3. *Effects on Lipid Accumulation in 3T3-L1 Adipocytes*

To assess the anti-obesity potential of GmO and GmOLb, their effects on lipid accumulation were assessed during 3T3-L1 adipocyte differentiation. As illustrated in Figure 1A, treatment with GmO and GmOLb at concentrations up to 300 μg/mL did not exhibit significant cytotoxicity in 3T3-L1 preadipocytes, as determined by the MTT assay. These results confirm the biocompatibility of both oils and support their suitability for further adipogenic studies. Oil Red O staining (Figure 1B) revealed significant intracellular lipid accumulation in the control group. In contrast, both GmO- and GmOLb-treated groups exhibited visibly reduced lipid droplet formation. Quantitative analysis of Oil Red O absorbance (Figure 1C) demonstrated a dose-dependent inhibition of lipid accumulation by GmO and GmOLb. Although GmO exhibited slightly lower lipid accumulation than that of GmOLb at 200 μg/mL, GmOLb exerted superior anti-lipogenic effects at most concentrations. Notably, at 200 μg/mL, GmOLb reduced lipid content by approximately 45% compared to that of the control group ($p < 0.01$), indicating enhanced efficacy through fermentation. Additionally, intracellular TG levels were significantly reduced in both GmO- and GmOLb-treated adipocytes (Figure 1D), further supporting their lipid-lowering capacity. GmOLb exhibited significant TG-lowering effects than that of GmO at all tested concentrations except 50 μg/mL, indicating enhanced bioactivity through fermentation. The superior anti-adipogenic effect of GmOLb may be attributed to its high levels of ω-3 PUFAs (EPA and DHA) and fat-soluble vitamins (A, D, E, and K2) (Tables 1 and 2). These components regulate adipogenesis and lipid metabolism by modulating peroxisome proliferator-activated receptor gamma (PPARγ) signaling, reducing oxidative stress, and enhancing mitochondrial function [50–52]. In summary, these findings indicate that lactic acid bacterial fermentation of marine oils, such as GmO, may enhance their efficacy in reducing adipocyte lipid accumulation. GmOLb is a promising candidate as a functional food or nutraceutical for targeting obesity and related metabolic disorders.

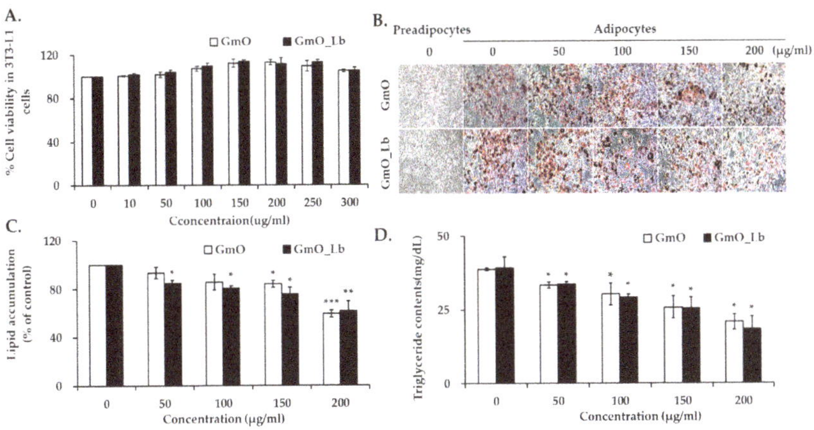

Figure 1. Effects of GmO and GmOLb on lipid accumulation in 3T3-L1 adipocytes. (**A**) Cell viability of 3T3-L1 preadipocytes treated with GmO or GmOLb at concentrations ranging from 10 to 300 μg/mL for 24 h, assessed by the MTT assay. (**B**) Representative images of Oil Red O staining to visualize intracellular lipid accumulation during adipocyte differentiation in response to GmO or GmO_Lb treatment. (**C**) Quantitative analysis of intracellular lipid content based on the absorbance measurements of eluted Oil Red O dye from stained adipocytes. (**D**) Intracellular TG content in differentiated 3T3-L1 adipocytes treated with GmO or GmO_Lb. All data are presented as the mean ± standard deviation ($n = 3$). Statistical significance is indicated as * $p < 0.05$, ** $p < 0.01$, and *** $p < 0.001$ vs. control group. GmO, Gwamegi oil; GmOLb, Gwamegi oil fermented with *L. brevis* KCCM 13538P; TG, triglyceride; MTT, 3-(4,5-dimethylthiazol-2-yl)-2,5-diphenyltetrazolium bromide.

3.4. Effects on Body Weight Gain and Adiposity in HFD-Fed Mice

To assess the anti-obesity effects of GmO and GmOLb, we measured body weight gain, cumulative feed intake, and visceral adiposity in an HFD-induced obesity mouse model (Figure 2A). Gm-O exhibited the highest feed intake among the HFD-fed groups; however, differences in body weight were observed between the treatment groups, indicating that variations in weight gain were not attributable to differences in feed consumption (Figure 2B). As anticipated, the HFD_L group exhibited significantly higher body weight gain than that of the ND group (Figure 2C). Both GmO- and GmOLb-supplemented HFD attenuated weight gain, with GmOLb producing a more significant suppression. Ovary fat pad mass was significantly increased in the HFD_L group than that in the ND group ($p < 0.001$). However, both the HFD_GmO and HFD_GmOLb groups exhibited significant reductions ($p < 0.05$), with no difference between them (Figure 2D). These findings indicate that GmO alleviates HFD-induced visceral fat accumulation and that fermentation with *L. brevis* augments the anti-obesity efficacy of the extract.

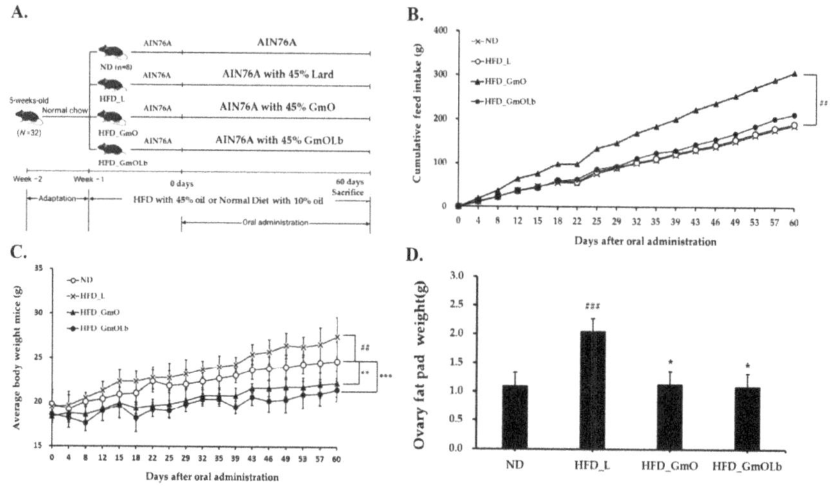

Figure 2. Effects of GmO and GmOLb on body weight gain and adipose tissue accumulation in HFD-fed mice. (**A**) Experimental design. Five-week-old female mice (n = 32) were assigned to four groups (n = 8 each): ND (normal diet), HFD_L (HFD with 45% lard), HFD_GmO (HFD with 45% Gwamegi oil), and HFD_GmOLb (HFD with 45% *L. brevis*-fermented Gwamegi oil). After a 1-week acclimation, diets were administered for 60 days. (**B**) Cumulative feed intake during the 60-day period. (**C**) Body weight alterations during the feeding period. (**D**) Ovary fat pad weight at sacrifice. Data are presented as the mean ± standard deviation (n = 8). Statistical significance: $^{\#\#}$ $p < 0.01$ and $^{\#\#\#}$ $p < 0.001$ vs. ND; * $p < 0.05$, ** $p < 0.01$, and *** $p < 0.001$ vs. HFD_L. ND, normal diet; HFD, high-fat diet; GmO, Gwamegi oil; GmOLb, Gwamegi oil fermented with *L. brevis* KCCM 13538P.

Fatty acid composition analysis revealed that both GmO and GmOLb contained significantly higher levels of PUFAs, specifically EPA and DHA (Table 1). Following fermentation with *L. brevis*, GmOLb exhibited a slight increase in overall ω-3 PUFAs, whereas ω-6 PUFA levels also increased modestly, resulting in only a minor enhancement in the ω-3/ω-6 PUFA ratio (from approximately 5.10 to 5.05). Despite this subtle shift, the increased absolute concentrations of ω-3 PUFAs—recognized for their anti-adipogenic and lipid-oxidizing effects—contributed to the favorable lipid profile of GmOLb. Notably, fermentation significantly reduced the SFA content in GmOLb. The co-occurrence of high PUFA and reduced SFA levels enhances mitochondrial β-oxidation and suppresses adipogenesis, processes associated with reduced adipose tissue mass [53,54]. These compositional alterations likely highlight the reduced fat accumulation observed in the GmOLb-treated group. In addition, fermentation was associated with increased levels of bioactive lipid components, including medium-chain triglycerides (MCTs) and conjugated linoleic acid (CLA). MCTs rapidly undergo hepatic β-oxidation and facilitate energy expenditure through thermogenesis [55], whereas CLA modulates lipid metabolism by activating PPARγ and uncoupling protein 1 in the adipose tissue [56,57]. Moreover, GmOLb exhibited increased levels of fat-soluble vitamins, including MK-4, MK-7, A, D, and E (Table 2). These micronutrients play crucial roles in enhancing insulin sensitivity, antioxidant defense, and metabolic regulation [58–60]. Specifically, MK-4 enhances insulin secretion and supports glucose homeostasis in both preclinical and clinical studies [61].

In summary, these findings indicate that GmOLb exhibits superior anti-obesity effects than that of unfermented GmO, likely because of compositional enhancements achieved through microbial fermentation. The altered fatty acid profile, increased levels of MCTs and CLA, and enrichment of functional lipid micronutrients highlight the potential of lactic acid bacteria-mediated fermentation as an effective strategy for enhancing the metabolic benefits of marine-derived oils.

3.5. Effects on Serum Lipid Profiles and Cardiovascular Risk in HFD-Fed Mice

After a 60-day intervention in HFD-fed mice, we measured serum lipid and cardiometabolic marker levels to assess the effects of GmO and GmOLb (Figure 3). As anticipated, HFD_L mice exhibited significant hyperlipidemia—TG, TC, and LDL-C, alongside increased HDL-C—confirming diet-induced dyslipidemia (Figure 3A–D). Supplementation with either GmO or GmOLb significantly lowered TG and TC levels while restoring HDL-C levels compared to HFD_L. Both the GmO and GmOLb groups exhibited a significant effect, with TG and LDL-C levels reduced to nearly those of the ND group, and HDL-C levels significantly increased. This indicates that GmO has a lipid-modulating efficacy. Serum leptin, an adipose-derived hormone implicated in energy regulation and leptin resistance, increased from approximately 4 ng/mL in ND mice to 7 ng/mL in HFD_L. Both GmO and GmOLb significantly reduced leptin levels (to approximately 0.57 and 0.49 ng/mL, respectively; *** $p < 0.001$ vs. HFD_L) (Figure 3E), indicating enhanced leptin sensitivity and adiposity reduction. These leptin-lowering effects align with previous studies demonstrating that vitamin K2 supplementation reduces circulating leptin levels in obese animal and clinical models [62].

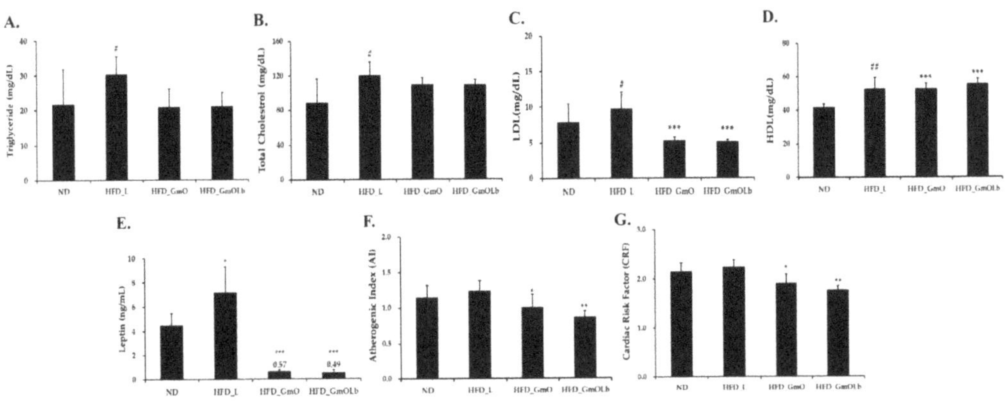

Figure 3. Effects of GmO and GmOLb on serum lipid profiles and cardiovascular risk markers in HFD-fed mice. (**A–D**) Serum levels of TG, TC, LDL-C, and HDL-C after 60 days of dietary intervention. (**E–G**) Leptin, AI, and CRF values in each treatment group. Data are presented as the mean ± standard deviation ($n = 8$ per group). Statistical significance: # $p < 0.05$ and ## $p < 0.01$ vs. ND; * $p < 0.05$, ** $p < 0.01$, and *** $p < 0.001$ vs. HFD_L. ND, normal diet; HFD, high-fat diet; GmO, Gwamegi oil; GmOLb, Gwamegi oil fermented with *L. brevis* KCCM13538P; TG, triglycerides; TC, total cholesterol; LDL-C, low-density lipoprotein cholesterol; HDL-C, high-density lipoprotein cholesterol; AI, atherogenic index; CRF, cardiac risk factor. AI = (TC–HDL-C)/HDL-C; CRF = TC/HDL-C.

To quantify cardiovascular risk, we calculated the atherogenic index (AI = [TC–HDL-C]/HDL-C) and cardiac risk factor (CRF = TC/HDL-C). HFD_L mice exhibited significantly increased AI and CRF, indicating increased cardiovascular disease (CVD) risk (Figure 3F,G). Both GmO and GmOLb significantly ameliorated these indices, with GmOLb yielding the largest reduction, indicating superior cardioprotection. This effect may stem from the fermentation-induced enhancement of GmOLb's fatty acid profile and fat-soluble micronutrients (MK-4 and MK-7). Vitamin K2 activates the γ-carboxylation of matrix Gla protein (MGP), a potent inhibitor of vascular calcification. Carboxylated MGP prevents the osteogenic transformation of vascular smooth muscle cells, thereby mitigating arterial calcification and stiffness, i.e., primary contributors to cardiovascular morbidity [63,64]. In addition to vascular health, vitamin K2 supports metabolic regulation. Carboxylated osteocalcin, another K2-dependent protein, enhances insulin secretion and adiponectin expression, thereby enhancing insulin sensitivity and lipid metabolism. Consistently, MK-7 supplementation in humans reduces fasting insulin and leptin levels, parallel to our observations in GmOLb-supplemented mice [65,66]. Furthermore, HFD-induced dyslipidemia contributes to leptin resistance and adiposity; however, the combination of PUFAs and vitamin K2 in GmOLb likely exerts synergistic effects by enhancing lipid metabolism, reducing inflammation, and supporting energy homeostasis.

3.6. *Effects on Hepatic Enzyme and Bilirubin Levels in HFD-Fed Mice*

To assess the hepatoprotective potential of GmO and GmOLb, the serum levels of AST, ALT, and total bilirubin were measured after 60 days of treatment in HFD-fed mice (Figure 4). Mice receiving HFD supplemented with HFD_L exhibited a significant increase in AST and ALT levels than those in the ND group, indicating liver injury associated with lipid accumulation and inflammation. A slight but significant increase in total bilirubin levels was observed, indicating mild cholestatic impairment. In contrast, both HFD_GmO and fermented HFD_GmOLb attenuated the increase in AST and ALT levels compared to HFD_L, indicating reduced hepatocellular damage. Notably, the GmOLb group demonstrated a significant reduction, with enzyme levels approaching those of the ND controls. Similarly, GmOLb significantly reduced total bilirubin levels, supporting enhanced hepatic function and bile excretion. The enhanced efficacy of GmOLb may stem from fermentation-induced increases in ω-3 PUFAs (EPA and DHA) that regulate AMP-activated protein kinase and PPARα pathways to alleviate inflammation and steatosis [67]. In addition, increased menaquinones (MK-4 and MK-7) in GmOLb exert anti-fibrotic and antioxidative effects by suppressing pro-inflammatory cytokines and stabilizing mitochondrial redox balance [68,69]. These findings indicate that GmOLb enhances systemic lipid metabolism and confers robust protection against HFD-induced hepatic dysfunction. Fermentation appears to synergistically enhance the bioactivity of GmO by enriching its lipid profile and micronutrient content. In this study, we assessed how fermentation with *L. brevis* KCCM13538P (GmOLb) modifies the fatty acid and micronutrient compositions of herring and saury oils. We also assessed the therapeutic efficacy of GmOLb in diet-induced obesity and metabolic dysfunction models. Through a series of in vivo and in vitro experiments, GmOLb con sistently demonstrated potent anti-obesity, lipid-lowering, hepatoprotective, and cardiometabolic effects. These findings highlight the potential of lactic acid bacterial fermentation to enhance the physiological activity of marine oils, thereby transforming them into functional foods with significant health-promoting potential.

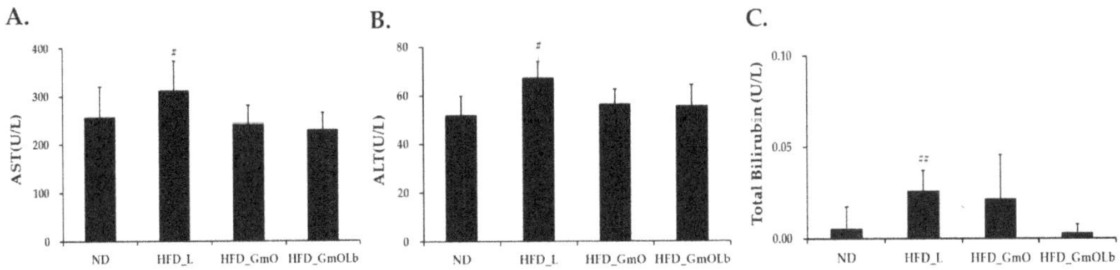

Figure 4. Effects of GmO and GmOLb on hepatic enzyme and bilirubin levels in HFD-fed mice. (**A**) AST, (**B**) ALT, and (**C**) total bilirubin levels after 60 days of dietary intervention. Data are expressed as the mean ± standard deviation (n = 8). Statistical significance: $^{\#}$ $p < 0.05$, $^{\#\#}$ $p < 0.01$ vs. ND. ND, normal diet; HFD, high-fat diet; GmO, Gwamegi oil; GmOLb, Gwamegi oil fermented with *L. brevis* KCCM13538P; AST, aspartate aminotransferase; ALT, serum alanine aminotransferase.

4. Discussion

The increasing global prevalence of obesity and its associated health complications, including metabolic syndrome and liver dysfunction, underscores the need for effective preventive and therapeutic strategies. In this study, we assessed the anti-obesity and hepatoprotective effects of GmO and GmOLb, produced using *L. brevis* KCCM13538P. Our findings revealed that microbial fermentation significantly enhances the bioactivity of fish oil, making GmOLb a promising candidate for treating obesity-associated metabolic disorders.

In vitro experiments using 3T3-L1 adipocytes demonstrated that GmOLb exhibited stronger anti-adipogenic effects than non-fermented GmO. GmOLb significantly suppressed lipid accumulation and intracellular TG content in differentiated adipocytes in a dose-dependent manner. These findings align with those of previous studies reporting that fermented products, particularly those involving lactic acid bacteria, inhibit adipogenesis by modulating key pathways related to lipid synthesis and storage [26,27]. The enhanced efficacy of GmOLb may be attributed to the enzymatic activity of *L. brevis*, including esterase and lipase, which can hydrolyze complex lipids and increase the release of bioavailable ω-3 PUFAs during fermentation [22]. Microbial fermentation is thus a promising strategy for enhancing the functional properties of marine-derived oils [70–72]. In this study, although the increases in ω-3 PUFAs (EPA and DHA) following *L. brevis* fermentation of GmO were not statistically significant, the fermented oil (GmOLb) exhibited superior physiological effects in vivo. GmOLb significantly improved serum lipid profiles, reduced visceral adiposity, and attenuated hepatic injury markers in HFD-induced obese mice. These findings suggest that fermentation may enhance the health benefits of fish oil not only through changes in fatty acid composition, but also by increasing bioavailability or generating bioactive metabolites. Further studies incorporating metabolomic profiling and dose–response analyses are warranted to elucidate the underlying mechanisms.

Our in vivo studies confirmed the superior anti-obesity effects of GmOLb. In HFD-induced obese mice, GmOLb significantly attenuated body weight gain and reduced epididymal white adipose tissue mass, a crucial indicator of visceral obesity. This aligns with studies demonstrating that ω-3-rich fish oils can mitigate diet-induced obesity by facilitating fatty acid oxidation and reducing adipogenesis [28,29]. However, the more significant effects of GmOLb indicate that fermentation amplifies these benefits. The enhanced efficacy of GmOLb can be attributed to the increased bioavailability of ω-3 PUFAs or the synergistic effects of other bioactive compounds produced during fermentation, which may affect satiety, energy expenditure, or lipid absorption [30,31].

Furthermore, GmOLb significantly enhanced serum lipid profiles in HFD-fed mice, characterized by reduced total cholesterol, TG, and LDL-C levels, and increased HDL-C levels. These changes significantly reduced the AI and CRF ratio, indicating a reduced risk of CVD. Obesity often leads to dyslipidemia owing to impaired lipid metabolism and increased synthesis of very LDL in the liver [32]. The ability of GmOLb to normalize these lipid parameters highlights its potential in ameliorating metabolic complications. The ω-3 PUFAs in fish oils are well-known for their hypolipidemic effects, primarily by suppressing hepatic TG synthesis and facilitating fatty acid oxidation [8]. Fermentation may further potentiate these effects by enhancing fatty acid absorption or producing metabolites that effectively regulate lipid metabolism. In this study, GmOLb supplementation significantly improved serum lipid profiles in HFD-fed mice, with significant reductions in total cholesterol, TG, and LDL-C, and a significant increase in HDL-C. These changes resulted in lower AI and CRF values, indicating enhanced cardiovascular protection.

A significant finding was the potent hepatoprotective effect of GmOLb. HFD-induced obesity often causes NAFLD, characterized by hepatic steatosis, inflammation, and potential progression to more severe liver pathologies [33]. Our results demonstrated that GmOLb significantly reduced the increased serum levels of ALT, AST, and total bilirubin and ameliorated hepatic steatosis, as confirmed by histopathological analysis. These findings indicate that GmOLb effectively protects the liver from HFD-induced damage. The hepatoprotective mechanism of ω-3 PUFAs involves reducing hepatic lipid accumulation, oxidative stress, and inflammatory responses within the liver [9,34]. The enhanced efficacy of GmOLb may stem from the increased bioavailability of ω-3 PUFAs because of fermentation or the production of specific anti-inflammatory or antioxidant compounds by *L. brevis*. For example, certain bacterial metabolites modulate the gut microbiota, thereby influencing liver health by affecting gut barrier integrity and reducing the translocation of bacterial endotoxins to the liver [35]. Despite the observed improvements in adiposity and lipid profiles, this study has some limitations. The short duration, single-dose design, and absence of dose optimization limit conclusions regarding long-term efficacy and safety. In addition, the inherent variability in individual responses and the use of a murine model restrict the direct applicability of the findings to human physiology. Future studies should incorporate extended intervention periods, dose–response analyses, and clinical validation to substantiate the therapeutic potential of GmOLb.

Our study demonstrates the promising effects of GmOLb; however, certain additional limitations warrant consideration. First, the mechanisms by which *L. brevis* fermentation enhances the bioactivity of GmO remain unclear. In this study, a small amount of yeast extract (0.1%) was added to support microbial growth and metabolic activity during lipid-based fermentation. Fish oil served as the primary substrate; however, this auxiliary nitrogen source may have influenced microbial metabolism and fermentation outcomes. In addition to improving lipid composition, *L. brevis* fermentation may also generate short-chain fatty acids (SCFAs), including acetate, propionate, and butyrate, through partial metabolism of available substrates. SCFAs are known to exert anti-obesity effects by modulating lipid metabolism, appetite regulation, and inflammation. While SCFA levels were not measured in this study, their potential contribution to the observed metabolic benefits cannot be excluded. Future studies should investigate the effects of different fermentation substrates, monitor SCFA production, and perform targeted metabolomic analyses to identify key fermentation-derived bioactives and elucidate their molecular mechanisms of action. Second, although the in vivo model provided valuable insights, long-term studies are needed to evaluate the sustained efficacy and safety of GmOLb. Finally, investigating the effects of GmOLb on gut microbiota composition may offer a more comprehensive understanding of its metabolic impact. While ω-3 PUFAs and fat-soluble vitamins likely play a central role, additional bioactive compounds—such as SCFAs, lipid mediators (e.g., resolvins), bioactive peptides, conjugated linoleic acids, and medium-chain triglycerides—may also contribute to the observed physiological benefits. These effects may be mediated by fermentation-derived SCFAs and modified lipids, both of which have been shown to regulate energy metabolism and inflammatory responses [70,71]. In addition to ω-3 PUFAs, fermented marine oils may be enriched with micronutrients such as vitamin K2 (menaquinones), which are known to exert beneficial effects on lipid metabolism and hepatic function. Menaquinone-4 (MK-4) and menaquinone-7 (MK-7), the bioactive forms of vitamin K2, have been shown to modulate insulin signaling, suppress hepatic lipogenesis, and reduce inflammation, thereby contributing to improved lipid profiles and reduced hepatic steatosis [14–16]. Microbial fermentation, particularly with *L. brevis*, may enhance the conversion of vitamin K1 or endogenous precursors into MK-4 and MK-7, thereby increasing the bioavailability of these fat-soluble micronutrients. Vitamin K levels were not quantified in this study; however, the potential enrichment of menaquinones during fermentation represents a promising mechanistic link that may explain the improved metabolic outcomes observed with GmOLb. Future studies should include the targeted quantification of MK vitamins, and investigate their synergistic interactions with ω-3 PUFAs and other fermentation-derived bioactives in regulating lipid and liver metabolism. While polyphenols and terpenes are not typically associated with fish-derived oils, these fermentation-related metabolites warrant further investigation.

5. Conclusions

This study demonstrates that fermenting marine oils—specifically herring and saury—with *L. brevis* KCCM13538P (GmOLb) significantly enhances their nutritional and functional profiles. Fermentation enhanced the fatty acid composition (notably increasing ω-3 PUFAs) and micronutrient levels (MK-4 and MK-7). In both in vivo and in vitro models of diet-induced obesity and metabolic dysfunction, GmOLb exhibited strong anti-obesity, lipid-lowering, hepatoprotective, and cardiometabolic effects. These consistent and multifaceted therapeutic effects support the conclusion that lactic acid fermentation significantly enhances the bioactivity of marine oils. Therefore, GmOLb is a promising candidate as a functional food supplement with significant potential for preventing or mitigating obesity-associated metabolic disorders and liver dysfunction. Future research on long-term safety, human clinical efficacy, and mechanistic pathways will be essential to fully use its nutraceutical potential.

Author Contributions: Conceptualization, S.-M.H. and T.-W.G.; methodology, S.-M.H. and T.-W.G.; validation, S.-M.H. and T.-W.G.; formal analysis, H.-S.J.; resources, S.-M.H. and H.-S.J.; writing—original draft preparation, S.-M.H., H.-S.J. and T.-W.G.; writing—review and editing, S.-M.H.; visualization, S.-M.H.; supervision, S.-M.H. and T.-W.G.; project administration, S.-M.H. All authors have read and agreed to the published version of the manuscript.

Funding: This study was supported by the Korea Institute of Planning and Evaluation for Technology in Food, Agriculture, and Forestry (IPET) through the Technology Commercialization Support Program, funded by the Ministry of Agriculture, Food and Rural Affairs (MAFRA; 122055-03) and Gyeongsangbuk-do (2025; Medicare BlueFood Development Project).

Institutional Review Board Statement: The animal study protocol was approved by the Institutional Animal Care and Use Committee (IACUC) of Dongguk University (protocol code: [IACUC-2024-04] and 25 April 2024). The 3T3-L1 cells were obtained from the American Type Culture Collection (ATCC, Manassas, VA, USA), commercially available.

Informed Consent Statement: Not applicable.

Data Availability Statement: The original contributions presented in this study are included in the article. Further inquiries can be directed to the corresponding authors.

Acknowledgments: The authors would like to thank Yeob-Ju Kim, the director of Hyundai Special Feed Ind. Co., Ltd., for providing the herring and saury oil used in this research.

Conflicts of Interest: The authors declare no conflicts of interest.

References

1. Monteiro, C.A.; Moubarac, J.C.; Cannon, G.; Ng, S.W.; Popkin, B. Ultra-processed products are becoming dominant in the global food system. *Obes. Rev.* **2013**, *14*, 21–28. [CrossRef]
2. Cordain, L.; Eaton, S.B.; Sebastian, A.; Mann, N.; Lindeberg, S.; Watkins, B.A.; O'Keefe, J.H.; Brand-Miller, J. Origins and evolution of the Western diet: Health implications for the 21st century. *Am. J. Clin. Nutr.* **2005**, *81*, 341–354. [CrossRef] [PubMed]
3. Hariri, N.; Thibault, L. High-fat diet-induced obesity in animal models. *Nutr. Res. Rev.* **2010**, *23*, 270–299. [CrossRef] [PubMed]
4. Buettner, R.; Schölmerich, J.; Bollheimer, L.C. High-fat diets: Modeling the metabolic disorders of human obesity in rodents. *Obesity* **2007**, *15*, 798–808. [CrossRef]
5. De Souza, R.J.; Mente, A.; Maroleanu, A.; Cozma, A.I.; Ha, V.; Kishibe, T.; Uleryk, E.; Budylowski, P.; Schünemann, H.; Beyene, J.; et al. Intake of saturated and trans unsaturated fatty acids and risk of all cause mortality, cardiovascular disease, and type 2 diabetes: Systematic review and meta-analysis of observational studies. *BMJ* **2015**, *351*, h3978. [CrossRef] [PubMed]
6. Calder, P.C. Marine omega-3 fatty acids and inflammatory processes: Effects, mechanisms and clinical relevance. *Biochim. Biophys. Acta Mol. Cell Biol. Lipids* **2015**, *1851*, 469–484. [CrossRef]
7. Swanson, D.; Block, R.; Mousa, S.A. Omega-3 fatty acids EPA and DHA: Health benefits throughout life. *Adv. Nutr.* **2012**, *3*, 1–7. [CrossRef]
8. Shahidi, F.; Ambigaipalan, P. Omega-3 polyunsaturated fatty acids and their health benefits. *Annu. Rev. Food Sci. Technol.* **2018**, *9*, 345–381. [CrossRef]
9. Serhan, C.N.; Levy, B.D. Resolvins in inflammation: Emergence of the pro-resolving superfamily of mediators. *J. Clin. Investig.* **2018**, *128*, 2657–2669. [CrossRef]
10. Jump, D.B.; Depner, C.M.; Tripathy, S. Omega-3 fatty acid supplementation and cardiovascular disease: Thematic review series: New lipid and lipoprotein targets for the treatment of cardiometabolic diseases. *J. Lipid Res.* **2012**, *53*, 2525–2545. [CrossRef]
11. Ross, A.C. Vitamin A and retinoic acid in T cell–related immunity. *Am. J. Clin. Nutr.* **2012**, *96*, 1166S–1172S. [CrossRef] [PubMed]
12. Holick, M.F. Vitamin D deficiency. *N. Engl. J. Med.* **2007**, *357*, 266–281. [CrossRef]
13. Brigelius-Flohé, R.; Traber, M.G. Vitamin E: Function and metabolism. *FASEB J.* **1999**, *13*, 1145–1155. [CrossRef]
14. Beulens, J.W.J.; Booth, S.L.; van den Heuvel, E.G.; Stoecklin, E.; Baka, A.; Vermeer, C. The role of menaquinones (vitamin K2) in human health. *Br. J. Nutr.* **2013**, *110*, 1357–1368. [CrossRef]
15. Shearer, M.J.; Newman, P. Metabolism and cell biology of vitamin K. *Thromb. Haemost.* **2008**, *100*, 530–559. [CrossRef]
16. Schurgers, L.J.; Teunissen, K.J.F.; Hamulyák, K.; Knapen, M.H.J.; Vik, H.; Vermeer, C. Vitamin K–containing dietary supplements: Comparison of synthetic vitamin K1 and natto-derived menaquinone-7. *Blood* **2007**, *109*, 3279–3283. [CrossRef]
17. Rai, A.K.; Jini, R.; Swapna, H.C.; Sachindra, N.M.; Bhaskar, N.; Baskaran, V. Application of native lactic acid bacteria (LAB) for fermentative recovery of lipids and proteins from fish processing wastes: Bioactivities of fermentation products. *J. Aquat. Food Prod. Technol.* **2011**, *20*, 32–44. [CrossRef]
18. Özyurt, G.; Durmuş, M.; Sakarya, Y.; Uslu, L.; Kuley, E. Improved oxidative stability of functional oils with spirulina enhanced by probiotic fermentation and encapsulation. *J. Appl. Phycol.* **2024**, *36*, 1837–1847. [CrossRef]
19. Eratte, D.; Gengenbach, T.R.; Dowling, K.; Barrow, C.J.; Adhikari, B. Survival, oxidative stability and surface characteristics of spray-dried co-microcapsules containing omega-3 fatty acids and probiotic bacteria. *Dry. Technol.* **2016**, *34*, 1926–1935. [CrossRef]
20. Tingö, L.; Hutchinson, A.N.; Bergh, C.; Stiefvatter, L.; Schweinlin, A.; Jensen, M.G.; Krüger, K.; Bischoff, S.; Brummer, R.J. Potential modulation of inflammation by probiotic and omega-3 polyunsaturated fatty acid supplementation in elderly with chronic low-grade inflammation—A randomized, placebo-controlled trial. *Nutrients* **2022**, *14*, 3998. [CrossRef]
21. Park, J.E.; Oh, S.H.; Cha, Y.S. *Lactobacillus brevis* OPK-3 from kimchi prevents obesity and modulates inflammatory gene expression in high-fat diet-induced obese mice. *Nutrients* **2020**, *12*, 604. [CrossRef]
22. Feyereisen, M.; Mahony, J.; Kelleher, P.; Roberts, R.J.; O'Sullivan, T.; Geertman, J.M.A.; van Sinderen, D. Comparative genome analysis of the *Lactobacillus brevis* species. *BMC Genom.* **2019**, *20*, 416. [CrossRef] [PubMed]
23. Lyberg, A.-M.; Adlercreutz, P. Lipase specificity towards Eicosapentaenoic acid and docosahexaenoic acid depends on substrate structure. *Biochim. Biophys. Acta* **2008**, *1784*, 343–350. [CrossRef]
24. Kuo, C.H.; Huang, C.Y.; Lee, C.L.; Kuo, W.C.; Hsieh, S.L.; Shieh, C.J. Synthesis of DHA/EPA ethyl esters via lipase-catalyzed acidolysis using Novozym® 435: A kinetic study. *Catalysts* **2020**, *10*, 565. [CrossRef]
25. Ichihara, K.; Shibahara, A.; Yamamoto, K.; Nakayama, T. An improved method for rapid analysis of the fatty acids of glycerolipids. *Lipids* **1996**, *31*, 535–539. [CrossRef]

26. Bošnir, J.; Bevardi, M.; Hećimović, I.; Budeč, M.; Cindrić, I.J.; Kober, R.; Jurak, G.; Lasić, D.; Brkić, D.; Racz, A. Optimization of sample preparation procedure for determination of fat-soluble vitamins in milk and infant food by HPLC technique. *Processes* **2024**, *12*, 1530. [CrossRef]
27. Ahmed, S.; Mahmoud, A.M. A novel salting-out assisted extraction coupled with HPLC-fluorescence detection for trace determination of vitamin K homologues in human plasma. *Talanta* **2015**, *144*, 480–487. [CrossRef]
28. Zebisch, K.; Voigt, V.; Wabitsch, M.; Brandsch, M. Protocol for effective differentiation of 3T3-L1 cells to adipocytes. *Anal. Biochem.* **2012**, *425*, 88–90. [CrossRef]
29. Gregoire, F.M.; Smas, C.M.; Sul, H.S. Understanding adipocyte differentiation. *Physiol. Rev.* **1998**, *78*, 783–809. [CrossRef]
30. Mosmann, T. Rapid colorimetric assay for cellular growth and survival: Application to proliferation and cytotoxicity assays. *J. Immunol. Methods* **1983**, *65*, 55–63. [CrossRef] [PubMed]
31. Ramirez-Zacarias, J.L.; Castro-Munozledo, F.; Kuri-Harcuch, W. Quantitation of adipose conversion and triglycerides by staining intracytoplasmic lipids with Oil red O. *Histochemistry* **1992**, *97*, 493–497. [CrossRef]
32. Smith, P.K.; Krohn, R.I.; Hermanson, G.T.; Mallia, A.K.; Gartner, F.H.; Provenzano, M.D.; Fujimoto, E.K.; Goeke, N.M.; Olson, B.J.; Klenk, D.C. Measurement of protein using bicinchoninic acid. *Anal. Biochem.* **1985**, *150*, 76–85. [CrossRef]
33. Reeves, P.G.; Nielsen, F.H.; Fahey, G.C., Jr. AIN-93 purified diets for laboratory rodents: Final report of the American Institute of Nutrition Ad hoc writing committee on the reformulation of the AIN-76A rodent diet. *J. Nutr.* **1993**, *123*, 1939–1951. [CrossRef] [PubMed]
34. Shahidi, F.; Wanasundara, U.N. Omega-3 fatty acid concentrates: Nutritional aspects and production technologies. *Trends Food Sci. Technol.* **1998**, *9*, 230–240. [CrossRef]
35. Calder, P.C. n-3 PUFA and Inflammation: From Membrane to Nucleus and from Bench to Bedside. *Proc. Nutr. Soc.* **2020**, *79*, 404–416. [CrossRef]
36. Simopoulos, A.P. Omega-3 fatty acids in inflammation and autoimmune diseases. *J. Am. Col. Nutr.* **2002**, *21*, 495–505. [CrossRef]
37. Innes, J.K.; Calder, P.C. Omega-6 fatty acids and inflammation. *Prostaglandins Leukot. Essent. Fat. Acids* **2018**, *132*, 41–48. [CrossRef]
38. Traber, M.G.; Stevens, J.F. Vitamins C and E: Beneficial effects from a mechanistic perspective. *Free. Radic. Biol. Med.* **2011**, *51*, 1000–1013. [CrossRef]
39. Lips, P. Vitamin D physiology. *Prog. Biophys. Mol. Biol.* **2006**, *92*, 4–8. [CrossRef] [PubMed]
40. Vermeer, C. Vitamin K: The effect on health beyond coagulation—An overview. *Food Nutr. Res.* **2012**, *56*, 5329. [CrossRef] [PubMed]
41. LeBlanc, J.G.; Milani, C.; De Giori, G.S.; Sesma, F.; Van Sinderen, D.; Ventura, M. Bacteria as vitamin suppliers to their host: A gut microbiota perspective. *Curr. Opin. Biotechnol.* **2013**, *24*, 160–168. [CrossRef] [PubMed]
42. Morishita, T.; Tamura, A.; Makino, T.; Kudo, S. Production of menaquinones by lactic acid bacteria. *J. Dairy Sci.* **1999**, *82*, 1897–1903. [CrossRef] [PubMed]
43. LeBlanc, J.G.; Laiño, J.E.; Del Valle, M.J.; Vannini, V.; van Sinderen, D.; Taranto, M.P.; de Valdez, G.F.; de Giori, G.S.; Sesma, F. B-Group vitamin production by lactic acid bacteria—Current knowledge and potential applications. *J. Appl. Microbiol.* **2011**, *111*, 1297–1309. [CrossRef]
44. Calvo, M.S.; Whiting, S.J.; Barton, C.N. Vitamin D intake: A global perspective of current status. *J. Nutr.* **2005**, *135*, 310–316. [CrossRef]
45. Bayat, F.; Danafar, H.; Aminzare, M.; Mohseni, M. Investigating the stability of vitamin D_3 and Bifidobacterium lactis nanoparticles coated with polycaprolactone-polyethylene glycol-polycaprolactone triblock copolymer in Iranian white cheese: Physicochemical, microbiological, and sensory properties. *J. Agric. Food Res.* **2024**, *15*, 101039. [CrossRef]
46. Traber, M.G. Vitamin E: Antioxidant activity, biokinetics, and bioavailability. *Annu. Rev. Nutr.* **1990**, *10*, 357–382. [CrossRef]
47. Yan, Q.; Zhang, T.; O'Connor, C.; Barlow, J.W.; Walsh, J.; Scalabrino, G.; Xu, F.; Sheridan, H. The Biological Responses of Vitamin K2: A Comprehensive Review. *Food Sci. Nutr.* **2023**, *11*, 1634–1656. [CrossRef]
48. Bonaldo, F.; Leroy, F. Bacterially produced vitamin K_2 and its potential to generate health benefits in humans. *Trends Food Sci. Technol.* **2024**, *147*, 104461. [CrossRef]
49. Brooijmans, R.J.W.; Smit, B.A.; dos Santos, F.; Van Riel, J.W.; de Vos, W.M.; Hugenholtz, J. Heme and menaquinone induced electron transport in lactic acid bacteria. *Microb. Cell Fact* **2009**, *8*, 28. [CrossRef] [PubMed]
50. Kim, S.H.; Plutzky, J. Brown fat and browning for the treatment of obesity and related metabolic disorders. *Diabetes Metab. J.* **2016**, *40*, 12–21. [CrossRef]
51. Kim, H.K.; Della-Fera, M.A.; Lin, J.; Baile, C.A. Docosahexaenoic acid inhibits adipocyte differentiation and induces apoptosis in 3T3-L1 preadipocytes. *J. Nutr.* **2006**, *136*, 2965–2969. [CrossRef]
52. Clarke, S.D. Polyunsaturated fatty acid regulation of gene transcription: A molecular mechanism to improve the metabolic syndrome. *J. Nutr.* **2001**, *131*, 1129–1132. [CrossRef] [PubMed]
53. Buckley, J.D.; Howe, P.R. Anti-obesity effects of long-chain omega-3 polyunsaturated fatty acids. *Obes. Rev.* **2009**, *10*, 648–659. [CrossRef]

54. Flachs, P.; Horáková, O.; Brauner, P.; Rossmeisl, M.; Pecina, P.; Franssen-van Hal, N.; Růžičková, J.; Šponarová, J.; Drahota, Z.; Vlček, C.; et al. Polyunsaturated fatty acids of marine origin upregulate mitochondrial biogenesis and induce beta-oxidation in white fat. *Diabetologia* **2005**, *48*, 2365–2375. [CrossRef]
55. St-Onge, M.P.; Jones, P.J. Physiological effects of medium-chain triglycerides: Potential agents in the prevention of obesity. *J. Nutr.* **2002**, *132*, 329–332. [CrossRef] [PubMed]
56. Shen, W.; Chuang, C.C.; Martinez, K.; Reid, T.; Brown, J.M.; Xi, L.; Hixson, L.; Hopkins, R.; Starnes, J.; McIntosh, M. Conjugated linoleic acid reduces adiposity and increases markers of browning and inflammation in white adipose tissue of mice. *J. Lipid Res.* **2013**, *54*, 909–922. [CrossRef]
57. Den Hartigh, L.J.; Han, C.Y.; Wang, S.; Omer, M.; Chait, A. 10E,12Z-conjugated linoleic acid impairs adipocyte triglyceride storage by enhancing fatty acid oxidation, lipolysis, and mitochondrial reactive oxygen species. *J. Lipid Res.* **2013**, *54*, 2964–2978. [CrossRef] [PubMed]
58. Booth, S.L. Roles for vitamin K beyond coagulation. *Annu. Rev. Nutr.* **2009**, *29*, 89–110. [CrossRef]
59. Aaseth, J.; Alexander, J.; Alehagen, U. The Importance of Vitamin K and the Combination with Vitamin D3 for Bone and Cardiovascular Health. *Nutrients* **2024**, *16*, 2420. [CrossRef]
60. Knapen, M.H.J.; Jardon, K.M.; Vermeer, C. Vitamin K-induced effects on body fat and weight: Results from a 3-year vitamin K_2 intervention study. *Eur. J. Clin. Nutr.* **2018**, *72*, 136–141. [CrossRef]
61. Ho, H.J.; Shirakawa, H.; Hirahara, K.; Sone, H.; Kamiyama, S.; Komai, M. Menaquinone-4 amplified glucose-stimulated insulin secretion in isolated mouse pancreatic islets and INS-1 rat insulinoma cells. *Int. J. Mol. Sci.* **2019**, *20*, 1995. [CrossRef]
62. Zhang, Y.; Liu, L.; Wei, C.; Wang, X.; Li, R.; Xu, X.; Zhang, Y.; Geng, G.; Dang, K.; Ming, Z.; et al. Vitamin K_2 supplementation improves impaired glycemic homeostasis and insulin sensitivity for type 2 diabetes through gut microbiome and fecal metabolites. *BMC Med.* **2023**, *21*, 174. [CrossRef]
63. Adeli, S.; Gargari, B.P.; Karamzad, N. The effects of vitamin K_2 (menaquinone-7) on leptin and adiponectin levels in overweight/obese type 2 diabetes patients: A randomized clinical trial. *Pharm. Sci.* **2023**, *29*, 346–354. [CrossRef]
64. Choi, H.J.; Yu, J.; Choi, H.; An, J.H.; Kim, S.W.; Park, K.S.; Jang, H.C.; Kim, S.Y.; Shin, C.S. Vitamin K_2 supplementation improves insulin sensitivity via osteocalcin metabolism: A placebo-controlled trial. *Diabetes Care* **2011**, *34*, e147. [CrossRef] [PubMed]
65. Van Ballegooijen, A.J.; Pilz, S.; Tomaschitz, A.; Grübler, M.; Verheyen, N. The synergistic interplay between vitamins D and K for bone and cardiovascular health: A narrative review. *Int. J. Endocrinol.* **2017**, *2017*, 7454376. [CrossRef] [PubMed]
66. Singh, V.; Jain, S.; Prakash, S.; Thakur, M. Studies on the synergistic interplay of vitamin D and K for improving bone and cardiovascular health. *Curr. Res. Nutr. Food Sci.* **2022**, *10*, 840–857. [CrossRef]
67. Zúñiga, J.; Cancino, M.; Medina, F.; Varela, P.; Vargas, R.; Tapia, G.; Videla, L.A.; Fernandez, V. n-3 PUFA supplementation triggers PPAR-α activation and PPAR-α/NF-κB interaction: Anti-inflammatory implications in liver ischemia-reperfusion injury. *PLoS ONE* **2011**, *6*, e28502. [CrossRef]
68. Kieronska-Rudek, A.; Kij, A.; Kaczara, P.; Tworzydlo, A.; Napiorkowski, M.; Sidoryk, K.; Chlopicki, S. Exogenous vitamins K exert anti-inflammatory effects dissociated from their role as substrates for synthesis of endogenous MK-4 in murine macrophages cell line. *Cells* **2021**, *10*, 1571. [CrossRef]
69. El-Sherbiny, M.; Atef, H.; Helal, G.M.; Al-Serwi, R.H.; Elkattawy, H.A.; Shaker, G.A.; Said, E.; Abulfaraj, M.; Albalawi, M.A.; Elsherbiny, N.M. Vitamin K_2 (MK-7) intercepts the Keap-1/Nrf-2/HO-1 pathway and attenuates inflammatory/apoptotic signaling and liver aging in naturally aging rats. *Antioxidants* **2022**, *11*, 2150. [CrossRef] [PubMed]
70. Kim, M.; Qie, Y.; Park, J.; Kim, C.H. Gut Microbial Metabolites Fuel Host Antibody Responses. *Cell Host Microbe* **2016**, *20*, 202–214. [CrossRef]
71. Marco, M.L.; Heeney, D.; Binda, S.; Cifelli, C.J.; Cotter, P.D.; Foligné, B.; Gänzle, M.; Kort, R.; Pasin, G.; Pihlanto, A.; et al. Health Benefits of Fermented Foods: Microbiota and Beyond. *Curr. Opin. Biotechnol.* **2017**, *44*, 94–102. [CrossRef] [PubMed]
72. Yin, F.; Sun, X.; Luo, X.; Zheng, W.; Yin, L.; Zhang, Y.; Fu, Y. A Review on Marine Microbial Docosahexaenoic Acid Production through Circular Economy, Fermentation Engineering, and Antioxidant Technology. *Mar. Drugs* **2025**, *23*, 256. [CrossRef] [PubMed]

foods

MDPI

Article

Soy Protein Isolate Supplementation Favorably Regulates the Fermentation Characteristics of *Debaryomyces hansenii* and Flavor Profile in a Sausage Model

Wenwen Duan [1], Qiujin Zhu [2] and Jing Wan [1,2,*]

1 School of Liquor and Food Engineering, Guizhou University, Guiyang 550025, China; dwwyfd@163.com
2 Key Laboratory of Agricultural and Animal Products Storage and Processing, Guizhou University, Guiyang 550025, China; ls.qjzhu@gzu.edu.cn
* Correspondence: jwan@gzu.edu.cn

Abstract: The metabolic activity of fermentative microorganisms plays a critical role in determining the flavor profile of fermented meat products. Modulating carbon and nitrogen sources represents a promising strategy for enhancing product quality. In this study, *Debaryomyces hansenii* strains isolated from dry-cured ham were assessed in a sterile sausage model to evaluate the effects of different carbon sources (sucrose, corn starch) and nitrogen sources (leucine, soy protein isolate) on colony growth, enzyme activity, and physicochemical properties. These nutritional factors significantly affected the fermentation performance of *D. hansenii*. Corn starch and soy protein isolate increased colony count by 14.94% and 90%, respectively, and enhanced protease activity by 2-fold and 4.5-fold. Both treatments maintained high lipase activity (>50 U/g). Both supplements improved the water-holding capacity and decreased the water activity. Carbon sources reduced the medium pH, whereas nitrogen sources contributed to the maintenance of pH stability. A further analysis indicated that corn starch promoted the accumulation of aldehydes and ketones, which intensified the sourness and suppressed the saltiness. In contrast, soy protein isolate increased the abundance of free amino acids associated with umami and sweetness, and stimulated the formation of esters, ketones, and pyrazines, thereby enhancing flavor richness and umami intensity. Both ingredients also reduced saturated fatty acid levels and increased the unsaturated to saturated fatty acid ratio. Soy protein isolate exhibited a more pronounced effect on *D. hansenii* fermentation. This study provides a technical reference for enhancing the flavor characteristics of fermented meat products via the adjustment of carbon and nitrogen sources to regulate *D. hansenii* fermentation.

Keywords: *Debaryomyces hansenii*; fermented sausage; carbon and nitrogen sources; protease activity; free amino acids; free fatty acids; volatile flavor compounds; electronic tongue; fermentation optimization

1. Introduction

Fermented sausage is a representative of traditional fermented meat products, with a unique flavor primarily resulting from the metabolic activities of microorganisms. In traditional natural fermentation, the random colonization of environmental microorganisms led to variability in product quality across batches. Modern fermentation technology ensures controlled flavor and stable quality through targeted inoculation with fermentation strains such as lactic acid bacteria, staphylococci, and yeasts [1]. Among these strains, *Debaryomyces hansenii* (*D. hansenii*) is a promising yeast strain for enhancing the flavor of fermented meat products. The inoculation of *D. hansenii* into low-sodium fermented sausages made from boar meat can effectively mask boar taint by generating volatile compounds with fruity and cheesy aromas [2]. In dry-cured pork belly, inoculation with *D. hansenii* has also been shown to promote muscle protein degradation, enhance amino acid accumulation, and facilitate the formation of aldehydes, esters, alcohols, ketones, and organic acids [3]. These flavor-enhancing effects are generally attributed to the yeast's growth performance, enzymatic activities, and metabolite production [4]. Accordingly, enhancing the fermentation characteristics of *D. hansenii* is considered a promising strategy for improving the sensory quality of fermented meat products.

Modern food biotechnology has improved microbial fermentation performance through strain enhancement approaches such as directed evolution, gene editing, and adaptive evolution targeting tolerance to high osmotic pressure or extreme pH conditions [5]. However, long development cycles and high validation costs often limit these strategies. In contrast, supplementing fermentation substrates with conventional carbon and nitrogen sources offers a practical alternative with lower technical barriers and shorter implementation timelines [6]. Carbon and nitrogen sources not only serve as essential sources of energy and nutrients for microorganisms but also regulate their metabolic pathways. Studies have shown that the microbial utilization of these nutrients is controlled by catabolite carbon repression (CCR) and nitrogen catabolite repression (NCR) mechanisms [7]. The CCR mechanism enables microorganisms to preferentially utilize high-efficiency carbon sources (e.g., glucose) while repressing the expression of genes involved in the metabolism of secondary carbon sources (e.g., lactose and maltose), thereby improving energy utilization efficiency. The NCR mechanism regulates nitrogen utilization by repressing the metabolic pathways of secondary nitrogen sources (e.g., proline and urea) when preferred nitrogen sources (e.g., ammonium ions and glutamine) are abundant, and by activating alternative nitrogen utilization pathways under nitrogen-limited conditions [8].

Furthermore, the properties of carbon and nitrogen sources can directly affect the physicochemical characteristics and sensory properties of fermented meat products. For instance, corn starch, owing to its strong water-binding capacity, can improve moisture retention and enhance the structural integrity of the final product [9]. Soy protein isolate, characterized by its high protein content and low fat composition, is widely used to improve water retention, reduce fat content, and facilitate the development of plant-based meat products [10]. In addition, carbon and nitrogen sources can indirectly influence product quality by modulating microbial metabolic pathways. For example, sucrose can increase oxidative stress in staphylococci, thereby promoting the formation of nitrosyl myoglobin, which enhances the color of dry-cured sausages [11]. Branched-chain amino acids (leucine, isoleucine, and valine) can be metabolized by microorganisms into aldehydes, alcohols, and organic acids, contributing to the malt-like and nutty aromas of fermented meat products [12]. The type of carbon and nitrogen sources can affect the accumulation of flavor precursors and final flavor compounds by modulating microbial metabolic pathways. For example, xylose has been shown to promote acetoin production by *Leuconostoc mesenteroides* while inhibiting diacetyl accumulation, whereas the metabolism of fructose and sucrose is closely associated with mannitol production by *L. mesenteroides* [13]. The addition of soy protein isolate and wheat gluten as nitrogen sources has been reported to increase ketone levels in fermented fish paste, thereby enhancing cheese-like flavor characteristics [14].

Current research on *D. hansenii* has primarily focused on its role in enhancing the flavor of fermented meat products, whereas studies investigating the effects of carbon and nitrogen sources on its fermentation performance remain limited. In this study, four carbon and nitrogen sources commonly used in fermented meat products were selected: sucrose and corn starch as simple and complex carbon sources, and leucine and soy protein isolate as small- and large-molecule nitrogen sources. A sterile sausage model was employed to evaluate their effects on the growth, enzymatic activity, and physicochemical properties of a local *D. hansenii* strain. Corn starch and soy protein isolate, which demonstrated the best overall performance, were subsequently selected for the further analysis of flavor compound accumulation. The objective of this research is to explore how carbon and nitrogen sources impact flavor formation by *D. hansenii* during fermented sausage production and to offer practical strategies for optimizing fermentation through formulation design and microbial nutrient regulation.

2. Materials and Methods

2.1. Preparation of Fermentation Agent

The *D. hansenii* strain used in this study was previously isolated from local dry-cured Panxian ham by the research group and was identified via 26S rDNA sequencing (GenBank accession number: PV249420). The strain was inoculated into Yeast Extract Peptone Dextrose (YPD) broth and incubated at 180 rpm for 48 h. The culture was then centrifuged at 5000× *g* for 5 min at 4 °C, and the harvested cells were washed twice with sterile 0.9% saline. The cell density was adjusted to 1×10^8 cells/mL using a hemocytometer and stored at 4 °C until use.

2.2. Preparation of Sterile Sausage Medium

The raw materials used in this study were fresh pork back fat and hind leg meat, obtained from Guizhou Tainong Xingwang Food Co., Ltd. (Guiyang, China). The back fat was sterilized by autoclaving at 121 °C for 15 min. The hind leg meat was sterilized by immersion in 75% ethanol for 10 min, followed by two rinses with sterile water and then exposure to ultraviolet light for 10 min to minimize protein denaturation [15]. Following sterilization, the back fat and hind leg meat were minced and blended at a 3:7 (*w*/*w*) ratio. A sterile solution of 3.0% NaCl, 0.015% $NaNO_2$, and 0.05% ascorbic acid (sterilized by 0.22 µm membrane filtration) was added to prepare the sausage medium. A control group (CK) was prepared without additional carbon or nitrogen sources. Sucrose (CS) and corn starch (CC) served as the carbon source treatments, while leucine (NL) and soy protein isolate (NP) were used as the nitrogen source treatments. The carbon and nitrogen sources were added at 5% (*w*/*w*) for all treatment groups, based on the studies by Prestes et al. and Huang et al., with slight modifications [16,17]. Each group was inoculated with 2% *D. hansenii* cell suspension and fermented at 28 ± 0.5 °C with 85% relative humidity for 5 days. After fermentation, the samples were designated as FCK, FCS, FCC, FNL, and FNP for subsequent analyses.

2.3. Colony Count

Samples (5 g) from each group were homogenized with 45 mL of 0.9% sterile saline using a YM-08X stomacher homogenizer (Shanghai Yuming Instrument Co., Ltd., Shanghai, China) for 2 min. After appropriate serial dilutions, 100 µL of each dilution was plated onto YPD agar plates. The plates were incubated at 28 ± 0.5 °C for 5 days, and the number of colony-forming units per gram of sample (CFU/g) was determined.

2.4. Determination of Enzyme Activity

Protease activity was measured according to the method described by Ding et al. [14]. Samples (5 g) were mixed with 25 mL of phosphate buffer (10 mM, pH 6.8) and homogenized in an ice-water bath for 60 s at 5000 rpm using an XHF-D homogenizer (Ningbo Xinzhisheng Biotechnology Co., Ltd., Ningbo, China). The homogenate was centrifuged at 10,000× *g* for 20 min at 4 °C, and the supernatant was collected as the crude enzyme extract. Subsequently, 1 mL of the crude enzyme was mixed with 1 mL of 2% (*w*/*v*) casein solution and incubated at 40 °C for 10 min. The reaction was terminated by adding 2 mL of 0.4 mol/L trichloroacetic acid. After centrifuging at 10,000× *g* for 20 min at 4 °C, 1 mL of the resulting supernatant was mixed with 5 mL of 0.4 mol/L Na_2CO_3 and 1 mL of Folin reagent. After a 20 min reaction at 40 °C, the absorbance was measured at 680 nm using a Multiskan SkyHigh microplate reader (Thermo Fisher Scientific, Waltham, MA, USA). A standard curve was constructed using L-tyrosine, and one unit of protease activity (U/g) was defined as the amount of enzyme required to catalyze the production of 1 μg of L-tyrosine per minute.

Lipase activity was determined using the G0902F Lipase Activity Assay Kit (Suzhou Greys Biotechnology Co., Ltd., Suzhou, China) according to the manufacturer's instructions. One unit (U/g) of lipase activity was defined as the amount of enzyme required to catalyze the production of 1 nmol of *p*-nitrophenol per minute.

2.5. Determination of Physicochemical Properties

The pH value of each sample was measured directly using a Testo 205 portable pH meter (Testo Inc., Lenzkirch, Germany), which was calibrated with standard buffer solutions (pH 4.01 and 7.01) before use. Water activity was measured using an SFY-30 water activity meter (Shanghai Guanya Precision Instruments Co., Ltd., Shanghai, China) after equilibrating 5 g of sample at 25 °C for 8 min. For the water-holding capacity (WHC) analysis, the sample (G_1) was wrapped in three layers of filter paper, placed in a 50 mL centrifuge tube, and centrifuged at 4 °C and 5000× *g* for 20 min. The sample was then quickly removed and weighed to obtain mass G_2, and WHC (%) was calculated as WHC (%) = $(G_2/G_1) \times 100$. Color parameters including lightness (L*), redness (a*), and yellowness (b*) were measured using an NH350 portable colorimeter (Shenzhen Sanenshi Technology Co., Ltd., Shenzhen, China).

2.6. Free Amino Acids

Free amino acids were determined according to the method of Xi et al., with minor modifications [18]. Each 2 g sample was homogenized with ultrapure water and allowed to stand for 24 h. The resulting supernatant was then mixed with an equal volume of 5% (*w*/*v*) sulfosalicylic acid solution to precipitate proteins. After centrifugation at 6000× *g* for 10 min, the supernatant was collected and concentrated using a rotary evaporator. The residue was dissolved in 1 mL of sodium citrate buffer and filtered through a 0.45 μm membrane filter. The filtrate was analyzed using a Biochrom30+ amino acid analyzer (Biochrom Ltd., Cambridge, UK) equipped with a post-column ninhydrin detector. Quantification was performed using the external standard method with a mixed standard solution of 16 amino acids (Sigma-Aldrich, St. Louis, MO, USA). Information on the purity and CAS numbers of the amino acid standards is summarized in Supplementary Table S1.

2.7. Free Fatty Acids

Free fatty acid levels were determined according to the method of Lin et al., with slight modifications [19]. Each 100 mg sample was mixed with 4 mL of chloroform, vortexed for 30 s, and centrifuged at 3500× *g* for 15 min at room temperature. The lower organic phase was collected, and the extraction was repeated twice using 2 mL of dichloromethane each time. The combined organic extracts were dried under a gentle stream of nitrogen with mild heating (~35 °C). Derivatization was then carried out by adding 2 mL of methylation reagent, vortexing for 30 s, and incubating in a water bath at 80 °C for 2 h. The container was loosely covered with gauze to minimize gas loss. After cooling to room temperature, 2 mL of n-hexane and 1 mL of water were added and thoroughly vortexed. The mixture was centrifuged at 2000× *g* for 5 min, and the upper organic phase was collected and evaporated under a gentle stream of nitrogen with mild heating. An appropriate volume of iso-octane was added based on the sample concentration, vortexed, and left to stand. The final solution was filtered through a 0.22 μm PTFE syringe filter before being transferred to a sample vial for analysis. Quantification was performed using the external standard method with a mixed standard solution containing 36 fatty acids (Sigma-Aldrich, St. Louis, MO, USA).

A chromatographic analysis was performed using an Agilent 6890 gas chromatograph (Agilent Technologies, Santa Clara, CA, USA) equipped with a CP-Sil 88 capillary column (100 m × 0.25 mm × 0.25 μm). A 1 μL sample was injected with a split ratio of 10:1, and helium was used as the carrier gas at a constant flow rate of 1.0 mL/min. The initial oven temperature was set at 100 °C and held for 5 min, then increased at a rate of 4 °C/min to 240 °C, and held at 240 °C for 15 min. Mass spectrometric analysis was conducted using an Agilent 5977 mass spectrometer (Agilent Technologies, Santa Clara, CA, USA) equipped with an electron ionization (EI) source and operated using the MassHunter Workstation software package. The analysis was conducted in selected ion monitoring (SIM) mode, with the ion source temperature set at 260 °C, quadrupole temperature at 150 °C, ionization voltage at 70 eV, and scan rate at 57.7 cycles per second. In SIM mode, only specific characteristic ions (m/z) for each target fatty acid were monitored, all of which fall within the scan window of m/z 30 to 550. Information on the purity, CAS numbers, monitored m/z values, and retention times of the fatty acid standards is summarized in Supplementary Table S2. Compound identification was performed by matching retention times and mass spectral data with those of authentic standards. Data pre-processing—including baseline correction, peak detection, alignment, and retention time calibration—was performed using the Quant-My-Way Agilent B.09.00 software program.

2.8. Volatile Compound

Volatile compounds were analyzed using the Pegasus GC HRT4D+ system (LECO Corporation, St. Joseph, MI, USA). Each 5 g sample was transferred into a 20 mL headspace autosampler vial. Solid-phase microextraction (SPME) was performed using a CTC 3-in-1 autosampler equipped with a 50/30 μm CAR/PDMS/DVB fiber (Supelco, Bellefonte, PA, USA). Extraction was performed at 60 °C for 30 min, followed by fiber conditioning at 250 °C for 3 min and analyte desorption at 250 °C for 2 min. Data acquisition was performed using the LECO Pegasus 4D workstation and analyzed with the integrated ChromaTOF® 4.61.1.0 software program. Chromatographic peak widths were set to 24 s and 0.2 s for the first and second dimensions, respectively. Peaks with a signal-to-noise ratio (S/N) greater than 200 were automatically detected, integrated, and deconvoluted. The linear retention index (LRI) was calculated based on the retention times of a C_5–C_{30} n-alkane series and matched against the NIST 2017 and Wiley 9 databases. Compounds with both similarity and reverse match scores $\geq$ 800 were selected for final identification.

2.9. Electronic Tongue Analysis

Electronic tongue analysis was conducted according to the method of Liu et al., with modifications [20]. Each 10 g sample was homogenized and heated in a water bath until the core temperature reached 40 °C. Subsequently, 80 mL of water at the same temperature was added and further mixed for 1 min. After centrifugation at 3000× g for 10 min at 4 °C, the supernatant was collected for electronic tongue testing. A SA402B taste sensing system (Insent, Kanagawa, Japan) was used to characterize the taste profile. The system was equipped with five lipid membrane sensors—SB2CA0 (sourness), SB2C00 (bitterness), SB2AE1 (astringency), SB2AAE (umami), and SB2CT0 (saltiness)—along with an Ag/AgCl reference electrode.

Before measurement, the taste sensors and reference electrode were activated for 24 h in a reference solution (30 mmol/L KCl + 0.3 mmol/L tartaric acid) and saturated KCl solution, respectively, to stabilize signal responses. Measurements were conducted at room temperature (25 ± 2 °C) following a standardized five-step procedure: cleaning, equilibration, sample measurement, re-cleaning, and aftertaste recording. Sensors were first cleaned with cation and anion cleaning solutions for 90 s, rinsed in two reference solutions for 120 s each, and the baseline potential (Vr) was recorded for 30 s. The sensors were then immersed in the sample solution to measure the sample potential (Vs, 30 s), followed by two brief rinses in reference solution (3 s), after which the aftertaste potential (Vr′, 30 s) was recorded. Each sample was tested in quadruplicate, with the first reading discarded and the mean of the remaining three used to ensure signal stability and reproducibility. Data were processed using the instrument's dedicated software and expressed as relative taste intensity values. A principal component analysis (PCA) was applied to identify taste distribution patterns and visualize differences among sample groups.

2.10. Data Statistics and Analysis

All experiments were independently performed in triplicate to ensure reliability, and the results were expressed as mean ± standard deviation (mean ± SD). Statistical analyses were conducted using the SPSS 25.0 software package (IBM, Armonk, NY, USA), with significance set at $p < 0.05$. To validate the statistical assumptions, normality was tested using the Shapiro–Wilk test, and the homogeneity of variance was assessed using Levene's test. If both conditions were met, a one-way analysis of variance (ANOVA) was performed, followed by Waller–Duncan's multiple range test for post hoc comparisons. When the assumptions were not satisfied, the Kruskal–Wallis nonparametric test was applied, and pairwise comparisons were conducted based on mean ranks. A principal component analysis (PCA) was performed using the Anaconda software program (Anaconda, Inc., Austin, TX, USA). Graphs were generated using Origin 2021 (OriginLab Corporation, Northampton, MA, USA) and the Chiplot platform (https://www.chiplot.online/, accessed on 25 April 2025).

3. Results and Discussion

3.1. Effects of Different Carbon and Nitrogen Sources on D. hansenii Growth and Enzymatic Activities

The growth and metabolic characteristics of microorganisms are strongly influenced by the type and concentration of carbon and nitrogen sources in the medium, a factor closely related to the metabolic regulation mechanisms of the fermenting strain [21]. As shown in Figure 1A, the supplementation of different exogenous carbon and nitrogen sources significantly affected the proliferation of *D. hansenii*. Under carbon source treatment, the FCC group (corn starch) exhibited a colony count of 1.00×10^{10} CFU/g, which was significantly higher than that of the control group, 8.70×10^{9} CFU/g ($p < 0.05$). This result was likely due to the slow-release property of corn starch, which may provide a stable energy supply to support sustained cell activity. In contrast, the FCS group (sucrose) showed a 40.23% decrease in colony count, which may be associated with osmotic stress. According to Clara Navarrete et al., under high-salt stress conditions, *D. hansenii* tends to allocate energy to maintaining cellular homeostasis rather than promoting rapid cell division [22]. Under nitrogen source treatment, both the FNL group (leucine) and the FNP group (soy protein isolate) significantly increased colony counts. The FNP group exhibited the highest colony count at 1.66×10^{10} CFU/g, representing a 90.23% increase compared to the control ($p < 0.05$). This enhanced proliferation may be attributed to the higher nitrogen content (~15%) and more complex nitrogenous composition of soy protein isolate, compared to leucine (~10.7%). Upon hydrolysis, soy protein isolate releases a broad spectrum of amino acids and peptides, which may more effectively support the synthesis of nucleic acids, proteins, and cofactors essential for cell proliferation. In contrast, the FNL group was supplemented with a single amino acid. Although leucine modestly enhanced growth, its limited nitrogen profile may not have met the broader metabolic requirements of the strain.

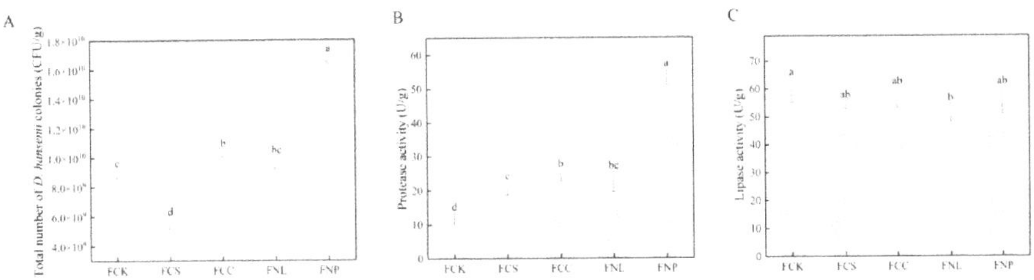

Figure 1. Total colony count of *D. hansenii* (**A**), protease activity (**B**), and lipase activity (**C**) in the sausage medium after 5 days of fermentation. FCK: control; FCS: sucrose; FCC: corn starch; FNL: leucine; FNP: soy protein isolates. Different letters indicate significant differences ($p < 0.05$).

Protease and lipase are two crucial enzymes produced by *D. hansenii* during fermentation; they are involved in the hydrolysis of proteins and fats, respectively, thereby enhancing the flavor profile of fermented meat products [23]. As shown in Figure 1B, protease activity was significantly higher in all treatment groups compared to the control group, with the FNP group exhibiting the highest activity at 55.38 U/g, approximately 5-fold higher than the control and 2-fold higher than the other treatment groups ($p < 0.05$). These results indicate that exogenous carbon and nitrogen supplementation enhanced the protease activity in *D. hansenii*, with various sources potentially affecting the strain through distinct nutritional pathways or physiological effects. Carbon sources can indirectly stimulate protease synthesis by providing carbon skeletons and an efficient energy supply to sustain high metabolic activity, while nitrogen sources can boost metabolic activity and enhance protease expression by supplying essential nitrogen [24]. In addition, soy protein isolate, as a high-molecular-weight protein, may serve as a substrate for extracellular proteases, potentially enhancing protease activity through substrate–enzyme recognition [25]. The elevated activity observed in the treatment groups may facilitate the release of amino acids and peptides—key precursors of flavor compounds—enhancing the umami and aromatic complexity of the final product.

Figure 1C shows that exogenous carbon and nitrogen sources did not significantly affect the lipase activity of *D. hansenii*. This may be because lipase is an inducible enzyme, and its synthesis typically depends on the presence of specific lipid substrates acting as inducers [26]. It is possible that the carbon and nitrogen sources used in this experiment lacked the specific components required to induce lipase expression. However, since the lipid profile of the carbon and nitrogen sources was not measured in this experiment, this inference remains to be validated. The *D. hansenii* strain used in this experiment exhibited inherently high baseline lipase activity, and all treatment groups had lipase activity exceeding 50 U/g.

3.2. Physicochemical Properties

3.2.1. pH, Water Activity, and Water-Holding Capacity

pH, water activity, and water-holding capacity are important indicators for evaluating the flavor and quality of fermented meat products. They also reflect microbial regulation of the substrate's microenvironment through metabolism. As shown in Figure 2A, the pH value of all treatment groups significantly decreased after fermentation ($p < 0.05$). Among the groups, the carbon source treatments showed a more pronounced decrease in pH, with the FCS group reaching the lowest pH of 5.33. This may be due to the rapid metabolism of sucrose by *D. hansenii*, producing acidic byproducts like lactic and acetic acids [27]. In contrast, the pH decrease in FCC group was slower, which can be attributed to the gradual degradation characteristics of corn starch, a complex polysaccharide. The pH in the nitrogen source treatment groups remained relatively stable, especially in the FNP group, where the pH was maintained between 5.90 and 5.81. This stability may be due to the buffering capacity of amino acids such as aspartic acid and glutamic acid, released during the microbial metabolism of soy protein isolates. These amino acids help neutralize some of the acidic metabolic byproducts, thereby maintaining a relatively stable pH [28].

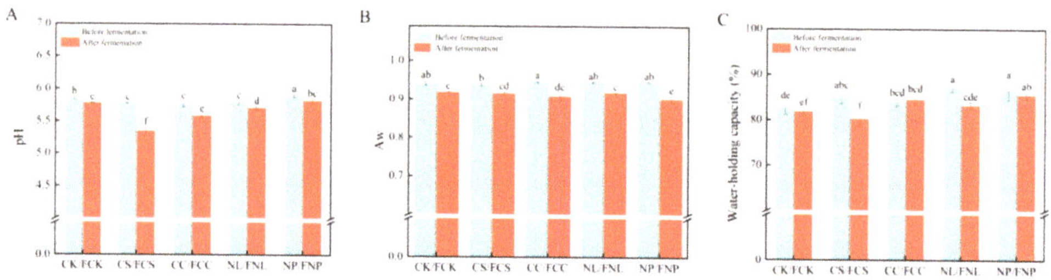

Figure 2. pH (**A**), Aw (**B**), and water-holding capacity (**C**) of the sausage medium before and after fermentation. CK: control; CS: sucrose; CC: corn starch; NL: leucine; NP: soy protein isolates. FCK, FCS, FCC, FNL, and FNP represent the samples after fermentation. Different letters indicate significant differences between groups ($p < 0.05$).

Figure 2B,C illustrate changes in water activity and water-holding capacity of the sausage matrix before and after fermentation. Both the FCC group (complex carbon source) and the FNP group (complex nitrogen source) significantly reduced water activity and increased water-holding capacity ($p < 0.05$). Among these, corn starch exhibited a strong water absorption capacity, continuously adsorbing free water and forming a stable protein–water network structure [29]. Soy protein isolates, possessing both water absorption and gel-forming properties, released degradation products, such as small peptides and hydrophilic amino acids, during microbial metabolism. These products further enhance water binding and improve the structural stability of the system [30]. In contrast, the simple carbon and nitrogen source treatment groups (FCS and FNL) showed no significant differences in water activity or water-holding capacity compared to the control group. This may be because neither the carbon nor the nitrogen sources were able to form stable protein–water networks, nor did they provide the spatial structure and hydrophilic groups required for effective water retention [31].

3.2.2. Color

Color is a key sensory indicator of the quality of fermented meat products and directly influences consumer acceptance. Figure 3A shows the appearance of the sausage matrix before and after fermentation, while Figure 3B, Figure 3C, and Figure 3D illustrate the specific changes in L*, a*, and b* values, respectively.

As shown in Figure 3B, fermentation led to a decrease in L* in the control group, while the addition of corn starch in the FCC group effectively slowed this decline. Corn starch, with its water absorption and expansion properties, reduced transparency and formed a uniform structure with proteins, enhancing light scattering and reflection, thereby maintaining a higher L* value [32]. The soy protein isolate in the FNP group, with its inherent yellowish color, may undergo browning during fermentation, leading to lower L* values before and after fermentation. Figure 3C shows that the a* value in the FNP group was significantly higher than in the other groups ($p < 0.05$). The relatively stable pH observed in the FNP group may help explain the preserved color. Previous studies have suggested that increased acidity may accelerate myoglobin oxidation and cause discoloration [33]. Figure 3D illustrates that fermentation caused a decrease in b* values for all groups. During fermentation, the increased NaCl concentration decreased oxygen solubility in the system, promoting the formation of high-iron myoglobin, which led to a decrease in b* [34]. The yellow color inherent in soy protein isolates resulted in higher b* values in the FNP group both before and after fermentation.

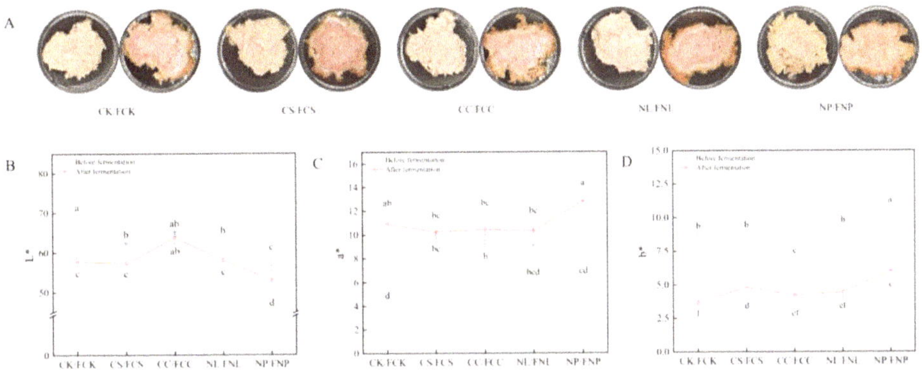

Figure 3. Appearance changes in the sausage medium before and after fermentation (**A**), and color changes: L* (**B**), a* (**C**), b* (**D**). CK: control; CS: sucrose; CC: corn starch; NL: leucine; NP: soy protein isolates. FCK, FCS, FCC, FNL, and FNP represent the samples after fermentation. Different letters indicate significant differences ($p < 0.05$).

In summary, supplementation with exogenous carbon and nitrogen sources had varying effects on the growth characteristics, enzyme activities, and physicochemical properties of *D. hansenii* in the fermented sausage model. Among the different treatments, the complex carbon source FCC group (corn starch) and the complex nitrogen source FNP group (soy protein isolates) exhibited the most significant improvements. Both groups significantly increased the colony count and protease activity of *D. hansenii*, while maintaining high lipase activity. Additionally, improvements in the color attributes and water-holding capacity of the fermented sausage were likely due to their macromolecular structures. In contrast, the simple carbon source FCS group (sucrose) and the simple nitrogen source FNL group (leucine) showed fewer improvements. The subsequent section of this manuscript focuses on the effects of corn starch and soy protein isolates on the accumulation of amino acids, fatty acids, and volatile flavor compounds during *D. hansenii* fermentation.

3.3. Amino Acid Analysis

Amino acids are important flavor precursors in fermented meat products, participating in enzymatic reactions and microbial metabolic processes. They are also essential nutrients for the human body and significantly contribute to improving the nutritional value of the product [35]. Table 1 shows the effect of different carbon and nitrogen sources on the amino acid profile. The total free amino acid (FAA) content in the FCC group was 258.37 mg/100 g, with no significant difference compared to the control group. This suggests that although corn starch increased protease activity, the hydrolysis products did not accumulate significantly in the substrate, which aligns with the findings of Ding et al. [14]. *D. hansenii* may have rapidly utilized the hydrolyzed amino acids for cell proliferation or anabolic metabolism, leading to limited accumulation in the medium. Moreover, previous studies suggest that glucose can induce carbon catabolite repression, potentially affecting microbial amino acid metabolism. Since corn starch is degraded into glucose, it is possible that this mechanism also played a role in the current fermentation system.

Table 1. Amino acid composition and content of the fermented sausage medium with different carbon and nitrogen sources.

Free Amino Acids	FCK	FCC	FNP
		mg/100 g	
Asp	3.30 ± 0.70 [b]	2.40 ± 0.61 [b]	6.47 ± 0.40 [a]
Glu	20.85 ± 1.34 [b]	22.63 ± 2.20 [b]	34.03 ± 1.72 [a]
Umami FAAs	24.30 ± 2.26 [b]	25.03 ± 2.07 [b]	40.50 ± 1.42 [a]
Thr *	7.93 ± 0.67 [b]	7.50 ± 0.75 [b]	12.53 ± 0.71 [a]
Ser	17.20 ± 0.71 [b]	16.43 ± 1.80 [b]	33.53 ± 0.72 [a]
Gly	6.75 ± 0.35 [b]	7.03 ± 0.61 [b]	11.47 ± 1.20 [a]
Ala	16.43 ± 0.81 [b]	15.03 ± 1.87 [b]	25.77 ± 2.89 [a]
Pro	9.50 ± 0.72 [c]	12.13 ± 0.76 [b]	14.70 ± 1.04 [a]
Sweet FAAs	59.30 ± 1.32 [b]	58.13 ± 1.52 [b]	98.00 ± 0.66 [a]
Val *	19.30 ± 0.85 [b]	20.33 ± 1.44 [b]	30.23 ± 1.06 [a]
Met *	8.37 ± 0.91 [a]	8.80 ± 1.00 [a]	8.37 ± 0.50 [a]
Ile *	14.83 ± 0.31 [b]	14.60 ± 1.47 [b]	23.17 ± 1.21 [a]
Leu *	19.80 ± 1.25 [b]	20.37 ± 0.71 [b]	37.17 ± 1.29 [a]
Tyr	15.80 ± 0.50 [b]	16.33 ± 0.99 [b]	25.37 ± 2.83 [a]
Phe *	24.03 ± 1.34 [b]	25.23 ± 1.92 [b]	34.73 ± 1.36 [a]
His	6.77 ± 0.49 [a]	7.33 ± 0.85 [a]	9.67 ± 2.31 [a]
Lys *	25.20 ± 2.56 [b]	27.03 ± 1.42 [b]	40.53 ± 1.04 [a]
Arg	10.87 ± 1.86 [b]	10.07 ± 1.53 [b]	17.43 ± 0.61 [a]
Bitter FAAs	145.57 ± 1.12 [b]	150.10 ± 3.72 [b]	226.67 ± 9.18 [a]
Cys	25.07 ± 0.06 [a]	25.10 ± 0.56 [a]	26.47 ± 1.10 [a]
Total FAAs	255.50 ± 3.91 [b]	258.37 ± 20.59 [b]	391.63 ± 10.72 [a]

"*" indicates essential amino acids. Different letters under the same parameter indicate significant differences ($p < 0.05$).

The total free amino acid (FAA) content in the FNP group was 391.63 mg/100 g, corresponding to a 53.28% increase compared to the control group. This increase is primarily due to the effective degradation of soy protein isolates by *D. hansenii* during fermentation, which provides a rich source of amino acids and short peptides, promoting amino acid accumulation. Additionally, as a complex nitrogen source, soy protein isolates avoided triggering nitrogen catabolite repression (NCR), which stimulated the strain to secrete more proteases, enhancing protein degradation and its conversion [36]. Furthermore, when nitrogen sources are limited, the inactivation of the TOR (target of rapamycin) signaling pathway may promote the degradation and utilization of non-preferred nitrogen sources, further enhancing the synthesis and accumulation of FAAs [37]. Therefore, the significant increase in free amino acid levels in the FNP group was likely due to the combined effects of raw material degradation, the strain's high baseline protease activity, and nitrogen metabolism regulation.

Amino acids not only serve as important flavor precursors but also undergo chemical reactions to contribute to the formation of complex flavors. In the FNP group, umami FAAs and sweet FAAs increased by 66.67% and 65.26%, respectively, likely contributing to the enhancement of umami and sweetness in the sausage. Furthermore, in the FNP group, the levels of all essential amino acids, except for methionine (which is naturally deficient in soy protein isolates), were significantly increased ($p < 0.05$). Essential amino acids improve the nutritional profile of food by providing the body with necessary amino acids, promoting protein synthesis, boosting immune function, and supporting nervous system health [38,39]. Therefore, they play a key role in improving the health benefits of food. Overall, the higher free amino acid levels in the FNP group suggest a stronger flavor potential and greater nutritional value.

3.4. Fatty Acid Analysis

The extent of fat hydrolysis and the composition of fatty acids are key factors influencing both the flavor and nutritional quality of fermented meat products. In fermented meat matrices, *D. hansenii* secretes lipase to catalyze fat hydrolysis, producing free fatty acids (FFAs), which are important flavor precursors [2]. In preliminary tests, the strain used in this study also exhibited notable lipolytic activity. Based on these findings, this section explores the effects of exogenous carbon and nitrogen sources on the fatty acid profile in a sausage model.

The free fatty acid content in each group after 5 days of fermentation is presented in Table 2. The total FFA content in the FCC group (1251.13 mg/100 g) and the FNP group (1342.92 mg/100 g) was significantly lower than that in the control group (1772.35 mg/100 g), corresponding to decreases of 29.4% and 24.2%, respectively ($p < 0.05$). Despite the reduction in total FFAs, the carbon and nitrogen source treatments retained sufficient levels of flavor precursors (>1000 mg/100 g). The fatty acid composition in all three groups was similar, with major components including oleic acid (cis) (30%), palmitic acid (25–26%), linoleic acid (16–17%), and stearic acid (13–14%), all of which are important precursors for aldehyde formation, such as nonanal and hexanal. Since there was no significant difference in lipase activity among the three groups, it suggests that the accumulation of free fatty acids may be influenced by factors such as raw material composition, lipid metabolism mechanisms, and the strain's energy supply strategy.

Table 2. Fatty acid composition and content of the fermented sausage medium with different carbon and nitrogen sources.

Free Fatty Acids	FCK	FCC	FNP
		mg/100 g	
Butyric Acid	0.21 ± 0.08 [a]	0.08 ± 0.01 [b]	0.07 ± 0.01 [b]
Caproic acid	0.04 ± 0.02 [a]	0.02 ± 0.02 [ab]	0.01 ± 0.01 [b]
Caprylic Acid	0.17 ± 0.02 [a]	0.10 ± 0.06 [a]	0.11 ± 0.01 [a]
Capric Acid	2.39 ± 0.23 [a]	1.73 ± 0.79 [a]	1.87 ± 0.09 [a]
Undecylic acid	0.08 ± 0.01 [a]	0.06 ± 0.03 [a]	0.07 ± 0.01 [a]
Lauric Acid	3.17 ± 0.28 [a]	2.28 ± 1.03 [a]	2.47 ± 0.11 [a]
Tridecylic acid	0.07 ± 0.01 [a]	0.05 ± 0.02 [a]	0.05 ± 0.01 [a]
Myristic Acid	42.36 ± 3.33 [a]	29.69 ± 1.28 [b]	32.11 ± 1.03 [b]
Pentadecanoic Acid	1.19 ± 0.09 [a]	0.80 ± 0.35 [a]	0.85 ± 0.02 [a]
Palmitic Acid	459.87 ± 16.92 [a]	327.85 ± 7.89 [b]	345.56 ± 10.99 [b]
Heptadecanoic Acid	6.06 ± 0.60 [a]	4.01 ± 1.85 [a]	4.39 ± 0.26 [a]
Stearic Acid	256.87 ± 14.02 [a]	175.67 ± 4.62 [b]	187.31 ± 5.16 [b]
Arachidic Acid	5.98 ± 0.52 [a]	3.75 ± 0.36 [b]	4.33 ± 0.21 [b]
Heneicosanoic acid	0.13 ± 0.02 [a]	0.07 ± 0.03 [b]	0.09 ± 0.01 [b]
Behenic Acid	0.39 ± 0.04 [a]	0.27 ± 0.07 [b]	0.32 ± 0.01 [ab]
Tricosanoic acid	0.08 ± 0.01 [a]	0.05 ± 0.01 [b]	0.06 ± 0.01 [b]
Lignoceric Acid	0.22 ± 0.04 [a]	0.14 ± 0.02 [a]	0.19 ± 0.09 [a]
Myristoleic Acid	0.50 ± 0.03 [a]	0.35 ± 0.20 [a]	0.35 ± 0.05 [a]
Palmitoleic Acid	61.95 ± 4.86 [a]	44.27 ± 2.91 [b]	48.74 ± 2.17 [b]
Margaroleic Acid	4.97 ± 1.96 [a]	3.22 ± 1.64 [a]	3.26 ± 0.09 [a]
Oleic Acid (trans)	3.88 ± 0.31 [a]	2.59 ± 1.23 [a]	2.85 ± 0.12 [a]
Oleic Acid (cis)	533.37 ± 22.61 [a]	376.65 ± 6.25 [b]	403.13 ± 13.20 [b]
Gadoleic Acid	25.23 ± 2.33 [a]	16.18 ± 1.27 [b]	17.63 ± 0.95 [b]

Table 2. *Cont.*

Free Fatty Acids	FCK	FCC	FNP
	mg/100 g		
Erucic Acid	1.52 ± 0.29 [a]	1.37 ± 0.14 [a]	1.36 ± 0.02 [a]
Nervonic Acid	0.41 ± 0.05 [a]	0.34 ± 0.14 [ab]	0.23 ± 0.03 [b]
Linoleic Acid	299.71 ± 7.21 [a]	216.81 ± 7.01 [c]	238.00 ± 9.29 [b]
Linoleic Acid (trans)	0.16 ± 0.01 [a]	0.11 ± 0.02 [b]	0.16 ± 0.01 [a]
γ-Linolenic Acid	1.16 ± 0.10 [a]	0.79 ± 0.37 [a]	0.88 ± 0.07 [a]
Alpha-Linolenic Acid	21.50 ± 2.19 [a]	15.06 ± 0.94 [b]	16.89 ± 0.92 [b]
Eicosadienoic Acid	20.38 ± 1.35 [a]	13.04 ± 2.45 [b]	14.47 ± 0.61 [b]
Dihomo-gamma-Linolenic Acid	2.69 ± 0.19 [a]	1.82 ± 0.68 [b]	2.09 ± 0.03 [ab]
Erucic acid	1.52 ± 0.29 [a]	1.37 ± 0.14 [a]	1.36 ± 0.02 [a]
Eicosatrienoic Acid	3.63 ± 0.38 [a]	2.13 ± 0.85 [b]	2.44 ± 0.28 [b]
Arachidonic Acid	9.93 ± 0.21 [a]	8.04 ± 1.27 [b]	8.81 ± 0.15 [ab]
Eicosapentaenoic Acid	0.26 ± 0.02 [a]	0.18 ± 0.02 [b]	0.19 ± 0.01 [b]
Docosadienoic Acid	0.29 ± 0.04 [a]	0.18 ± 0.08 [b]	0.19 ± 0.01 [ab]
SFAs	779.29 ± 34.73 [a]	546.63 ± 15.60 [b]	579.88 ± 17.63 [b]
MUFAs	631.83 ± 29.89 [a]	444.96 ± 10.67 [b]	477.55 ± 16.33 [b]
PUFAs	361.23 ± 10.92 [a]	259.54 ± 12.69 [c]	285.49 ± 10.64 [b]
UFAs	993.06 ± 40.61 [a]	704.50 ± 23.07 [b]	763.04 ± 26.89 [b]
UFAs/SFAs	1.27 ± 0.01 [b]	1.29 ± 0.02 [ab]	1.32 ± 0.01 [a]
Total FFAs	1772.35 ± 75.34 [a]	1251.13 ± 36.90 [b]	1342.92 ± 44.36 [b]

SFAs: saturated fatty acids; MUFAs: monounsaturated fatty acids; PUFAs: polyunsaturated fatty acids; UFAs: unsaturated fatty acids. Different letters under the same parameter indicate significant differences ($p < 0.05$).

The low free fatty acid levels in the FCC group may be attributable to the energy-providing function of corn starch. The additional energy supply reduced the strain's reliance on fat as an energy source, thereby lowering the fat hydrolysis rate. Moreover, glucose, a product of corn starch degradation, may activate the CCR mechanism in *D. hansenii*, thereby inhibiting fat catabolism [40]. The reduced level of free fatty acids in the FNP group may also result from the energy-yielding function of soy protein isolate, which could reduce the strain's dependence on lipid catabolism. In addition, serine—a degradation product of soy protein—has been identified as a key precursor for phospholipid synthesis [41]. Based on the observation of the highest colony count in the FNP group, it is speculated that lipid-derived metabolites may have been preferentially directed toward membrane construction and biomass production, rather than accumulating in the medium.

The saturated fatty acid (SFA) content in the FCC and FNP groups decreased by 29.86% and 25.58%, respectively, compared to the control. The corresponding UFA/SFA ratios increased to 1.29 and 1.32, respectively. Given that a high SFA intake has been linked to increased cardiovascular risk, this compositional change may be of potential nutritional interest [42]. High levels of free fatty acids (FFAs) are generally associated with increased risks of lipid oxidation and rancidity [43]. The elevated FFA content observed in the control group may therefore have implications for product shelf life and flavor stability, although this would require further verification through TBARS and peroxide value measurements. Overall, the addition of exogenous carbon and nitrogen sources (corn starch and soy protein isolate) modulated the fatty acid profile of *D. hansenii*, potentially improving lipid composition while preserving flavor precursor availability, and indicating a favorable nutritional profile.

3.5. Volatile Compound Analysis

Volatile compounds play a crucial role in determining the flavor characteristics and consumer acceptance of fermented sausages. In this study, SPME-GC-MS was employed to analyze the differences in volatile compounds among the groups, and heatmaps of their relative contents, along with stacked bar charts for each compound category, were generated (Figure 4A,B). A total of 54 volatile compounds were detected, including 12 alcohols, 16 aldehydes, 6 acids, 5 ketones, 8 esters, 6 pyrazines, and 1 furan compound. Specifically, 45 compounds were detected in the control group, 44 in the FCC group, and 38 in the FNP group. Significant differences in the composition of volatile flavor compounds were observed among the groups ($p < 0.05$), suggesting that the addition of carbon and nitrogen sources significantly affected the accumulation of volatile compounds in *D. hansenii*-fermented sausages.

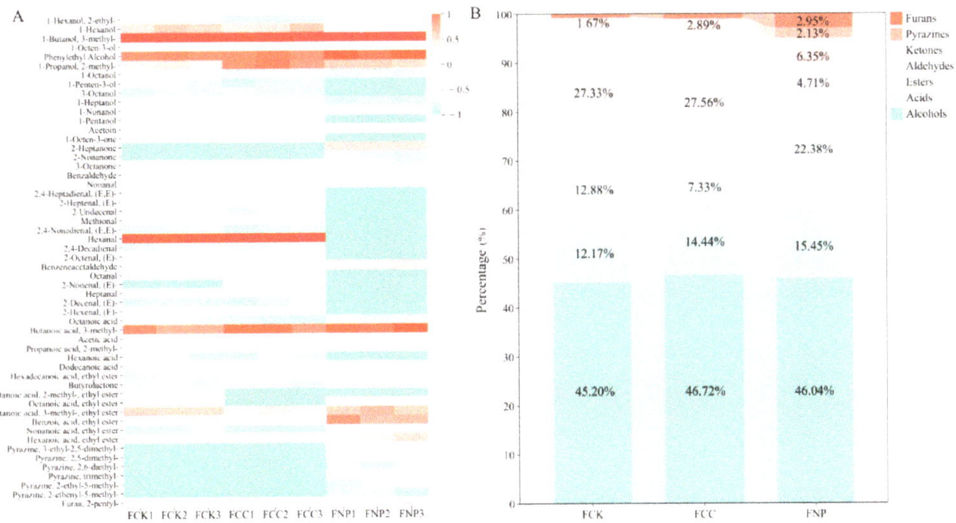

Figure 4. Abundance heatmap of volatile compounds (**A**) and stacked bar chart of relative content of each category of volatile compounds (**B**). The heatmap is color-coded from red to green, with red indicating relatively high content and green indicating relatively low content.

The relative contents of alcohols and acids were similar among the three groups, with alcohols being the most abundant volatile compounds in each group. These compounds accumulated in large quantities during the early stages of fermentation and served as precursors for key flavor compounds, such as esters and aldehydes [44]. Alcohols and acids primarily originated from yeast lipid oxidation, amino acid metabolism, and carbohydrate metabolism, with typically high thresholds, contributing to alcoholic, grassy, acidic, and cheesy aromas in fermented meat products [45]. The differences in alcohol content among the treatment groups may be attributed to the metabolic regulation characteristics of *D. hansenii*. In the FCC group, due to the lack of exogenous nitrogen sources, the strain may have relied more on sugar metabolism for energy, resulting in the production of alcohols, such as isobutanol [46]. In contrast, the high levels of FAAs in the FNP group may have triggered the Ehrlich pathway, which could convert amino acids into higher alcohols [47]. For instance, phenylethanol, a metabolic product of phenylalanine, was significantly increased in the FNP group ($p < 0.05$), contributing to rose and honey-like aroma characteristics.

Relatively high levels of aldehydes were observed in the FCC group (27.56%) and control group (27.33%). Aldehydes generally result from the oxidation and degradation of unsaturated fatty acids [1]. The high free fatty acid content in the control group served as a sufficient substrate for aldehyde accumulation. Although the free fatty acid content in the FCC group was lower, aldehyde intermediates might still have been generated through metabolic pathways such as sugar degradation [48]. In contrast, the aldehyde content in the FNP group was significantly lower (4.71%) than that in the other two groups ($p < 0.05$). The observed effect may be associated with the antioxidant activity of soy protein isolate hydrolysates. According to previous studies, these hydrolysates can contribute to lipid stabilization through mechanisms such as metal ion chelation, hydrogen donation, and free radical scavenging [49]. It is worth noting that aldehydes have a low sensory threshold, and even low concentrations can impart a fresh, herbal aroma to fermented meat products, whereas higher concentrations may result in greasy or rancid odors [50].

Esters are key flavor components in fermented meat products, characterized by low sensory thresholds and playing a crucial role in enhancing aroma. They are typically formed through the esterification of alcohols and acids, and yeasts can also catalyze the formation of ethyl esters via the alcohol acetyltransferase pathway, utilizing acetyl-CoA and higher alcohols [51]. In the FCC group, the relative content of esters decreased to 7.33%, which may be related to the CCR effect triggered by glucose, degraded from corn starch. This effect is suggested to further inhibit metabolic pathways in *D. hansenii* associated with the synthesis of higher alcohols and the expression of alcohol acetyltransferase, thereby limiting the generation and conversion of ester precursors and reducing ester accumulation, as reported in previous studies [7]. In contrast, the FNP group, by alleviating the NCR effect, may have enhanced the activity of metabolic pathways related to ester synthesis, leading to a significant increase in ester content, which reached 22.38%. The FNP group showed the substantial accumulation of volatile compounds such as ethyl hexanoate and ethyl benzoate, which are commonly associated with "fruity" and "floral" aromas in previous studies.

The relative content of ketones in the FCC and FNP groups increased to 2.89% and 6.35%, respectively. Ketones, which have a low sensory threshold, contribute to fruity, floral, and creamy aromas. They primarily originate from β-keto acid decarboxylation, incomplete β-oxidation of free fatty acids, and amino acid degradation. Some studies have found that carbon sources are converted into pyruvate through glycolysis, which can then be transformed into acetyl-CoA, and under specific metabolic conditions, short-chain ketones may be produced via an atypical pathway [52]. In the FNP group, the abundant nitrogen source may have maintained metabolic balance in *D. hansenii*, reducing its dependence on fat for energy. This increase in incomplete β-oxidation flux could have promoted the accumulation of medium-chain ketones, such as 2-heptanone, which have contributed to fruity aromas [53].

The FNP group also exhibited a significant accumulation of 2-pentylfuran (2.95%), a low-threshold compound known to contribute to plant-like aromas such as beany and green notes. Additionally, pyrazine compounds (2.13%) were detected exclusively in the FNP group. Pyrazines require carbohydrates as carbon source precursors and amino acids as nitrogen source precursors [54]. The elevated FAA levels in the FNP group may have provided sufficient substrates for non-enzymatic pyrazine synthesis, promoting the early accumulation of pyrazines during fermentation. Pyrazines have an extremely low flavor threshold and can contribute unique roasted and nutty aromas at low concentrations.

In summary, the FCC group generated aldehydes and ketones under conditions of low free fatty acid levels, contributing fresh, grassy flavor characteristics to the sausages. The relative content of esters, ketones, and pyrazines increased in the FNP group, with reduced aldehyde contents, leading to a complex and rich flavor profile with fruity, nutty, and other distinctive characteristics.

3.6. Taste Characteristics of Sausage Model

Electronic tongue technology was used to further examine the impact of corn starch and soy protein isolates on the flavor characteristics of a fermented sausage model. The electronic tongue simulates the human tongue's ability to perceive key taste attributes, such as sourness, sweetness, saltiness, bitterness, and umami, offering an objective reflection of taste differences between samples. As shown in Figure 5A, the FCC group (corn starch) exhibited significantly increased sourness and decreased saltiness, which can be attributed to the accumulation of organic acids produced by an active glycolysis pathway and the lower degree of fat hydrolysis. The FNP group (soy protein isolates) exhibited a prominent umami response, attributable to its high content of glutamic acid and flavor peptides, which synergistically interact with ester compounds to enhance the overall flavor. As shown in Figure 5B, a principal component analysis (PCA) explained 87.46% of the total variance, effectively distinguishing the taste distribution characteristics among the different treatment groups. PC1 (71.80%) primarily differentiates between carbon source and nitrogen source treatments, while PC2 (15.66%) reflects differences in fermentation levels. The FNP and FCC groups were positioned on either side of the control group in the PCA space, suggesting that carbon and nitrogen sources may induce distinct taste clustering patterns through different metabolic regulation mechanisms. When combined with analyses of FAAs, free fatty acids, and volatile flavors, it can be inferred that the enhanced sourness in the FCC group was related to the acidic byproducts from active sugar metabolism, while the decreased saltiness was likely due to lower fat hydrolysis and limited salt-soluble protein degradation. An electronic tongue analysis indicated enhanced umami and taste richness in the FNP group, which may be associated with the accumulation of short peptides and FAAs, as well as abundant esters, pyrazines, and furans.

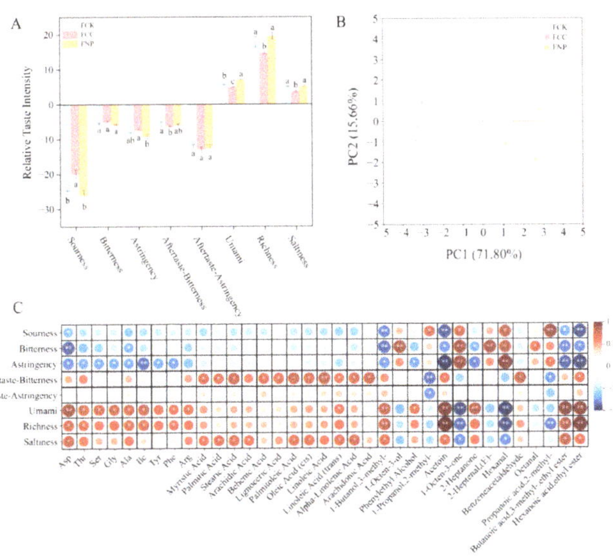

Figure 5. Taste characteristics of fermented sausage model, including bar chart of the electronic tongue (**A**), PCA plot of the electronic tongue (**B**), and correlation heatmap with major amino acids, fatty acids, and volatile compounds (**C**). Red indicates a positive correlation, blue indicates a negative correlation, and white indicates no correlation. Different letters under the same parameter indicate significant differences ($p < 0.05$); asterisks indicate significance levels: * $p < 0.05$, ** $p < 0.01$.

To further explore the relationship between flavor compounds and sensory taste attributes, a Spearman correlation analysis was performed to construct a correlation heatmap between 44 major volatile and non-volatile flavor compounds and 8 electronic tongue sensory characteristics (Figure 5C). The results showed that compounds such as isoamyl alcohol, 3-hydroxy-2-butanone, isobutyric acid, and ethyl isoamyl ester were positively correlated with umami, richness, and saltiness, suggesting their role in enhancing these taste attributes. Phenylethanol, 2-heptanone, and the umami amino acid glutamic acid also significantly contributed to the umami characteristic. Additionally, while free fatty acids enhance saltiness, they may also impart bitterness and a bitter aftertaste. Compounds such as 1-octen-3-one, hexanal, and octanal were positively correlated with astringency, bitterness, and sourness, indicating that these compounds could contribute to unpleasant sensory experiences at high concentrations.

In summary, the analysis of both the electronic tongue data and flavor compounds revealed a significant association between key metabolites and taste attributes. The FCC group exhibited a fresh, acidic taste with reduced saltiness, while the FNP group exhibited prominent umami and richness.

4. Conclusions

This study initially assessed the effects of simple and complex carbon and nitrogen sources on the fermentation characteristics of local *D. hansenii*. The results demonstrated that the complex carbon and nitrogen sources, including corn starch and soy protein isolates, were particularly effective. Both sources exhibited good water retention properties. Corn starch, serving as a slow-release carbon source, maintained a stable energy supply, boosting colony counts, protease activity, and product brightness. Soy protein isolates, providing a stable nitrogen source and alleviating the NCR effect, led to enhanced growth and protease secretion in *D. hansenii*. A further analysis revealed that corn starch accumulated aldehydes and ketones, contributing fresh flavor characteristics. Soy protein isolates, through raw material degradation, the synergistic action of *D. hansenii*'s high protease activity, and nitrogen metabolism regulation, increased the total free amino acid content, particularly enhancing umami and sweetness components. It also enriched esters, ketones, and pyrazines, while reducing aldehydes, resulting in a more intense and complex aroma profile. Notably, both carbon and nitrogen sources optimized the fatty acid profile, contributing to the improved nutritional value of the final product. In summary, supplementation with different exogenous carbon and nitrogen sources can regulate the fermentation characteristics of *D. hansenii* through distinct metabolic pathways, producing differentiated effects in promoting growth and enzyme activity, improving physicochemical properties, and optimizing flavor compound accumulation. The rational selection of carbon and nitrogen sources provides technical support for improving the quality and flavor of fermented sausages, as well as for achieving controlled flavor regulation in fermented meat products. In future studies, the fermentation characteristics of the strain can be further optimized by combining carbon and nitrogen sources and adjusting the carbon-to-nitrogen ratio.

Supplementary Materials: The following supporting information can be downloaded at: https://www.mdpi.com/article/10.3390/foods14111840/s1, Table S1. Free Amino Acid Standard Information; Table S2. Free Fatty Acid Standard Information; Table S3. Volatile compound composition and content of fermented sausage culture medium with different carbon and nitrogen sources; Figure S1. Representative chromatograms of free fatty acids in fermented samples from different treatment groups: A, Standard sample; B, control (FCK); C, corn starch (FCC); D, soy protein isolate (FNP).

Author Contributions: Conceptualization, methodology, investigation, writing—original draft, W.D.; supervision, formal analysis, Q.Z.; conceptualization, supervision, writing—review and editing, funding acquisition, J.W. All authors have read and agreed to the published version of the manuscript.

Funding: This study was funded by the National Natural Science Foundation of China (No. 32160607).

Data Availability Statement: The original contributions presented in the study are included in the article, and further inquiries can be directed to the corresponding author.

Conflicts of Interest: The authors declare no conflicts of interest.

References

1. Hu, Y.; Tian, Y.; Zhu, J.; Wen, R.; Chen, Q.; Kong, B. Technological characterization and flavor-producing potential of lactic acid bacteria isolated from traditional dry fermented sausages in northeast China. *Food Microbiol.* **2022**, *106*, 104059. [CrossRef] [PubMed]
2. Corral, S.; Belloch, C.; Lopez-Diez, J.J.; Flores, M. Lipolysis and aroma generation as mechanisms involved in masking boar taint in sodium reduced fermented sausages inoculated with *Debaryomyces hansenii* yeast. *J. Sci. Food Agric.* **2018**, *98*, 2121–2130. [CrossRef]
3. Gong, X.; Chen, X.; Mi, R.; Qi, B.; Xiong, S.; Li, J.; Zhu, Q.; Wang, S. Two *Debaryomyces hansenii* strains as starter cultures for improving the nutritional and sensory quality of dry-cured pork belly. *Food Res. Int.* **2024**, *183*, 114227. [CrossRef] [PubMed]
4. Cano-Garcia, L.; Rivera-Jimenez, S.; Belloch, C.; Flores, M. Generation of aroma compounds in a fermented sausage meat model system by *Debaryomyces hansenii* strains. *Food Chem.* **2014**, *151*, 364–373. [CrossRef]
5. Yang, D.; Zhang, Y.; Zhao, L.; Wang, Y.; Rao, L.; Liao, X. Pressure-resistant acclimation of lactic acid bacteria from a natural fermentation product using high pressure. *Innov. Food Sci. Emerg. Technol.* **2021**, *69*, 102660. [CrossRef]
6. Olivares-Marin, I.K.; Madrigal-Perez, L.A.; Canizal-Garcia, M.; García-Almendárez, B.E.; González-Hernández, J.C.; Regalado-Gonzalez, C. Interactions between carbon and nitrogen sources depend on RIM15 and determine fermentative or respiratory growth in Saccharomyces cerevisiae. *Appl. Microbiol. Biotechnol.* **2018**, *102*, 4535–4548. [CrossRef]
7. Nair, A.; Sarma, S.J. The impact of carbon and nitrogen catabolite repression in microorganisms. *Microbiol. Res.* **2021**, *251*, 126831. [CrossRef]
8. Lleixa, J.; Martin, V.; Giorello, F.; Portillo, M.C.; Carrau, F.; Beltran, G.; Mas, A. Analysis of the NCR Mechanisms in *Hanseniaspora vineae* and *Saccharomyces cerevisiae* During Winemaking. *Front. Genet.* **2019**, *9*, 747. [CrossRef] [PubMed]
9. Sa Junior, P.L.S.; Silva, L.J.; Andrade, H.A.; Sucupira Maciel, M.I.; Sakugawa Shinohara, N.K.; Abreu Gloria, M.B.; Oliveira Filho, P.R.C. Optimization of mechanically separated meat washing cycles and of corn starch addition in saramunete (*Pseudupeneus maculatus*) sausages. *J. Food Process. Preserv.* **2021**, *45*, 16093. [CrossRef]
10. Jiang, L.; Zhang, H.; Zhang, J.; Liu, S.; Tian, Y.; Cheng, T.; Guo, Z.; Wang, Z. Improve the fiber structure and texture properties of plant-based meat analogues by adjusting the ratio of soy protein isolate (SPI) to wheat gluten (WG). *Food Chemistry-X* **2024**, *24*, 101962. [CrossRef]
11. Huang, P.; Shao, X.; Zhu, M.; Xu, B.; Chen, C.; Li, P. Sucrose enhances colour formation in dry sausages by up-regulating gene expression of nitric oxide synthase in *Staphylococcus vitulinus*. *Int. J. Food Microbiol.* **2020**, *315*, 108419. [CrossRef] [PubMed]
12. Liu, R.; Ma, Y.; Chen, L.; Lu, C.; Ge, Q.; Wu, M.; Xi, J.; Yu, H. Effects of the addition of leucine on flavor and quality of sausage fermented by *Lactobacillus fermentum* YZU-06 and *Staphylococcus saprophyticus* CGMCC 3475. *Front. Microbiol.* **2023**, *13*, 1118907. [CrossRef]
13. Baek, J.H.; Han, D.M.; Choi, D.G.; Jeon, C.O. Unraveling the carbohydrate metabolic characteristics of *Leuconostoc mesenteroides* J18 through metabolite and transcriptome analyses. *Food Chem.* **2024**, *435*, 137594. [CrossRef] [PubMed]
14. Ding, Y.; He, W.; Dai, W.; Xie, X.; Pan, Y.; Tang, X.; Zheng, R.; Zhou, X. Quality and flavor development of solid-state fermented surimi with *Actinomucor elegans*: A perspective on the impacts of carbon and nitrogen sources. *Food Chem.* **2024**, *447*, 139053. [CrossRef]
15. Cho, A.J.; Han, S.; Nahar, S.; Her, E.; Kang, J.G.; Ha, S.-D. Synergistic effects of ε-poly-l-lysine and lysozyme against *Pseudomonas aeruginosa* and *Listeria monocytogenes* biofilms on beef and food contact surfaces. *Meat Sci.* **2024**, *214*, 109534. [CrossRef]
16. Prestes, R.C.; Silva, L.B.; Torri, A.M.P.; Kubota, E.H.; Rosa, C.S.; Roman, S.S.; Kempka, A.P.; Demiate, I.M. Sensory and physicochemical evaluation of low-fat chicken mortadella with added native and modified starches. *J. Food Sci. Technol.* **2015**, *52*, 4369–4377. [CrossRef] [PubMed]
17. Huang, Q.H.; Xiong, K.X.; Wang, Y.Y.; Xiong, Z.M.; Lu, H.Y.; Peng, L.J.; Jin, W.Z.; Wang, W.; Ni, L.X.; Wang, H.B. Physical and flavor characteristics of reduced sodium sausages containing plant proteins and blends of chloride salts. *Lwt-Food Sci. Technol.* **2025**, *215*, 117268. [CrossRef]

18. Xi, X.; Ke, J.; Ma, Y.; Liu, X.; Gu, X.; Wang, Y. Physiochemical and taste characteristics of traditional Chinese fermented food sufu. *J. Food Process. Preserv.* **2022**, *46*, 16845. [CrossRef]
19. Lin, Y.; Yuan, D.D.; Zhang, S.J.; Wu, T.X. Differences in flavor characteristics of rounds of Sauce-based Baijiu based on GC-IMS, GC-MS, and electronic tongue. *J. Food Compos. Anal.* **2024**, *135*, 106696. [CrossRef]
20. Liu, B.-Y.; Zhu, K.-X.; Guo, X.-N.; Peng, W.; Zhou, H.-M. Effect of deamidation-induced modification on umami and bitter taste of wheat gluten hydrolysates. *J. Sci. Food Agric.* **2017**, *97*, 3181–3188. [CrossRef]
21. Egli, T.; Zinn, M. The concept of multiple-nutrient-limited growth of microorganisms and its application in biotechnological processes. *Biotechnol. Adv.* **2003**, *22*, 35–43. [CrossRef] [PubMed]
22. Navarrete, C.; Frost, A.T.; Ramos-Moreno, L.; Krum, M.R.; Martinez, J.L. A physiological characterization in controlled bioreactors reveals a novel survival strategy for *Debaryomyces hansenii* at high salinity. *Yeast* **2021**, *38*, 302–315. [CrossRef]
23. Gong, X.; Mi, R.; Chen, X.; Zhu, Q.; Xiong, S.; Qi, B.; Wang, S. Evaluation and selection of yeasts as potential aroma enhancers for the production of dry-cured ham. *Food Sci. Hum. Wellness* **2023**, *12*, 324–335. [CrossRef]
24. Sun, Y.; Qian, Y.C.; Zhang, J.X.; Wang, Y.F.; Li, X.H.; Zhang, W.C.; Wang, L.S.; Liu, H.; Zhong, Y.H. Extracellular protease production regulated by nitrogen and carbon sources in Trichoderma reesei. *J. Basic Microbiol.* **2021**, *61*, 122–132. [CrossRef] [PubMed]
25. Zhang, Y.Z.; Zhang, W.X.; Chen, X.L. Mechanisms for Induction of Microbial Extracellular Proteases in Response to Exterior Proteins. *Appl. Environ. Microbiol.* **2020**, *86*, e01036-20. [CrossRef]
26. Akhtar, M.W.; Mirza, A.Q.; Chughtai, M.I. Lipase Induction in Mucor hiemalis. *Appl. Environ. Microbiol.* **1980**, *40*, 257–263. [CrossRef]
27. Jayaram, V.B.; Cuyvers, S.; Lagrain, B.; Verstrepen, K.J.; Delcour, J.A.; Courtin, C.M. Mapping of *Saccharomyces cerevisiae* metabolites in fermenting wheat straight-dough reveals succinic acid as pH-determining factor. *Food Chem.* **2013**, *136*, 301–308. [CrossRef]
28. Brown, J.; Grimaud, A. Proton-donating and chemistry-dependent buffering capability of amino acids for the hydrogen evolution reaction. *Phys. Chem. Chem. Phys.* **2023**, *25*, 8005–8012. [CrossRef]
29. Lopez-Silva, M.; Agama-Acevedo, E.; Bello-Perez, L.A.; Alvarez-Ramirez, J. Influence of gelatinization degree and octenyl succinic anhydride esterification on the water sorption characteristics of corn starch. *Carbohydr. Polym.* **2021**, *270*, 118378. [CrossRef]
30. Webb, D.; Dogan, H.; Li, Y.; Alavi, S. Physico-Chemical Properties and Texturization of Pea, Wheat and Soy Proteins Using Extrusion and Their Application in Plant-Based Meat. *Foods* **2023**, *12*, 1586. [CrossRef]
31. Bellissent-Funel, M.-C.; Hassanali, A.; Havenith, M.; Henchman, R.; Pohl, P.; Sterpone, F.; van der Spoel, D.; Xu, Y.; Garcia, A.E. Water Determines the Structure and Dynamics of Proteins. *Chem. Rev.* **2016**, *116*, 7673–7697. [CrossRef] [PubMed]
32. Zhang, F.; Fang, L.; Wang, C.; Shi, L.; Chang, T.; Yang, H.; Cui, M. Effects of starches on the textural, rheological, and color properties of surimi-beef gels with microbial tranglutaminase. *Meat Sci.* **2013**, *93*, 533–537. [CrossRef] [PubMed]
33. Hoa, V.-B.; Cho, S.-H.; Seong, P.-N.; Kang, S.-M.; Kim, Y.-S.; Moon, S.-S.; Choi, Y.-M.; Kim, J.-H.; Seol, K.-H. The significant influences of pH, temperature and fatty acids on meat myoglobin oxidation: A model study. *J. Food Sci. Technol.* **2021**, *58*, 3972–3980. [CrossRef] [PubMed]
34. Bedia, M.; Méndez, L.; Bañón, S. Evaluation of different starter cultures (*Staphylococci* plus Lactic Acid Bacteria) in semi-ripened Salami stuffed in swine gut. *Meat Sci.* **2011**, *87*, 381–386. [CrossRef]
35. Ardo, Y. Flavour formation by amino acid catabolism. *Biotechnol. Adv.* **2006**, *24*, 238–242. [CrossRef]
36. Zhang, W.; Du, G.; Zhou, J.; Chen, J. Regulation of Sensing, Transportation, and Catabolism of Nitrogen Sources in *Saccharomyces cerevisiae*. *Microbiol. Mol. Biol. Rev.* **2018**, *82*, 10-1128. [CrossRef]
37. Heitman, J.; Movva, N.R.; Hall, M.N. Targets for cell cycle arrest by the immunosuppressant rapamycin in yeast. *Science* **1991**, *253*, 905–909. [CrossRef]
38. Bifari, F.; Ruocco, C.; Decimo, I.; Fumagalli, G.; Valerio, A.; Nisoli, E. Amino acid supplements and metabolic health: A potential interplay between intestinal microbiota and systems control. *Genes Nutr.* **2017**, *12*, 27. [CrossRef]
39. Xiao, F.; Guo, F.F. Impacts of essential amino acids on energy balance. *Mol. Metab.* **2022**, *57*, 101393. [CrossRef]
40. Simpson-Lavy, K.; Kupiec, M. Carbon catabolite repression: Not only for glucose. *Curr. Genet.* **2019**, *65*, 1321–1323. [CrossRef]
41. Kleetz, J.; Vasilopoulos, G.; Czolkoss, S.; Aktas, M.; Narberhaus, F. Recombinant and endogenous ways to produce methylated phospholipids in *Escherichia coli*. *Appl. Microbiol. Biotechnol.* **2021**, *105*, 8837–8851. [CrossRef] [PubMed]
42. Ruiz-Nunez, B.; Dijck-Brouwer, D.A.J.; Muskiet, F.A.J. The relation of saturated fatty acids with low-grade inflammation and cardiovascular disease. *J. Nutr. Biochem.* **2016**, *36*, 1–20. [CrossRef] [PubMed]
43. Rey, A.I.; De Cara, A.; Segura, J.F.; Martí, P.; Hechavarría, T.; Calvo, L. Dietary oleuropein extract supplementation and its combination with α-tocopheryl acetate and selenium modifies the free fatty acid profile of pork and improves its stability. *J. Sci. Food Agric.* **2021**, *101*, 2337–2344. [CrossRef] [PubMed]
44. Sallan, S.; Kaban, G.; Kaya, M. The effects of nitrite, sodium ascorbate and starter culture on volatile compounds of a semi-dry fermented sausage. *Lwt-Food Sci. Technol.* **2022**, *153*, 112540. [CrossRef]

45. Wen, R.; Li, X.-A.; Han, G.; Chen, Q.; Kong, B. Fungal community succession and volatile compound dynamics in Harbin dry sausage during fermentation. *Food Microbiol.* **2021**, *99*, 103764. [CrossRef]
46. Xu, Z.; Lin, L.; Chen, Z.; Wang, K.; Sun, J.; Zhu, T. The same genetic regulation strategy produces inconsistent effects in different *Saccharomyces cerevisiae* strains for 2-phenylethanol production. *Appl. Microbiol. Biotechnol.* **2022**, *106*, 4041–4052. [CrossRef]
47. Matheis, K.; Granvogl, M.; Schieberle, P. Quantitation and Enantiomeric Ratios of Aroma Compounds Formed by an Ehrlich Degradation of L-Isoleucine in Fermented Foods. *J. Agric. Food Chem.* **2016**, *64*, 646–652. [CrossRef]
48. Hiltunen, J.K.; Mursula, A.M.; Rottensteiner, H.; Wierenga, R.K.; Kastaniotis, A.J.; Gurvitz, A. The biochemistry of peroxisomal beta-oxidation in the yeast Saccharomyces cerevisiae. *FEMS Microbiol. Rev.* **2003**, *27*, 35–64. [CrossRef]
49. Huang, L.; Jia, S.; Wu, R.; Chen, Y.; Ding, S.; Dai, C.; He, R. The structure, antioxidant and antibacterial properties of thiol-modified soy protein isolate induced by allicin. *Food Chem.* **2022**, *396*, 133713. [CrossRef]
50. Dominguez, R.; Purrinos, L.; Perez-Santaescolastica, C.; Pateiro, M.; Barba, F.J.; Tomasevic, I.; Bastianello Campagnol, P.C.; Lorenzo, J.M. Characterization of Volatile Compounds of Dry-Cured Meat Products Using HS-SPME-GC/MS Technique. *Food Anal. Methods* **2019**, *12*, 1263–1284. [CrossRef]
51. Yoshimoto, H.; Bogaki, T. Mechanisms of production and control of acetate esters in yeasts. *J. Biosci. Bioeng.* **2023**, *136*, 261–269. [CrossRef] [PubMed]
52. Yuzawa, S.; Mirsiaghi, M.; Jocic, R.; Fujii, T.; Masson, F.; Benites, V.T.; Baidoo, E.E.K.; Sundstrom, E.; Tanjore, D.; Pray, T.R.; et al. Short-chain ketone production by engineered polyketide synthases in *Streptomyces albus*. *Nat. Commun.* **2018**, *9*, 4569. [CrossRef] [PubMed]
53. Flores, M.; Moncunill, D.; Montero, R.; Javier Lopez-Diez, J.; Belloch, C. Screening of *Debaryomyces hansenii* Strains for Flavor Production under a Reduced Concentration of Nitrifying Preservatives Used in Meat Products. *J. Agric. Food Chem.* **2017**, *65*, 3900–3909. [CrossRef] [PubMed]
54. Corral, S.; Leitner, E.; Siegmund, B.; Flores, M. Determination of sulfur and nitrogen compounds during the processing of dry fermented sausages and their relation to amino acid generation. *Food Chem.* **2016**, *190*, 657–664. [CrossRef]

foods

MDPI

Article

Physicochemical Properties, Antioxidant Activities, and Aromatic Profile of Yogurt Co-Fermented by *Weissella cibaria* G232 with Traditional Starters

Qian Huang, Haixiao Ye, Yangyang Yang, Chenglin Zhu * and Junni Tang *

College of Pharmacy and Food, Southwest Minzu University, Chengdu 610225, China; huangqian0221@163.com (Q.H.); yehaixiao@stu.swun.edu.cn (H.Y.); yang124856@163.com (Y.Y.)

* Correspondence: chenglin.zhu@swun.edu.cn (C.Z.); junneytang@swun.edu.cn (J.T.)

Abstract: To improve the quality and functional properties of yogurts, a multi-starters co-fermentation system was used during yogurt preparation. In this work, *Weissella cibaria* G232 (added at 0%, 3%, 5%, and 7%) was involved as a co-fermenter with a traditional starter (*Lactobacillus delbrueckii* subsp. *bulgaricus* G119 and *Streptococcus thermophilus* Q019). The results showed that *W. cibaria* G232 co-fermentation could shorten the fermentation time and significantly enhance the viable counts of yogurt ($p < 0.05$). Moreover, the incorporation of *W. cibaria* G232 improved the water holding ability, viscosity, and texture of yogurt. Notably, the highest levels of firmness, consistency, and cohesiveness of yogurt were observed at the 5% addition level of *W. cibaria* G232. Furthermore, co-fermentation with *W. cibaria* G232 significantly enhanced the antioxidant activity of yogurt, as evidenced by increased free radical scavenging capacity and ferric ion reducing antioxidant power (FRAP) value. The intelligent sensory technology and Gas Chromatography-Ion Mobility Spectrometry (GC-IMS) indicated that co-fermentation with *W. cibaria* G232 and a traditional starter notably altered the accumulation of aldehydes, ketones, and alcohols in yogurt. These findings suggest that co-fermentation of *W. cibaria* G232 with a traditional starter present the potential for the quality and functionality improvement of yogurt and also lay the foundation for the application of *W. cibaria* G232.

Keywords: *Weissella cibaria*; co-fermentation; yogurt; antioxidant; aromatic profile

1. Introduction

Yogurt is a semi-solid fermented product produced by fermenting cow's milk with a starter culture consisting of *Streptococcus thermophilus* and *Lactobacillus delbrueckii* subsp. *bulgaricus* [1]. Yogurt, known for its distinctive flavor and nutritional benefits, offers multiple health advantages, including maintaining the balance of intestinal microbiota, improving the intestinal barrier, scavenging free radicals, and increasing the body's resistance to disease [2]. Other benefits are also associated with the consumption of yogurt, such as stimulating peristaltic movements [3], promoting the colonization of the gastrointestinal tract by beneficial microorganisms [4], improving the digestibility of nutrients, and relieving irritable bowel syndrome [5]. However, the quality of plain yogurt could alter during fermentation, storage, and transportation, mainly due to whey separation, low bacterial viability, poor viscosity, and rough texture, which negatively impact the properties of the yogurt [4]. Starter cultures are the key ingredients for the production of yogurt. To address the aforementioned issues, researchers have used probiotic co-fermentation with traditional starters to improve yogurt quality. For example, fermentation by *Lactiplantibacillus*

https://doi.org/10.3390/foods14091607

plantarum MTCC 25432 and 25433 could enhance the rheological, texture, and antimicrobial properties of yogurt [6]. *Lactiplantibacillus plantarum* ZFM55 fermentation also distributed a desirable flavor profile of yogurt; at the same time, the water holding capacity, cohesiveness, and viscosity of yogurt improved [7]. *Lactobacillus fermentum* HY01, as an adjunct starter, increased the abundance of volatile flavor compounds in yak yogurt, and 36 volatile substances, including aldehydes, esters, and alcohols, were detected by GC-MS [8]. Although *Lactiplantibacillus plantarum* has been extensively utilized as an adjunct starter culture in yogurt production, persistent acid production during refrigerated storage and mismatched acidification rates with the primary starter cultures may compromise the acid stability and textural properties of the final product. The development of compatible synergistic starter cultures is urgently needed to overcome key challenges in dairy fermentation.

Weissella spp. are facultatively anaerobic, Gram-positive, chemoorganotrophs bacteria that do not form spores or produce lactic acid via the hexose-monophosphate and phosphoketolase pathways [9]. The genus *Weissella* belongs to the family *Leuconostocaceae*, and has been frequently found in human and animal intestines, saliva, and traditional fermented foods, such as pickles, sourdough, and cheese [10,11], suggesting it is a lactic acid bacterial genus with Generally Recognized as Safe status [12]. As natural emulsifying stabilizers in yogurt fermentation, exopolysaccharides (EPS) produced by *Weissella* spp. effectively enhance milk gel texture and reduce syneresis. Notably, EPS from *W. cibaria* SP213 exhibits enhanced interfacial activity under acidic conditions (pH 3), making it particularly suitable for fermented dairy products like yogurt. This EPS contains abundant functional groups (e.g., carboxyl groups), conferring strong hydrophilicity and superior interfacial stabilization [13]. Furthermore, in mixed-culture fermentations, *Weissella* spp. contribute to flavor complexity through heterofermentative metabolism. For instance, Sichuan pickles co-inoculated with *Lactiplantibacillus plantarum* LP067 and *W. cibaria* WC018 showed accelerated acetic and lactic acid production, along with increased volatile flavor compounds (e.g., esters, aldehydes, and ketones) [14]. Due to their prevalence in fermented foods and putative probiotic properties, *Weissella* species are promising starter cultures and potential probiotics. However, their application in yogurt production remains underexplored.

Given the functional potential of *Weissella* spp. in fermented foods—particularly their EPS-producing capacity and flavor-modulating properties—this study investigated the potential of using *W. cibaria* G232 as an adjunct culture to enhance yogurt quality. Building on our laboratory's prior discovery of its strong EPS production, we systematically evaluated the effects of co-culturing *W. cibaria* G232 with traditional starters (*Lactobacillus delbrueckii* subsp. *bulgaricus* G119 and *Streptococcus thermophilus* Q019) on three critical parameters: physicochemical properties, antioxidant activities, and aromatic profile. To objectively characterize aroma variations, we employed intelligent sensory techniques coupled with gas chromatography-ion mobility spectrometry (GC-IMS). The findings offer novel insights into industrial application of *W. cibaria* G232 in fermented dairy production.

2. Materials and Methods

2.1. Preparation of Strains

Weissella cibaria G232 was isolated from traditional Sichuan fermented pickles. *Lactobacillus delbrueckii* subsp. *bulgaricus* G119 and *Streptococcus thermophilus* Q019 were obtained from commercial yogurt and identified via 16S rRNA sequencing. All strains were cryopreserved at −80 °C in 50% (v/v) glycerol until use. Prior to experimentation, the cultures were revived by two successive subcultures in de Man, Rogosa, and Sharpe (MRS) broth under aerobic conditions at 30 °C for 24 h.

2.2. Yogurt Fermentation

Initial preliminary experiments confirmed vigorous growth of *W. cibaria* G232, *L. bulgaricus* G119, and *S. thermophilus* Q019 on the MRS medium. Based on these findings, the three strains were inoculated into 10 mL of fresh MRS broth and incubated at 30 °C for 24 h. Cells were harvested through centrifugation (7200× *g*, 30 min, 4 °C, Centrifuge 5920 R, Eppendorf, Hamburg, Germany). Pelleted cells were PBS-washed twice and standardized to 5 × 10^6 CFU/mL for subsequent yogurt fermentation. Yogurt fermentation was performed according to Figure 1: sterilized milk inoculated with 5% (*v*/*v*) starter culture was incubated at 42 °C. The control group (YC) contained conventional starter cultures (*L. bulgaricus* G119 and *S. thermophilus* Q019, 1:1 ratio). Moreover, yogurt samples were co-inoculated with a traditional starter and a different ratio of *W. cibaria* G232 (3%, Y1, 5%, Y2, and 7%, Y3) while maintaining total inoculum volume at 5% (*v*/*v*), as detailed in Table 1. Fermentation was terminated when the pH decreased below 4.5, after which the coagulated product was immediately refrigerated at 4 °C overnight.

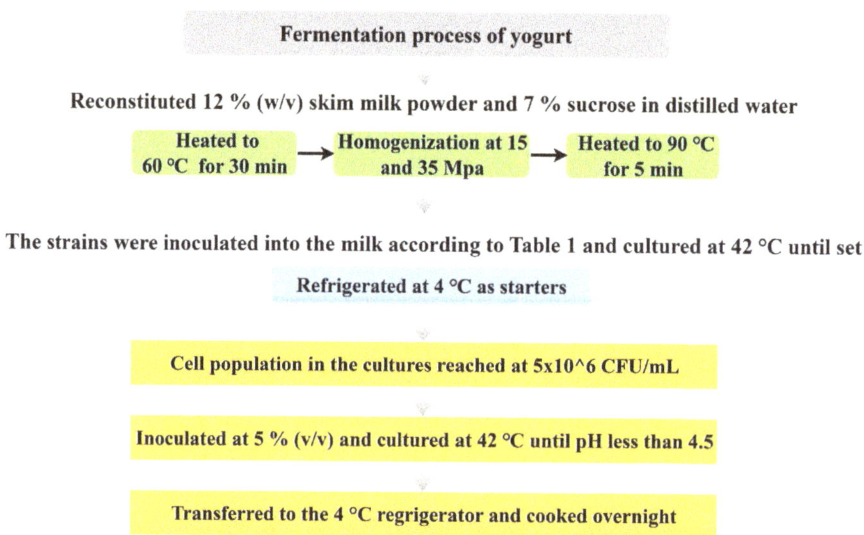

Figure 1. Fermentation process of yogurt.

Table 1. Combinations of strains.

Group	Different Proportional Combinations
YC	*L. bulgaricus* G119: *S. thermophilus* Q019 is 1:1
Y1	YC + *W. cibaria* G232 (3%)
Y2	YC + *W. cibaria* G232 (5%)
Y3	YC + *W. cibaria* G232 (7%)

2.3. Physicochemical Characteristics of Yogurt

2.3.1. Determination of pH and Titratable Acidity (TA)

The pH of yogurt was measured hourly during the fermentation with a digital pH meter (DGB-402A, Inesa, Shanghai, China). TA of yogurt was determined by titration with sodium hydroxide (NaOH). Specifically, 10 g of yogurt was accurately weighed into tapered bottles and mixed with 20 mL of distilled water. The mixture was titrated with

0.1 mol/L NaOH in the presence of phenolphthalein as an indicator. The TA was calculated using Equation (1).

$$\text{Titratable acidity} = \frac{V \times C \times 100}{m \times 0.1} \quad (1)$$

where C is the concentration of the NaOH standard solution (0.1 mol/L), V is the volume of NaOH (mL), and m is the quantity of the yogurt sample (g).

2.3.2. Determination of Viable Counts

Samples (1 mL) were diluted with 9 mL of normal saline. Serial dilutions were plated onto an MRS agar medium and cultured at 37 °C for 48 h. Colonies were counted on the plate, each containing 30 to 300 colonies.

2.3.3. Determination of Water-Holding Capacity (WHC)

A total of 10 g of yogurt sample was weighed into a 15 mL centrifuge tube and centrifuged at 5000× *g* at 4 °C for 20 min. The supernatant was collected for antioxidant capacity analysis. WHC was calculated following Equation (2).

$$\text{WHC} = \frac{W}{W_0} \times 100 \quad (2)$$

where W is the weight of precipitation after centrifugation, and W_0 is the weight of the sample.

2.3.4. Determination of Rheological Property

According to the method by Fan et al. [15], the rheological properties of the samples were determined using the HR-1 rheometer (TA-WATERS Ltd., New Castle, DE, USA). The apparent viscosity of the yogurts was measured at 25 °C, with a fixed frequency of 1 HZ and a share rate increasing from 0.1 s^{-1} to 100 s^{-1}, followed by a decrease from 100 s^{-1} to 0.1 s^{-1}. Modulus (G') and loss modulus (G'') were measured at 25 °C under a constant strain fixed at 0.5%; the frequency of scanning ranged from 0.1 to 10 HZ in the shear mode.

2.3.5. Texture Analysis

Hardness, consistency, and cohesiveness of yogurt were measured using a texture analyzer (TA. XT Plus, Stable Micro System, Godalming, Surrey, UK). A cake probe of 35 mm was selected with a test distance of 10 mm and a trigger point of 10.0 g. The pre-test speed was 6 mm/s, and the test speed was 2 mm/s [16]. The yogurt was tested immediately after the removal from the refrigerator, and the temperature in the testing environment was maintained at 25 °C.

2.4. Determination of Yogurt's Antioxidant Activity

A total of 10 g of the yogurt sample was weighed into 15 mL centrifuge tube and centrifuged at 5000× *g*, at 4 °C, for 20 min. The supernatant was mixed with 15 mL of acidified methanol (containing 0.05 mL of hydrochloric acid), thoroughly vortexed, and incubated at −20 °C for 1 h to ensure complete protein precipitation, then centrifuged at 5000× *g* and 4 °C for 10 min.

2.4.1. Determination of ABTS Free Radical Scavenging Ability

The ABTS solution was mixed with 7 mM of 2, 2-azino-bis-(3-ethylbenzo-thiazoline-6-sulfonic acid) diammonium salt and 2.45 mM of potassium persulfate, reacted in darkness overnight. The solution was diluted with 80% ethanol to achieve an absorbance of 0.7 ± 0.02 at 734 nm and stocked until use. Then, 50 μL of the sample was mixed with 1 mL of the diluent ABTS solution and incubated at 25 °C for 30 min in darkness. The absorbance at 734 nm was recorded [17] and the results were calculated following Equation (3).

$$\text{Scavenging activity (\%)} = [1 - (A2 - A1)/A0] \times 100 \quad (3)$$

where A0 is blank and represents the absorbance of ABTS and ethanol, A1 is control and represents the absorbance of supernatant and ethanol, and A2 represents the absorbance of ABTS and supernatant.

2.4.2. Determination of DPPH Free Radical Scavenging Ability

DPPH (2,2-diphenyl-1-picrylhydrazyl) solution (0.05 mmol/L) was prepared in anhydrous ethanol and stored in light-protected amber vials at 4 °C. For radical scavenging assays, 400 μL test samples were combined with 600 μL DPPH solution (final volume 1 mL) and incubated in darkness at 25 °C for 30 min [17]. Absorbance measurements were conducted at 517 nm (TU-1901 spectrophotometer, Persee General Instrument Co., Ltd., Beijing, China), with anhydrous ethanol serving as the blank control. Radical scavenging activity (%) was calculated according to Equation (4).

$$\text{Scavenging activity (\%)} = [1 - (B2 - B1)/B0] \times 100 \quad (4)$$

B0 as blank represents the absorbance of DPPH and ethanol, B1 as control represents the absorbance of supernatant and ethanol, and B2 represents the absorbance of DPPH and supernatant.

2.4.3. Ferrous Ion Chelating Ability (FRAP)

Fresh FRAP reagent was mixed with 300 mM acetate buffer, 40 mM 2,4,6-Tripyridyl-s-Triazine (TPTZ) solution, and 20 Mm ferric chloride solution in a volume ratio of 10:1:1. The reaction system consisted of 0.4 mL of the sample and 0.6 mL of the FRAP reagent, incubated in the dark at 37 °C for 10 min. The absorbance of mixture was recorded at 593 nm. $FeSO_4 \cdot 7H_2O$ solutions were used to graph standard curve, and the ferric-reducing ability of the yogurt was expressed as mmol/mL of Fe^{2+} [18].

2.5. Determination of Yogurt's Sensory Properties by Intelligent Technology

2.5.1. Electronic Nose and Tongue Analysis

A commercial electronic nose system (FOX 4000, Alpha MOS, Toulouse, France) equipped with 18 sensor chambers, an injection system, a mass flow controller, and a microcontroller acquisition board, was used to measure the different yogurt samples. The information detected by each sensor is presented in Table 2 [19]. A total of 3 mL of the yogurt sample was put into a 10 mL headspace bottle, then incubated at 40 °C for 5 min and injected into system immediately. The measurement process was maintained for 120 s. Each sample was observed five times. E-tongue System (Astree tongue, Alpha MOS, Toulouse, France) was equipped with 7 different taste sensors, including sourness (AHS), saltiness (CTS), umami (NMS), sweetness (ANS), bitterness (SCS), and two reference electrodes (PKS and CPS). Mixed with 150 mL yogurt sample and 200 mL deionized water, and filtered under normal pressure, the filtrate was collected for electronic tongue analysis [20]. The collection time was set to 120 s, cleaning time to 30 s, and the stirring speed to 60 rpm. Each sample was analyzed five times.

Table 2. Performance description of the E-nose sensors.

Sensors	Performance Description
LY2/LG	Sensitive to oxidizing gas
LY2/G	Sensitive to ammonia, carbon monoxide
LY2/AA	Sensitive to ethanol
LY2/Gh	Sensitive to ammonia/organic amines
LY2/gCT1	Sensitive to hydrogen sulfide
LY2/gCT	Sensitive to propane/butane
T30/1	Sensitive to organic solvents
P10/1	Sensitive to hydrocarbons
P10/2	Sensitive to methane
P40/1	Sensitive to fluorine
T70/2	Sensitive to aromatic compounds
PA/2	Sensitive to ethanol, ammonia/organic amines
P30/1	Sensitive to polar compounds (ethanol)
P40/2	Sensitive to heteroatom/chloride/aldehydes
P30/2	Sensitive to alcohol
T40/2	Sensitive to aldehydes
T40/1	Sensitive to chlorinated compounds
TA/2	Sensitive to air quality

2.5.2. GC-IMS Analysis

The GC-IMS instrument (Flavorspec, G.A.S. Instrument, Munich, Germany) was equipped with an automatic sampling system (CTC Analytics AG, Zwingen, Switzerland) and MXT-WAX capillary column. Accurately weighed 2.0 g sample into a headspace bottle and incubated at 60 °C for 15 min. Then, 500 μL of the headspace sample was injected at 65 °C using a heated syringe, with 99.99% nitrogen as the carrier gas. The injector temperature was maintained at 60 °C, and the GC program flow was set as follows: 2 mL/min for 2 min; 20 mL/min for 8 min; 130 mL/min for 10 min. The headspace samples were eluted and separated at 60 °C [21]. N-ketone (C4–C9) was used as external reference material to calculate the retention index (RI) of volatile compounds. The retention index (RI) and ions' drift time were compared with the standard in the GC-IMS database for identification.

2.6. *Statistical Analysis*

Data means were calculated from triplicate samples. Significant differences were determined by one-way ANOVA followed by Duncan's multiple comparison test. Non-parametric correlations between variables were assessed using Spearman's rank correlation analysis, with correlation coefficients and p-values calculated via IBM SPSS Statistics 27.0 (SPSS Inc., Chicago, IL, USA). PCA and clustering heatmap visualizations were generated using Origin 2022 (OriginLab, Northampton, MA, USA). All statistical analyses, including ANOVA, post hoc tests, and Spearman correlation, were conducted using SPSS.

3. Results

3.1. *Changes in Acidity of Yogurt During Fermentation*

Figure 2 exhibited the changes in pH and TA values of different groups. Before fermentation, the pH of milk was 6.5. In the first 4 h during fermentation, the acidity of yogurt drops dramatically, which corresponds to a significant increase in TA values. After that, the acidity levels among the four groups showed little variation. At 6 h, both Y2 and Y3 groups reached the end point, and their pH values were 4.47 and 4.36, respectively—lower than those of the control groups (4.59). Compared to the control group, 5% and 7% *W. cibaria* G232 addition reduced the fermentation time from 7 h to 6 h.

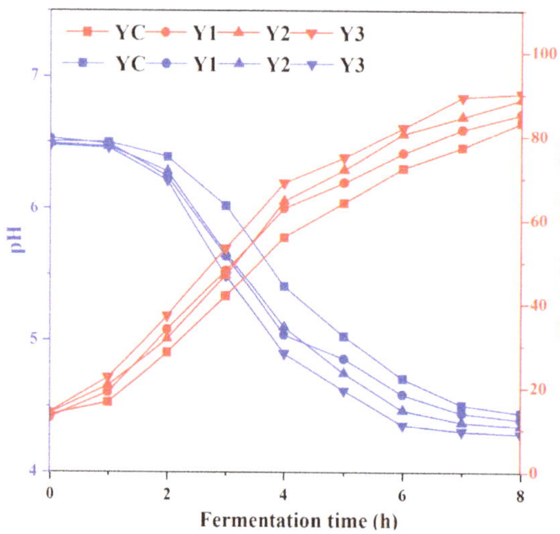

Figure 2. Changes in the pH and titratable acidity during fermentation. (Note: YC: *L. bulgaricus* G119: *S. thermophilus* Q019 (1:1); Y1: YC + 3% *W. cibaria* G232; Y2: YC + 5% *W. cibaria* G232; Y3: YC + 7% *W. cibaria* G232.).

3.2. Changes in Viable Counts and WHC of Yogurt During Fermentation

The changes in viable counts and water holding capacity of the fermented milk are shown in Figures 3A and 3B, respectively. With the incremental elevation in the proportion of *W. cibaria* G232, the viable counts in the yogurt increased significantly ($p < 0.05$), demonstrating a dose-dependent relationship with the proportion of *W. cibaria* G232. The experimental group Y3 demonstrated a marked increase from 8.41 log CFU/mL in the control group (YC) to 8.63 log CFU/mL ($p < 0.05$). Concurrently, water holding capacity (WHC) exhibited a concentration-dependent enhancement pattern, rising from 57.35% in YC to 67.38% in Y2 ($p < 0.05$). Although WHC experienced a marginal reduction to 65.72% at the maximal supplementation level (7% G232), this value remained significantly elevated compared to the control ($p < 0.05$). The enhanced WHC values indicated a compact tissue state and improved the retention of aromas. These findings collectively demonstrate that *W. cibaria* G232 functions as an effective co-starter culture, simultaneously enhancing both microbial viability and water holding capability in yogurt fermentation systems.

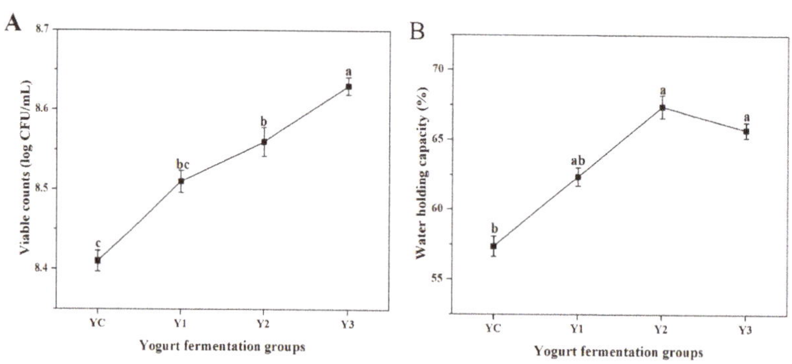

Figure 3. Changes in the viable counts (**A**) and water holding capability (**B**) in different yogurt groups. (Note: YC: *L. bulgaricus* G119: *S. thermophilus* Q019 (1:1); Y1: YC + 3% *W. cibaria* G232; Y2: YC + 5% *W. cibaria* G232; Y3: YC + 7% *W. cibaria* G232. Averages labeled with different lowercase letters indicate a significant difference ($p < 0.05$).).

3.3. Rheological Properties of Yogurt

The shear stability of yogurt was assessed in Figure 4A,B. The apparent viscosity of all yogurt showed the same change trends and decreased with the increasing of shear rate during the scan, until it finally stabilized. The apparent viscosity of yogurt co-fermented with *W. cibaria* G232 and traditional starter was higher than control group YC; the highest apparent viscosity was observed in Y2. However, when the addition of *W. cibaria* G232 reached 7% (Y3), the effect on the viscosity of yogurt is not obvious.

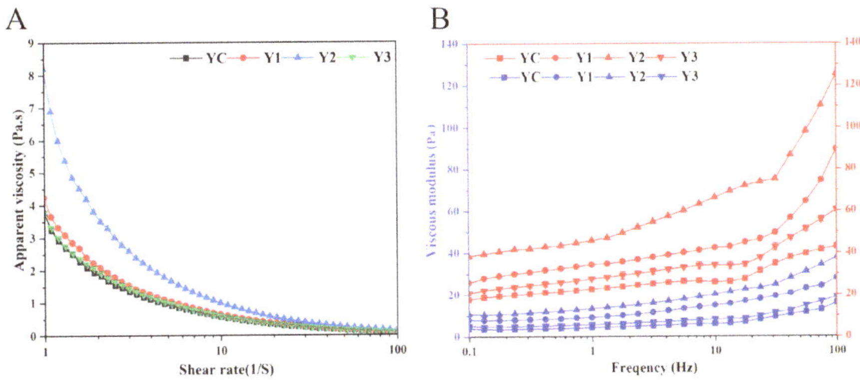

Figure 4. Changes in apparent viscosity (**A**), the viscous modulus G'' and elastic modulus G' (**B**) of yogurts supplemented with different proportions of *W. cibaria* G232. (Note: YC: *L. bulgaricus* G119: *S. thermophilus* Q019 (1:1); Y1: YC + 3% *W. cibaria* G232; Y2: YC + 5% *W. cibaria* G232; Y3: YC + 7% *W. cibaria* G232).

G' (elastic module) and G'' (viscous module) of the four groups of yogurts showed the same variation trends during test processing. $G' > G''$ on all frequency indicated that yogurt has solid-like characteristics; both indicators increased gradually when frequency increased from 0.1 to 100 Hz. On the lower frequency (0.1–10 Hz), G' and G'' increased slowly; however, the higher frequency (>10 Hz) also increased the rise rate of G' and G''. Furthermore, the value of G' and G'' of yogurt with *W. cibaria* G232 (Y1, Y2 and Y3) was higher than control group YC. When the addition of *W. cibaria* G232 was 5%, the highest values of G' and G'' were observed in the Y2 group.

3.4. Texture Change in Yogurt

Texture properties of yogurt are shown in Table 3. Firmness serves as an indicator of coagulation quality and resistance to deformation, and cohesiveness serves as an index to characterize the structural integrity and smoothness of yogurt. The most favorable textural properties were observed at a 5% addition of *W. cibaria* G232 (Y2), with the yogurt exhibiting the highest levels of firmness, consistency and cohesiveness. When the addition of G232 reached 7%, the three indicators of texture showed varying degrees of decline; however, the three groups Y1, Y2, and Y3 still showed improvement compared to the control group YC. Table 3 showed that the yogurt co-fermented with *W. cibaria* G232 had comprehensive advantages in firmness, consistency, and cohesiveness compared to the control group, indicating that *W. cibaria* G232 could greatly improve the texture and quality of the yogurt.

Table 3. Texture changes in yogurt supplemented with different proportions of G232.

Parameters	YC	Y1	Y2	Y3
Firmness (g)	212.85 ± 4.87 ^{c}	236.16 ± 7.31 ^{b}	259.12 ± 4.44 ^{a}	245.91 ± 3.82 ^{b}
Consistency (g·s)	936.02 ± 15.26 ^{d}	1778.28 ± 13.84 ^{b}	1986.75 ± 18.13 ^{a}	1324.94 ± 8.16 ^{c}
Cohesiveness (g)	−56.06 ± 3.94 ^{b}	−46.95 ± 4.73 ^{ab}	−40.41 ± 4.82 ^{a}	−50.18 ± 7.22 ^{ab}

Note: YC: *L. bulgaricus* G119: *S. thermophilus* Q019 (1:1); Y1: YC + 3% *W. cibaria* G232; Y2: YC + 5% *W. cibaria* G232; Y3: YC + 7% *W. cibaria* G232. The data are presented as average values ± SD (n = 3). Averages featuring varying lowercase letters within the same row show a notable difference ($p < 0.05$).

3.5. Antioxidant Activity Change in Yogurt

The addition of *W. cibaria* G232 significantly enhanced the antioxidant activity of yogurt, which also demonstrated a dose-dependent relationship with *W. cibaria* G232 in Figure 5. When the addition of *W. cibaria* G232 was at 7%, the highest antioxidant activity of yogurt was observed. Compared to the control group, the ABTS radical scavenging ability significantly increased from 33.54 ± 3.26% to 45.74 ± 1.03% ($p < 0.01$), and the value of DPPH radical scavenging ability increased from 44.37 ± 3.69% to 57.08 ± 6.48% ($p < 0.01$). Additionally, the FRAP value of yogurt significantly increased from 0.44 ± 0.09 to 1.13 ± 0.10 mM at the 7% addition of G232 ($p < 0.001$), showing a remarkable improvement in antioxidant activity compared to control group YC ($p < 0.001$).

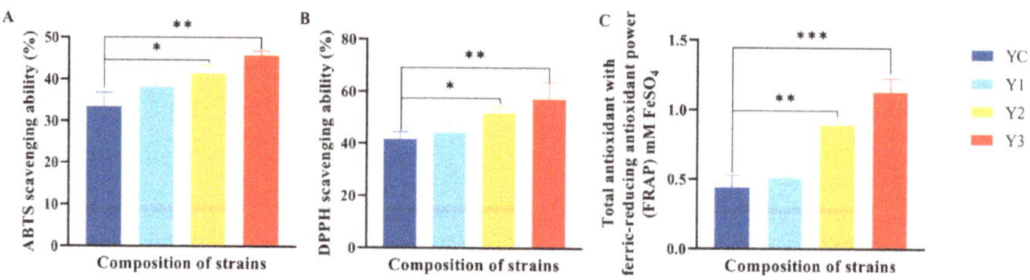

Figure 5. Antioxidant activity of yogurt with different proportions of *W. cibaria* G232. (Note: (**A**) is the ABTS scavenging ability; (**B**) is the DPPH scavenging ability; (**C**) is the total antioxidant with ferric-reducing antioxidant power; where * stands for $p < 0.05$, ** stands for $p < 0.01$, *** stands for $p < 0.001$. YC: *L. bulgaricus* G119: *S. thermophilus* Q019 (1:1); Y1: YC + 3% *W. cibaria* G232; Y2: YC + 5% *W. cibaria* G232; Y3: YC + 7% *W. cibaria* G232.).

3.6. E-Tongue and E-Nose Analysis of Yogurt

The electronic tongue analyzer was equipped with seven sensors. Sensors of PKS, ANS, and SCS showed significant responses when analyzing the four different groups of yogurts. Figure 6A presents the results of PCA, which was based on the response values of different sensors. The horizontal axis represents the contribution rate of PC1 (82.0%), meanwhile the vertical axis represents the contribution rate of PC2 (13.7%). The cumulative contribution rate of the two principal components reached 95.7%, indicating an effective dimensionality reduction while preserving essential sample characteristics. As shown in the graphic, the two principal components retained the main characteristics of the samples, with the horizontal axis (PC1) contributing more significantly than the vertical axis (PC2). The position of the YC sample was further along the PC1 axis compared to the other groups, suggesting statistically significant organoleptic differentiation in taste profile parameters.

The electronic nose analyzer was equipped with 18 different odor sensors to identify the flavor profile of fermented milks. Figure 6B shows that the results of PCA, YC, Y1, Y2, and Y3 were distinguished well by the E-nose, indicating that fermentation generated the distinct, characteristic, gas flavor profile. The contribution rate of horizontal axis (PC1) interpreted 81.9% of the data information, while the vertical axis (PC2) accounted for 12.1%. From a horizontal coordinate perspective, YC was more distant from the other samples, which indicates that YC differs greatly from the other samples in terms of PC1. In addition, Y1 and Y2 were closer together, indicating similar odor compositions. According to the loading diagram, the sensors of P10/1 and T70/2 contributed more significant values to the PCA. These results demonstrate that the sensory characteristics of the yogurt can be effectively differentiated using the E-nose and E-tongue. In the E-nose sensor response profiles, Sensors LY2/LG and P30/2 exhibited the highest loadings on PC1, suggesting that the volatile organic compounds detected by these sensors are strongly associated with the floral and delicate aromatic profile of fermented yogurt.

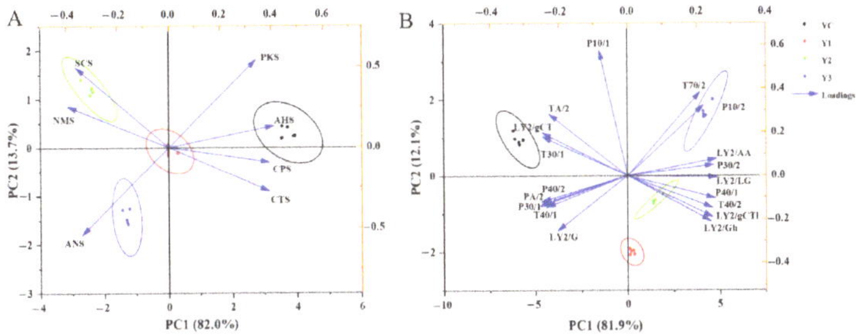

Figure 6. PCA model and loading plot based on the response data of E-tongue (**A**) and E-nose (**B**) of different yogurt groups. (Note: YC: *L. bulgaricus* G119: *S. thermophilus* Q019 (1:1); Y1: YC + 3% *W. cibaria* G232; Y2: YC + 5% *W. cibaria* G232; Y3: YC + 7% *W. cibaria* G232.).

3.7. Volatile Aroma Compound Analysis of Yogurt (GC-IMS)

The volatile metabolic profiles of fermented yogurts with different concentration of *W. cibaria* G232 were analyzed using GC-IMS. Figure 7A shows the 3D visualization plot where the transverse axis represents the ion migration time, the longitudinal axis represents the gas chromatography retention time, and the vertical axis refers to the intensity of the signal peak.

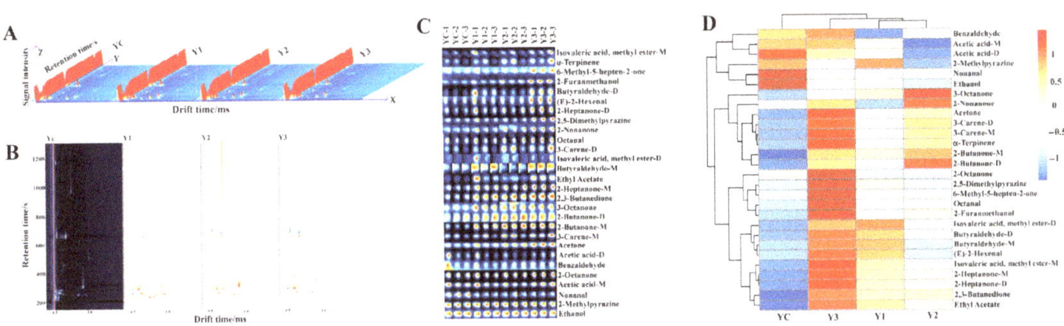

Figure 7. Observation results of GC-IMS on fermented yogurts with different proportions of *W. cibaria* G232. (Note: (**A**) is the three-dimensional representation; (**B**) is the bird's-eye view representation, with spectra from the control group employed as a reference and the corresponding spectra from yogurt fermented with *W. cibaria* G232 represented as differences from the control group YC; (**C**) is the representation of ion migration spectra, where the ions are numbered and then listed as gallery plots, in which the color was brighter, the content was higher; (**D**) is the clustering map of yogurts with different proportions of *W. cibaria* G232 metabolites. YC: *L. bulgaricus* G119: *S. thermophilus* Q019 (1:1); Y1: YC + 3% *W. cibaria* G232; Y2: YC + 5% *W. cibaria* G232; Y3: YC + 7% *W. cibaria* G232.).

In the 3D topographic map, the visualization patterns of different samples were similar, but the peak heights of different compounds varied slightly. When yogurt was fermented without the *W. cibaria* G232 used as the spectral background, the other samples exhibited distinct red and blue spots in the spectrogram (Figure 7B). The red spots represented higher concentration of volatile aroma compounds compared to the control group YC, while the blue spots represented lower concentrations. Compared to the control group YC, the yogurt supplemented with *W. cibaria* G232 showed increased diversity and abundance of volatile aroma substances, indicating that *W. cibaria* G232 facilitated the formation and accumulation of volatile aroma compounds.

Figure 7C showed the fingerprint of the volatile aroma substances in fermented yogurt, which exhibited the influence of the different *W. cibaria* G232 concentrations on these compounds. A total of twenty-two aroma compounds were detected in the four different yogurt groups, including six aldehydes, seven ketones, two alcohols, one acid, two esters, and four other aroma substances. As shown in Table 4, compared to control group YC, the content of acetone, 2-heptanone, 2-butanone, 2,3-butanedione, ethyl acetate, butyraldehyde, and other substance increased significantly in yogurt co-fermented with *W. cibaria* G232 and traditional starter cultures ($p < 0.05$). On the contrary, the content of nonanal and ethanol, which were the main aroma substances in the control group YC, significantly decreased ($p < 0.05$). The results showed that with the addition of different concentrations of *W. cibaria* G232, the content of the volatile aroma gradually increased, and when the concentration of *W. cibaria* G232 reached 7%, the content of the aroma substances reached maximum.

Table 4. Molecules' peak areas (mean ± sd) characterized by GC–IMS in fermented yogurt.

Volatile Flavor Compounds	CAS Registry Number	Molecule Formula	YC	Y1	Y2	Y3
Acetic acid-M	C64197	$C_2H_4O_2$	$4.25 \times 10^3 \pm 4.79 \times 10^2$ a	$3.96 \times 10^3 \pm 6.94 \times 10^2$ a	$3.50 \times 10^3 \pm 4.40 \times 10^2$ a	$4.27 \times 10^3 \pm 2.43 \times 10^2$ a
Acetic acid-D	C64197	$C_2H_4O_2$	$3.85 \times 10^2 \pm 1.04 \times 10^2$ a	$3.12 \times 10^2 \pm 1.06 \times 10^2$ a	$2.50 \times 10^2 \pm 43.02$ a	$3.44 \times 10^2 \pm 40.66$ a
Nonanal	C124196	$C_9H_{18}O$	$3.81 \times 10^3 \pm 2.34 \times 10^2$ a	$2.54 \times 10^3 \pm 1.81 \times 10^2$ b	$2.84 \times 10^3 \pm 3.20 \times 10^2$ b	$2.58 \times 10^3 \pm 49.62$ b
2,5-Dimethylpyrazine	C123320	$C_6H_8N_2$	$9.17 \times 10^2 \pm 51.61$ b	$1.03 \times 10^3 \pm 1.32 \times 10^2$ b	$9.24 \times 10^2 \pm 66.22$ b	$1.29 \times 10^3 \pm 58.32$ a
2-Octanone	C111137	$C_8H_{16}O$	$9.25 \times 10^3 \pm 9.17 \times 10^2$ a	$8.89 \times 10^3 \pm 1.56 \times 10^2$ a	$8.92 \times 10^3 \pm 7.39 \times 10^2$ a	$1.04 \times 10^4 \pm 3.02 \times 10^2$ a
2-Methylpyrazine	C109080	$C_5H_6N_2$	$8.74 \times 10^3 \pm 3.31 \times 10^2$ a	$8.78 \times 10^3 \pm 3.46 \times 10^2$ a	$8.12 \times 10^3 \pm 2.91 \times 10^2$ a	$8.31 \times 10^3 \pm 47.33$ a
Octanal	C124130	$C_8H_{16}O$	$8.51 \times 10^2 \pm 1.33 \times 10^2$ a	$8.29 \times 10^2 \pm 1.37 \times 10^2$ a	$8.04 \times 10^2 \pm 95.03$ a	$8.63 \times 10^2 \pm 27.26$ a
2-Nonanone	C821556	$C_9H_{18}O$	$1.11 \times 10^3 \pm 36.43$ ab	$1.05 \times 10^3 \pm 98.29$ b	$1.33 \times 10^3 \pm 1.26 \times 10^2$ a	$1.27 \times 10^3 \pm 20.43$ ab
Acetone	C67641	C_3H_6O	$4.01 \times 10^2 \pm 20.15$ c	$4.54 \times 10^2 \pm 37.53$ bc	$4.90 \times 10^2 \pm 22.22$ b	$5.48 \times 10^2 \pm 15.08$ a
2-Heptanone-M	C110430	$C_7H_{14}O$	$1.16 \times 10^3 \pm 98.33$ b	$1.38 \times 10^3 \pm \times 1.41 \times 10^2$ ab	$1.33 \times 10^3 \pm 1.22 \times 10^2$ ab	$1.51 \times 10^3 \pm 20.52$ a
2-Heptanone-D	C110430	$C_7H_{14}O$	$2.75 \times 10^2 \pm 34.11$ b	$3.63 \times 10^2 \pm 59.44$ ab	$3.34 \times 10^2 \pm 38.18$ ab	$4.21 \times 10^2 \pm 9.48$ a
3-Carene-M	C13466789	$C_{10}H_{16}$	$5.09 \times 10^2 \pm 1.06 \times 10^2$ a	$5.94 \times 10^2 \pm 1.74 \times 10^2$ a	$7.13 \times 10^2 \pm 1.64 \times 10^2$ a	$8.08 \times 10^2 \pm 1.44 \times 10^2$ a
3-Carene-D	C13466789	$C_{10}H_{16}$	$1.81 \times 10^2 \pm 43.73$ a	$2.18 \times 10^2 \pm 68.89$ a	$2.88 \times 10^2 \pm 90.04$ a	$3.24 \times 10^2 \pm 82.34$ a
α-Terpinene	C99865	$C_{10}H_{16}$	$1.40 \times 10^2 \pm 19.07$ b	$1.67 \times 10^2 \pm 46.86$ ab	$2.16 \times 10^3 \pm 50.74$ ab	$2.56 \times 10^2 \pm 53.92$ a
2-Butanone-M	C78933	C_4H_8O	$9.56 \times 10^2 \pm 91.59$ c	$1.79 \times 10^3 \pm 1.76 \times 10^2$ b	$2.21 \times 10^3 \pm 1.45 \times 10^2$ a	$2.09 \times 10^3 \pm 24.04$ a
2-Butanone-D	C78933	C_4H_8O	$5.21 \times 10^2 \pm 21.61$ c	$5.52 \times 10^2 \pm 51.26$ b	$5.97 \times 10^2 \pm 26.89$ a	$5.78 \times 10^2 \pm 11.93$ a
Ethanol	C64175	C_2H_6O	$2.79 \times 10^3 \pm 13.18$ a	$2.47 \times 10^3 \pm 7.23$ b	$2.41 \times 10^3 \pm 28.16$ c	$2.51 \times 10^3 \pm 14.21$ b
2,3-Butanedione	C431038	$C_4H_6O_2$	$1.98 \times 10^3 \pm 80.30$ c	$3.50 \times 10^3 \pm 1.18 \times 10^2$ b	$3.50 \times 10^3 \pm 3.04 \times 10^2$ b	$4.20 \times 10^3 \pm 98.20$ a
Ethyl Acetate	C141786	$C_4H_8O_2$	$1.00 \times 10^2 \pm 6.51$ b	$2.50 \times 10^2 \pm 85.55$ a	$2.10 \times 10^2 \pm 36.09$ a	$3.02 \times 10^2 \pm 7.57$ a
Butyraldehyde-M	C123728	C_4H_8O	$2.10 \times 10^2 \pm 11.29$ b	$3.60 \times 10^2 \pm 39.97$ a	$2.24 \times 10^2 \pm 16.91$ b	$3.78 \times 10^2 \pm 7.81$ a
Butyraldehyde-D	C123728	C_4H_8O	81.01 ± 7.85 b	$2.30 \times 10^2 \pm 47.81$ a	$1.31 \times 10^2 \pm 5.61$ b	$2.57 \times 10^2 \pm 5.51$ a
2-Furanmethanol	C98000	$C_5H_6O_2$	$1.53 \times 10^3 \pm 1.09 \times 10^2$ c	$1.74 \times 10^3 \pm 2.23 \times 10^2$ bc	$1.86 \times 10^3 \pm 91.15$ b	$2.38 \times 10^3 \pm 6.59$ a
Benzaldehyde	C100527	C_7H_6O	$2.21 \times 10^2 \pm 58.58$ a	$1.94 \times 10^2 \pm 16.22$ a	$2.09 \times 10^2 \pm 10.15$ a	$2.26 \times 10^2 \pm 8.72$ a
(E)-2-Hexenal	C6728263	$C_6H_{10}O$	$2.04 \times 10^2 \pm 11.26$ b	$2.38 \times 10^2 \pm 17.68$ a	$2.08 \times 10^3 \pm 10.04$ b	$2.45 \times 10^2 \pm 5.51$ a
Isovaleric acid, methyl ester-M	C556241	$C_6H_{12}O_2$	$1.80 \times 10^2 \pm 23.57$ a	$2.68 \times 10^2 \pm 97.63$ a	$2.31 \times 10^2 \pm 48.19$ a	$3.21 \times 10^2 \pm 61.19$ a
Isovaleric acid, methyl ester-D	C556241	$C_6H_{12}O_2$	70.98 ± 3.09 a	90.02 ± 24.14 a	$76.94 \times 10^3 \pm 7.79$ a	89.10 ± 12.06 a
3-Octanone	C106683	$C_8H_{16}O$	$1.36 \times 10^2 \pm 11.26$ ab	$1.51 \times 10^2 \pm 3.68$ ab	$1.67 \times 10^2 \pm 17.14$ a	$1.44 \times 10^2 \pm 0.66$ ab
6-Methyl-5-hepten-2-one	C110930	$C_8H_{14}O$	$2.58 \times 10^2 \pm 12.64$ b	$2.71 \times 10^2 \pm 20.57$ b	$2.60 \times 10^2 \pm 4.46$ b	$3.16 \times 10^3 \pm 18.04$ a

Note: The data are presented as average values ± SD (n = 3). Averages featuring varying lowercase letters within the same row show a notable difference ($p < 0.05$). YC: *L. bulgaricus* G119: *S. thermophilus* Q019 (1:1); Y1: YC + 3% *W. cibaria* G232; Y2: YC + 5% *W. cibaria* G232; Y3: YC + 7% *W. cibaria* G232.

The volatile aroma substances in yogurt are primarily produced through the degradation of milk fat and bioconversion by lactic acid bacteria. The heat map was generated based on the relative contents of volatile aroma substances in the four different yogurt groups, and cluster analysis was also performed in Figure 7D. The yogurt samples were divided into two clusters based on their volatile aroma profiles. YC formed one cluster and

Y1, Y2, and Y3 were gathered in other clusters. The heatmap visually represents the differences in volatile aroma substance profiles between YC and the other groups. Variations in the contents of different aroma substances contributed to the distinct aroma differences in YC and Y3, which aligns with the results of the PCA obtained from the E-nose analysis.

3.8. Correlation Between E-Nose and GC-IMS

As illustrated in Figure 8, the sensors LY2/LG, P10/2, and LY2/AA, associated with coffee, fruity, and special aromas, exhibited a significant positive correlation with volatile compounds of 2-heptanone, α-terpinene, 2-butanone, 2,3-butanedione, and ethyl acetate ($p < 0.01$). These compounds were present in higher concentration in yogurts co-fermented with *W. cibaria* G232. On the contrary, sensors LY2/gCT, PA/2 showed a significant positive correlation with nonanal, while sensors p40/1 was correlated with ethanol. Nonanal and ethanol were found in greater quantities in control group fermented without *W. cibaria* G232 (YC). This finding indicates that both E-nose and GC-IMS are effective in distinguishing aroma compounds in fermented milk at different levels. Moreover, the yogurt co-fermented with *W. cibaria* G232 exhibited a greater richness in aroma substances, highlighting its dominance in aroma profile development.

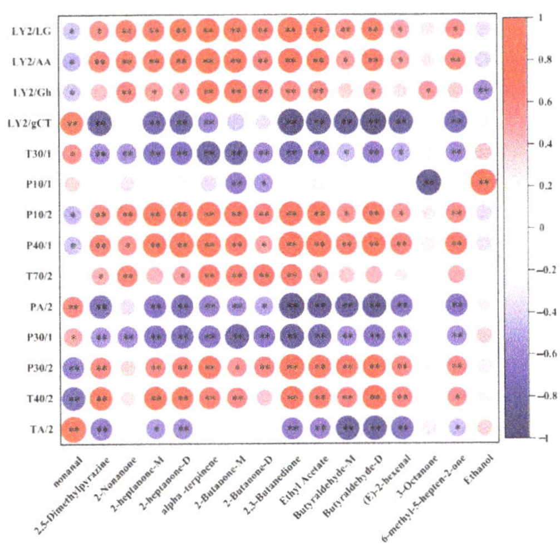

Figure 8. Spearman's correlation heatmap of the significantly altered volatile compound levels and the E-nose sensor responses. (Colors represent the correlation coefficients, with red and blue indicating positive and negative correlation, respectively. Asterisks * and ** stand for significance at $p < 0.05$ and $p < 0.01$, respectively).

4. Discussion

Fermented dairy products are widely consumed around the world, with LAB playing a crucial role in the fermentation of milk. Previous studies have highlighted several advantages of *Weissella* as a fermenting agent [22], such as reducing fermentation time and inhibiting the growth of pathogenic microorganisms through acid production [23]. Additionally, *Weissella* has been shown to improve the quality and flavor of fermented products by breaking down proteins, lactose, and fats [24]. Although *Weissella* spp. are recognized as excellent probiotics with significant potential in food fermentation, their application in yogurt production remains underexplored, and previous studies mainly focused on their effects on yogurt texture, rheological properties, and bacterial viability. In this study, gas chromatography-ion mobility spectrometry (GC-IMS) and intelligent sensory technologies (E-nose and E-tongue) were employed to systematically investigate the dynamic changes in volatile aroma compounds during yogurt fermentation with *W. cibaria* G232. Furthermore, the impact of co-fermentation with *W. cibaria* G232 on the antioxidant properties of yogurt was comprehensively evaluated. This research emphasizes three critical aspects (physicochemical properties, antioxidant activity, and aromatic profiles) of yogurt quality enhancement, providing valuable theoretical insights and experimental data to support the application of *W. cibaria* G232 as a co-starter in yogurt production.

Coagulation time and acidification rate of yogurt is significantly influenced by pH and titratable acidity (TA) values. In this study, the addition of 5%–7% *W. cibaria* G232 (Y2 and Y3) shortened the fermentation time from 7 h to 6 h compared to the control group YC. The reduction was attributed to the accelerated acidification process, driven by the synergistic interaction between *W. cibaria* G232 and the traditional starters (*L. bulgaricus* G119 and *S. thermophilus* Q019). *Weissella cibaria*, as a typical heterofermentative bacterium, produces lactic acid, CO_2, and acetate via the hexose monophosphate and phosphoketolase pathways [25]. *L. bulgaricus* and *S. thermophilus* are homofermentative lactic acid bacteria; however, the available literature indicates that the coexistence of homo- and hetero-lactic bacteria can accelerate the fermentation process [13]. Notably, the pH decline was most pronounced during the first four hours of fermentation; after that, the acidity levels among the four groups showed little variation. This phenomenon may be due to the promoting effect of *W. cibaria* G232 on the growth environment of *Streptococcus thermophilus* Q019. During the initial stage of fermentation, the rapid growth and acid production by *S. thermophilus* significantly shortened the fermentation time. These results align with the findings of Zou et al. [24], who observed that co-fermentation with *W. confusa* SW1 reduced the curd time of yogurt and promoted the growth of *S. thermophilus* ST51. Moreover, synergistic effects on glycolysis and lactate production may further enhance the fermentation process.

Water-holding capacity and rheological and texture properties are determined by the network arrangement structure of yogurt and influenced by the manufacturing process [26]. In this study, co-fermentation of traditional starters with *W. cibaria* G232 promoted protein aggregation in milk, resulting in higher WHC and apparent viscosity in the yogurt. At a 5% addition level of *W. cibaria* G232, both the elastic modulus (G') and viscous modulus (G'') reached their maximum, indicating optimal stability and smoothness of the yogurt. In addition, the co-fermentation of traditional starters with *W. cibaria* G232 improved the hardness, consistency, and cohesiveness of the yogurt to varying degrees. These quality enhancements may be attributed to the exopolysaccharides (EPS) produced by *W. cibaria* G232 during fermentation. Our previous studies have shown that *W. cibaria* G232 has a strong ability to produce extracellular polysaccharides (EPS). Also, the literature showed that the *Weissella* species, including *W. cibaria*, *W. confusa*, and *W. hellenica*, are well-known EPS producers [27]. The hydroxyl and negatively charged groups in the side chains of EPS can bind water molecules, thereby reducing whey separation and enhancing the stability of yogurt [28]. Moreover, the interaction between EPS and casein (CAS) via intermolecular hydrogen bonding and electrostatic and hydrophobic contacts enhances the charge density and electrostatic potential of molecules. This improves the viscidity and elasticity of the fermented system [29], which is consistent with the findings of our study.

Even though the positive effects of EPS on fermented products are well-documented, excessive EPS in a yogurt fermentation system can strengthen the steric barrier [30], reducing electrostatic interactions between particles and potentially weakening the stability of the fermented system. The finding is also consistent with our results. The addition of *W. cibaria* G232 at a concentration of 5% resulted in optimal fermentation outcomes, including improved texture and rheological properties, whereas a concentration of 7% was less effective, likely due to the steric hindrance caused by excessive EPS production. Antioxidant potential (AP) is one of the most important nutritional properties of yogurt, as it not only extends the shelf life but also helps to prevent oxidative damage. ABTS and DPPH are widely used to evaluate free radical scavenging capacity due to their rapid, simple, and cost-effective feature. FRAP assay measures the reducing power of antioxidants by quantifying the reduction in the Fe^{3+}-TPTZ complex to Fe^{2+}-TPTZ, resulting in strong absorption at specific wavelengths [31]. Our results reveal that mixed fermentation significantly enhanced the antioxidant activity of yogurt ($p < 0.05$) in a dose-dependent manner, with respect to the concentration of *W. cibaria* G232. *Weissella* spp. are well known for their antioxidant properties, particularly from the exopolysaccharides (EPS) they produce. These EPS hydrolyze in acidic environments, exposing active hemiacetal hydroxyl groups that transfer electrons to free radicals, facilitating the conversion of free radicals into stable compounds and reducing oxidative damage [32]. Moreover, the antioxidant capacity of fermented products also closely relates to the viable cell counts. Simionescu and Petrovici [33] demonstrated that the antioxidant activity in fermented media increased with the number of CFU/mL of *W. confusa* PP29, which is consistent with our results. The addition of *W. cibaria* G232 at 7% resulted in the highest viable cell count (8.63 log CFU/mL) among the four groups, accompanied by the highest ABTS and DPPH radical scavenging activities and ferric-reducing antioxidant power, which can be attributed to the synergistic effects of various antioxidant components, including casein, peptides, antioxidant enzymes [34].

E-nose and GC-IMS were employed to distinguish and analyze the aromatic profiles of fermented yogurt from two different dimensions. Previous studies have shown that the E-nose, equipped with a variety of odor sensors, enables it to effectively extract and reflect the overall information of flavor compounds [35]. In comparison, GC-IMS allows for both qualitative and quantitative analysis of volatile flavor compounds, providing more precise and detailed results. To obtain comprehensive flavor information, we combined both techniques in this study. The generation of volatile compounds in yogurt is highly dependent on the strains used during fermentation. According to previous studies, mixed fermentation tends to be more conducive to the production and accumulation of aroma substances due to the synergistic interactions between different microbial species [8,36]. The aroma of fermented dairy products is mainly derived from microbial transformation and chemical reactions, including glycolysis, proteolysis, and lipolysis [37]. The aromatic profile of yogurt is the result of the combined effects of various aroma compounds, including alcohols, acids, aldehydes, ketones, and esters. However, not all the aroma compounds in yogurt play a decisive role. In our study, the content of ethanol and nonanal in yogurt co-fermented with *W. cibaria* G232 significantly decreased compared to the control group ($p < 0.05$). Conversely, the content of ketones, including acetone, 2-heptanone, 2-butanone, and 2- and 3-butanedione was significantly elevated ($p < 0.05$). Ethanol probably provides a complementary flavor to yogurt accrued through decomposition of glucose and amino acids [38]. Nonanal is characterized by a sweet, floral, citrus, and fatty aroma [39]. Both compounds were more abundant in the control group and served as primary aroma compounds in yogurt. In this study, the concentrations of six ketone compounds in yogurt increased variably after co-fermentation. Ketones, primarily formed through the oxidation of free fatty acids and the degradation of amino acids, can directly influence the aroma of fermented dairy products [40]. 2,3-butanedione, an important aroma compound, is produced during the fermentation of citrate in milk. The increase in 2,3-butanedione enhanced the rate of citrate metabolism and contributed to a buttery flavor [41]. Additionally, 2-heptanone can help mitigate the pungent flavor in yogurt [42]. Other compounds such as acetone and butanedione considerably provide a pleasant floral, fruity, and creamy flavor [43]. The incorporation of *W. cibaria* G232 potentially augmented the metabolic pathways associated with fatty acids and amino acids, culminating in the biosynthesis of a substantial quantity of ketones, which contributed notably to positive aromatic profiles such as fruity, sweet, milky, and creamy-cheesy aromas, thereby imparting the complexity and balance of the flavor profile. Moreover, six different aldehyde aroma compounds were also detected. Butyraldehyde, which offers a cocoa-like flavor, showed significantly increased accumulation in the co-fermentation yogurt compared to the control group ($p < 0.05$). Hexanal and heptanal, which provide grassy aromas [44], were present in low concentrations across the fermentation groups without significant differences ($p > 0.05$). Although each flavor compound constitutes a recognizable aroma, the flavor of yogurt is determined by the balanced mixture of key volatile compounds. The synergistic effects of multiple volatile flavor compounds contribute to the production of superior yogurt products. Yogurt fermented with traditional strains contained higher levels of nonanal and ethanol, contributing to a rich and balanced oily flavor. In contrast, the co-fermented yogurt with *W. cibaria* G232 increased the concentration of 2,3-butanedione, enhancing the buttery flavor. Additionally, the accumulation of acetone, butanone, and ethyl acetate enhanced the complexity of the aroma profile, with distinct layers of fruity, creamy, and cocoa notes. The differences in the concentrations of typical volatile compounds were consistent with the results obtained from the E-nose analysis. There was a significant improvement in the aroma of yogurt co-fermented with *W. cibaria* G232, as evidenced by both intelligent sensory techniques and GC-IMS. These findings highlight the potential of mixed fermentation to enhance the aroma profile of yogurt through the synergistic effects of microbial metabolism.

5. Conclusions

In conclusion, this study investigated the effects of co-fermentation using *W. cibaria* G232 and traditional starter. The results demonstrate that *W. cibaria* G232 exhibits a positive effect on the physicochemical properties, antioxidant capacities, and volatile profiles of fermented milk. Co-fermentation with *W. cibaria* G232 resulted in a shorter fermentation duration, which was attributed to the synergistic acidification by *Weissella* spp. Furthermore, *W. cibaria* G232 and traditional starter enhanced the vitality of the strains through species interactions. The metabolites produced by *W. cibaria* G232 enhanced stability of yogurt, contributing to improvements in water-holding capacity, rheological properties, and texture. Antioxidant potential of yogurt was significantly enhanced by co-fermentation, showing a dose-dependent relationship with the concentration of *W. cibaria* G232. Intelligent sensory technology combined with GC-IMS revealed that compared to the control group, co-fermentation with *W. cibaria* G232 notably altered the accumulation of crucial aroma compounds, including aldehydes, ketones, and alcohols in the fermented milk. Optimized at 5% *W. cibaria* G232 supplementation, the yogurt exhibited superior quality through balanced physicochemical properties, enhanced antioxidant capacity, and refined aromatic profiles. Excessive supplementation (7%) triggered EPS overproduction, destabilizing casein gels and compromising structural integrity. Overall, the results support the quality improvement of yogurt with the addition of *W. cibaria* G232. However, the specific mechanism underlying the molecular mechanisms of interspecies interactions between *W. cibaria* G232 and traditional starters—particularly through multi-omic approaches (e.g., metabolomics and proteomics) to map cross-species metabolic networks associated with this enhancement—remain to be further explored.

Author Contributions: Q.H.: writing—original draft, methodology, investigation, formal analysis. H.Y.: writing—review, formal analysis. Y.Y.: writing—review and editing. C.Z.: writing—review and editing, investigation, methodology. J.T.: writing—review and editing, writing—original draft, project administration, funding acquisition, data curation, conceptualization. All authors have read and agreed to the published version of the manuscript.

Funding: This work was supported by "the Fundamental Research Funds for the Central Universities", Southwest Minzu University (Grant No. ZYN2025119) and the Natural Science Foundation of Sichuan Province (Grant No. 2024NSFSC0364).

Institutional Review Board Statement: Not applicable.

Informed Consent Statement: Not applicable.

Data Availability Statement: The original contributions presented in the study are included in the article, further inquiries can be directed to the corresponding authors.

Conflicts of Interest: All authors declare that there is no financial/personal interest. No conflict of interest exits in the submission of this manuscript, and the manuscript is approved by all authors for publication. All the authors listed have approved the manuscript that is enclosed.

References

1. Fan, X.; Zhang, A.; Zhang, T.; Tu, M.; Du, Q.; Ling, N.; Wu, J.; Zeng, X.; Wu, Z.; Pan, D. Effects of Semen Ziziphi Spinosae Extract and Binary Probiotics Co-Fermentation on the Quality of Yogurt and Their Underlying Molecular Mechanisms. *Food Chem.-X* **2024**, *21*, 101191. [CrossRef] [PubMed]
2. Rizzoli, R.; Biver, E. Role of Fermented Dairy Products in the Health Benefits of a Mediterranean Diet. *Aging Clin. Exp. Res.* **2024**, *36*, 75. [CrossRef] [PubMed]
3. Pyo, Y.; Kwon, K.H.; Jung, Y.J. Probiotic Functions in Fermented Foods: Anti-Viral, Immunomodulatory, and Anti-Cancer Benefits. *Foods* **2024**, *13*, 2386. [CrossRef]
4. de Souza, M.; Drunkler, D.A.; Colla, E. Probiotic Functional Yogurt: Challenges and Opportunities. *Fermentation* **2024**, *10*, 6. [CrossRef]
5. Wang, Y.; Liu, H.; Zhao, J. Macrophage Polarization Induced by Probiotic Bacteria: A Concise Review. *Probiotics Antimicrob. Proteins* **2020**, *12*, 798–808. [CrossRef] [PubMed]
6. Kumar, V.; Amrutha, R.; Ahire, J.J.; Taneja, N.K. Techno-Functional Assessment of Riboflavin-Enriched Yogurt-Based Fermented Milk Prepared by Supplementing Riboflavin-Producing Probiotic Strains of *Lactiplantibacillus plantarum*. *Probiotics Antimicrob. Proteins* **2024**, *16*, 152–162. [CrossRef]
7. Meng, X.; Huang, Y.; Xiong, J.; Cheng, Z.; Yang, T.; Li, Z.; Tuo, R.; Zhang, Z.; Wang, G.; Gu, Q.; et al. *Lactiplantibacillus plantarum* ZFM55 Improves Texture and Flavor of Yogurt, Increases Beneficial Metabolites, and the Co-Fermented Yogurt Promotes Human Gut Microbiota Health. *LWT-Food Sci. Technol.* **2024**, *198*, 115929. [CrossRef]
8. Zhang, X.; Yang, J.; Zhang, C.; Chi, H.; Zhang, C.; Zhang, J.; Li, T.; Liu, L.; Li, A. Effects of *Lactobacillus fermentum* HY01 on the Quality Characteristics and Storage Stability of Yak Yogurt. *J. Dairy Sci.* **2022**, *105*, 2025–2037. [CrossRef]
9. Fusco, V.; Chieffi, D.; Fanelli, F.; Montemurro, M.; Rizzello, C.G.; Franz, C.M.A.P. The *Weissella* and *Periweissella* Genera: Up-to-Date Taxonomy, Ecology, Safety, Biotechnological, and Probiotic Potential. *Front. Microbiol.* **2023**, *14*, 1289937. [CrossRef]
10. Singh, J.K.; Devi, P.B.; Reddy, G.B.; Jaiswal, A.K.; Kavitake, D.; Shetty, P.H. Biosynthesis, Classification, Properties, and Applications of *Weissella* Bacteriocins. *Front. Microbiol.* **2024**, *15*, 1406904. [CrossRef]
11. Xia, Y.; Qin, S.; Shen, Y. Probiotic Potential of *Weissella* Strains Isolated from Horse Feces. *Microb. Pathog.* **2019**, *132*, 117–123. [CrossRef] [PubMed]
12. Wang, Q.; Wang, W.; Zhang, Y.; Yao, Y.; Dong, B.; Zhao, G. Effects of New Fermentation Strains: *Weissella confusa* GTA22 and *Weissella koreensis* QAY6 on the Quality of Dough and Steamed Bread. *Food Biosci.* **2025**, *66*, 106223. [CrossRef]
13. Zammouri, A.; Ziadi, M.; Gharsallaoui, A.; Fguiri, I.; Sbissi, I.; Hammadi, M.; Khorchani, T. Characterization of Novel Exopolysaccharides from *Weissella cibaria* and *Lactococcus lactis* Strains and Their Potential Application as Bio-Hydrocolloid Agents in Emulsion Stability. *Fermentation* **2024**, *10*, 532. [CrossRef]
14. Xiong, H.; Lei, D.; Xiong, J.; Liu, J.-Y.; Yao, J. Exploring the Mechanism of *Weissella cibaria* Cooperating with *Lactiplantibacillus plantarum* to Improve Volatile Flavor of Sichuan Pickle Using Dialysis Membranes. *Food Biosci.* **2024**, *60*, 104520. [CrossRef]
15. Fan, X.; Shi, Z.; Xu, J.; Li, C.; Li, X.; Jiang, X.; Du, L.; Tu, M.; Zeng, X.; Wu, Z.; et al. Characterization of the Effects of Binary Probiotics and Wolfberry Dietary Fiber on the Quality of Yogurt. *Food Chem.* **2023**, *406*, 135020. [CrossRef]
16. Ahmed, S.A.; Helmy, W.A.; Ibrahim, O.A. Evaluation of Lupine Seeds (*Lupinus albus* L.) Neutral Extract as a Texture Improver in Low-Fat Yogurt Production. *Int. J. Biol. Macromol.* **2024**, *263*, 130303. [CrossRef]
17. Huo, Y.; Zhang, D.; Wang, X.; Xu, G.; Dai, M.; Zhang, S. Biofunctional Attributes and Storage Study of Milk Fermented by *Enterococcus italicus*. *Int. J. Food Microbiol.* **2024**, *423*, 110844. [CrossRef]
18. Shori, A.B. Proteolytic Activity, Antioxidant, and α-Amylase Inhibitory Activity of Yogurt Enriched with Coriander and Cumin Seeds. *LWT-Food Sci. Technol.* **2020**, *133*, 109912. [CrossRef]

19. Wen, R.; Kong, B.; Yin, X.; Zhang, H.; Chen, Q. Characterisation of Flavour Profile of Beef Jerky Inoculated with Different Autochthonous Lactic Acid Bacteria Using Electronic Nose and Gas Chromatography-Ion Mobility Spectrometry. *Meat Sci.* **2022**, *183*, 108658. [CrossRef]
20. Li, X.; Yang, Y.; Zhu, Y.; Ben, A.; Qi, J. A Novel Strategy for Discriminating Different Cultivation and Screening Odor and Taste Flavor Compounds in Xinhui Tangerine Peel Using E-Nose, E-Tongue, and Chemometrics. *Food Chem.* **2022**, *384*, 132519. [CrossRef]
21. Wu, T.; Guo, S.; Liu, K.; Yang, Y.; Wang, J.; Zhang, H. Comparison of Volatile Metabolic Profiles in Fermented Milk of *Streptococcus thermophilus* during the Postripening Period at Different Incubation Temperatures. *J. Dairy Sci.* **2023**, *106*, 2303–2313. [CrossRef] [PubMed]
22. Kim, E.; Yang, S.-M.; Kim, H.-Y. *Weissella* and the Two Janus Faces of the Genus. *Appl. Microbiol. Biotechnol.* **2023**, *107*, 1119–1127. [CrossRef] [PubMed]
23. Xu, B.; Mi, T.; Ma, S.; Yi, X.; Huang, P.; Huang, P.; Wu, C. Insight into the Autochthonous Lactic Acid Bacteria as Starter Culture for Improving the Quality of Sichuan Radish Paocai: Changes in Microbial Diversity and Metabolic Profiles. *Int. J. Food Microbiol.* **2024**, *425*, 110877. [CrossRef] [PubMed]
24. Zou, F.; Huo, Y.; Gao, W.; Dai, M.; Zhao, G.; Zhang, S. Physicochemical, Microbiological and Sensory Characterization of Yogurt Fermented by *Weissella confusa* SW1 and Traditional Starters. *LWT-Food Sci. Technol.* **2024**, *201*, 116229. [CrossRef]
25. Lee, S.H.; Whon, T.W.; Roh, S.W.; Jeon, C.O. Unraveling Microbial Fermentation Features in Kimchi: From Classical to Meta-Omics Approaches. *Appl. Microbiol. Biotechnol.* **2020**, *104*, 7731–7744. [CrossRef]
26. Pan, L.H.; Liu, F.; Luo, S.Z.; Luo, J. Pomegranate Juice Powder as Sugar Replacer Enhanced Quality and Function of Set Yogurts: Structure, Rheological Property, Antioxidant Activity and in Vitro Bioaccessibility. *LWT-Food Sci. Technol.* **2019**, *115*, 108479. [CrossRef]
27. Kavitake, D.; Devi, P.B.; Shetty, P.H. Overview of Exopolysaccharides Produced by Weissella Genus—A Review. *Int. J. Biol. Macromol.* **2020**, *164*, 2964–2973. [CrossRef]
28. London, L.E.E.; Chaurin, V.; Auty, M.A.E.; Fenelon, M.A.; Fitzgerald, G.F.; Ross, R.P.; Stanton, C. Use of *Lactobacillus mucosae* DPC 6426, an Exopolysaccharide-Producing Strain, Positively Influences the Techno-Functional Properties of Yoghurt. *Int. Dairy J.* **2015**, *40*, 33–38. [CrossRef]
29. Zhang, M.; Lai, T.; Yao, M.; Zhang, M.; Yang, Z. Interaction of the Exopolysaccharide from *Lactobacillus plantarum* YW11 with Casein and Bioactivities of the Polymer Complex. *Foods* **2021**, *10*, 1153. [CrossRef]
30. Payens, T.A.J. Association of Caseins and Their Possible Relation to Structure of the Casein Micelle1. *J. Dairy Sci.* **1966**, *49*, 1317–1324. [CrossRef]
31. Feng, Y.; Cheng, H.; Cheng, Y.; Zhao, J.; He, J.; Li, N.; Wang, J.; Guan, J. Chinese Traditional Pear Paste: Physicochemical Properties, Antioxidant Activities and Quality Evaluation. *Foods* **2023**, *12*, 187. [CrossRef] [PubMed]
32. Zhou, Y.; Cui, Y.; Qu, X. Exopolysaccharides of Lactic Acid Bacteria: Structure, Bioactivity and Associations: A Review. *Carbohydr. Polym.* **2019**, *207*, 317–332. [CrossRef] [PubMed]
33. Simionescu, N.; Petrovici, A.-R. Enhancing the Antioxidant Potential of *Weissella confusa* PP29 Probiotic Media through Incorporation of *Hibiscus sabdariffa* L. Anthocyanin Extract. *Antioxidants* **2024**, *13*, 165. [CrossRef]
34. Wang, W.; Li, S.; Heng, X.; Chu, W. *Weissella confusa* CGMCC 19,308 Strain Protects Against Oxidative Stress, Increases Lifespan, and Bacterial Disease Resistance in Caenorhabditis Elegans. *Probiotics Antimicrob. Proteins* **2022**, *14*, 121–129. [CrossRef]
35. Yang, X.; Li, M.; Ji, X.; Chang, J.; Deng, Z.; Meng, G. Recognition Algorithms in E-Nose: A Review. *IEEE Sens. J.* **2023**, *23*, 20460–20472. [CrossRef]
36. Zhou, B.; Wang, C.; Yang, Y.; Yu, W.; Bin, X.; Song, G.; Du, R. Structural Characterization and Biological Properties Analysis of Exopolysaccharides Produced by *Weisella cibaria* HDL-4. *Polymers* **2024**, *16*, 2314. [CrossRef]
37. Yang, S.; Bai, M.; Kwok, L.-Y.; Zhong, Z.; Sun, Z. The Intricate Symbiotic Relationship between Lactic Acid Bacterial Starters in the Milk Fermentation Ecosystem. *Crit. Rev. Food Sci. Nutr.* **2025**, *65*, 728–745. [CrossRef]
38. Cheng, H. Volatile Flavor Compounds in Yogurt: A Review. *Crit. Rev. Food Sci. Nutr.* **2010**, *50*, 938–950. [CrossRef]
39. Yu, X.; Sun, Y.; Shen, X.; Li, W.; Cai, H.; Guo, S.; Sun, Z. Effect of Different Isolation Sources of *Lactococcus lactis* subsp. *lactis* on Volatile Metabolites in Fermented Milk. *Food Chem. X* **2024**, *21*, 101224. [CrossRef]
40. Beltrán-Barrientos, L.M.; Garcia, H.S.; Reyes-Díaz, R.; Estrada-Montoya, M.C.; Torres-Llanez, M.J.; Hernández-Mendoza, A.; González-Córdova, A.F.; Vallejo-Cordoba, B. Cooperation between *Lactococcus lactis* NRRL B-50571 and NRRL B-50572 for Aroma Formation in Fermented Milk. *Foods* **2019**, *8*, 645. [CrossRef]
41. Papaioannou, G.; Kosma, I.; Badeka, A.V.; Kontominas, M.G. Profile of Volatile Compounds in Dessert Yogurts Prepared from Cow and Goat Milk, Using Different Starter Cultures and Probiotics. *Foods* **2021**, *10*, 3153. [CrossRef] [PubMed]
42. Ott, A.; Germond, J.E.; Baumgartner, M.; Chaintreau, A. Aroma Comparisons of Traditional and Mild Yogurts: Headspace Gas Chromatography Quantification of Volatiles and Origin of Alpha-Diketones. *J. Agric. Food Chem.* **1999**, *47*, 2379–2385. [CrossRef] [PubMed]

43. Tian, H.; Huang, N.; Yao, W.; Yu, H.; Yu, B.; Chen, X.; Chen, C. Comparative Transcriptomic Analysis of the Flavor Production Mechanism in Yogurt by Traditional Starter Strains. *J. Dairy Sci.* **2024**, *107*, 5402–5415. [CrossRef]
44. Vazquez-Landaverde, P.A.; Velazquez, G.; Torres, J.A.; Qian, M.C. Quantitative Determination of Thermally Derived Off-Flavor Compounds in Milk Using Solid-Phase Microextraction and Gas Chromatography. *J. Dairy Sci.* **2005**, *88*, 3764–3772. [CrossRef] [PubMed]

foods

MDPI

Review

Biosynthesis of Edible Terpenoids: Hosts and Applications

Mengyu Wang *, Zhengyi Zhang, Xinyu Liu, Zhixuan Liu and Ruirui Liu

Institute of Environmental Systems Biology, College of Environmental Science and Engineering, Dalian Maritime University, Dalian 116026, China; zhengyi-zhang@dlmu.edu.cn (Z.Z.); dmulxy2022@dlmu.edu.cn (X.L.); swxuan1224@163.com (Z.L.); 15840681379@163.com (R.L.)
* Correspondence: wangmengyu@dlmu.edu.cn; Tel.: +86-15140482161

Abstract: Microbial foods include microbial biomass, naturally fermented foods, and heterologously synthesized food ingredients derived from microbial fermentation. Terpenoids, using isoprene as the basic structure, possess various skeletons and functional groups. They exhibit diverse physicochemical properties and physiological activities, such as unique flavor, anti-bacterial, anti-oxidant, anti-cancer, and hypolipemic, making them extensively used in the food industry, such as flavor, fragrance, preservatives, dietary supplements, and medicinal health food. Compared to traditional strategies like direct extraction from natural species and chemical synthesis, microbial cell factories for edible terpenoids have higher titers and yields. They can utilize low-cost raw materials and are easily scaling-up, representing a novel green and sustainable production mode. In this review, we briefly introduce the synthetic pathway of terpenoids and the applications of microbial cell factories producing edible terpenoids. Secondly, we highlight several typical and non-typical microbial chassis in edible terpenoid-producing cell factories. In addition, we reviewed the recent advances of representative terpenoid microbial cell factories with a gram-scale titer in food flavor, food preservation, nutritional enhancers, and medicinal health foods. Finally, we predict the future directions of microbial cell factories for edible terpenoids and their commercialization process.

Keywords: microbial foods; cell factories; terpenoids; biosynthesis; fermentation

1. Introduction

Terpenoids, natural products formed by the head-to-tail linkage of isoprene units, are widely found in organisms, including plants, animals, microorganisms, and marine life. These compounds function as critical secondary metabolites within living organisms. The diversity in their skeleton structures and functional groups confers a broad spectrum of biological activities and physicochemical properties to these compounds, ranging from simple monoterpenes to complex polyterpenes (Figure 1). Terpenoids have substantial application prospects and market demand in fields such as medicine, food, and daily chemicals [1–3]. Particularly in the food industry, terpenoid compounds have diverse applications [4]. For example, limonene and menthol are employed as edible fragrances due to their unique flavors and tastes. Carnosic acid and α-pinene exhibit potential as food preservatives owing to their antioxidative and antibacterial properties. Retinol and vitamin K are frequently used as nutritional enhancers in the food industry due to their roles in vision, hemagglutination, and bone metabolism. *Ganoderma* triterpenoids and sclareol are widely used as medicinal health food for their capabilities in hypolipemic and antitumor activities. With the increasing attention to food safety, terpenoids are increasingly applied in the food industry due to their natural origin [5]. Hence, natural terpenoids hold immense potential for application.

https://doi.org/10.3390/foods14040673

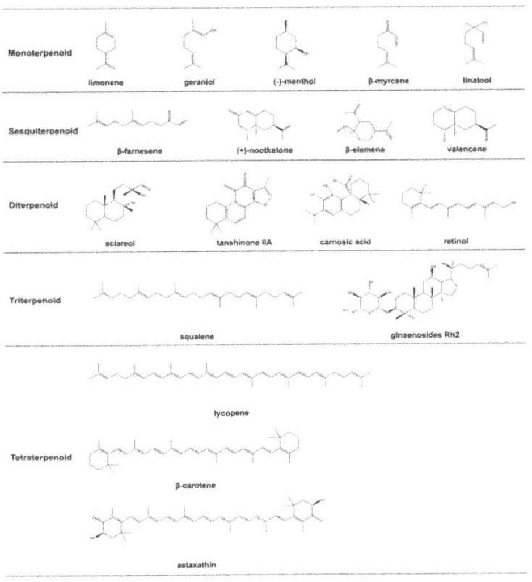

Figure 1. Chemical structures of some terpenoids.

However, the low content and uneven distribution of terpenoids in natural plants require the use of complex extraction and separation techniques, resulting in higher extraction costs. Furthermore, their synthesis and accumulation are influenced by various environmental factors such as geographical location and season, making it challenging to produce on a large scale continuously. While chemical synthesis of terpenoids offers scalability, it is encumbered with high raw material costs, intricate processes, and issues related to chiral recognition and stereoselectivity [6]. Microbial cell factories represent an innovative production approach that employs synthetic biology techniques to integrate key enzyme genes of the target biosynthetic pathway into heterologous hosts, such as *Escherichia coli* and *Saccharomyces cerevisiae*. It allows for the direct and efficient production of valuable natural or non-natural chemicals through microbial fermentation, such as amino acids, organic acids, fatty acid derivatives, polyketides, terpenoid compounds, and so on [7]. This production approach has advantages such as a broad spectrum of raw materials, environmentally friendly processing, stereospecific products with high yield potential, and stable year-round production to meet market demands, irrespective of seasons or regions. *Phaffia rhodozyma* naturally produces astaxanthin. Through chemical and ultraviolet radiation, the resulting strain *P. rhodozyma* CBS 6938 achieved an astaxanthin yield 50 times higher than the wild type, thereby reaching commercial levels of production [8]. Meadows et al. redesigned the central carbon metabolism to improve the carbon efficiency and cofactor balance in an engineered farnesene-producing *S. cerevisiae*. The titer of farnesene reached 130 g/L, achieving industrial production [9]. Therefore, the production of edible terpenes using microbial cell factories has become a hot topic in current research [10].

The biosynthetic pathway of terpenoids is a complex and sophisticated process, including precursor synthesis, skeleton synthesis, and subsequent modification steps [11] (Figure 2). There are two isoprenoid precursor biosynthetic pathways in nature: the mevalonate pathway (MVA pathway), which utilizes acetyl-CoA in the cytoplasm as the starting substrate, and the methylerythritol 4-phosphate pathway (MEP pathway), which uses 3-phosphoglycerate and pyruvate in plastids as the starting substrates (Figure 2a) [12,13]. Through these two pathways, isopentenyl pyrophosphate (IPP) and its isomer dimethylallyl pyrophosphate (DMAPP), which are the basic building blocks of terpenoids, can be synthesized. They condensed in a head-to-tail manner to form monoterpenes (C_{10}), sesquiterpenes (C_{15}), diterpenes (C_{20}), sesterterpenes (C_{20}), triterpenes (C_{30}), and tetraterpenes (C_{40}) (Figure 2b). Subsequently, the bioactive terpenoid compounds can be produced by a series of modification processes, including cyclization, redox reactions, methylation, acylation, and glycosylation (Figure 2c).

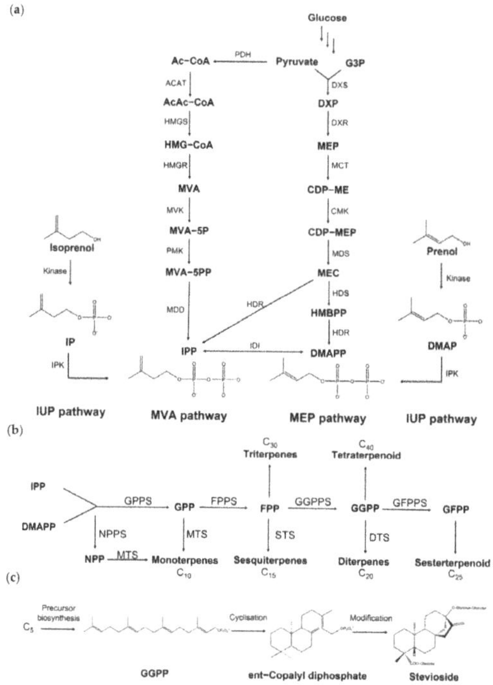

Figure 2. Biosynthetic pathway of terpenoids. (**a**) Natural and non-natural upstream pathways for the C_5 unit precursors isopentenyl diphosphate (IPP) and dimethylallyl diphosphate (DMAPP). Mevalonate pathway (MVA pathway) in yeast and animal cells utilizes acetyl-CoA (Ac-CoA) as the starting substrate, while methylerythritol 4-phosphate pathway (MEP pathway) in gram-negative bacteria and eukaryotic organelles uses 3-phosphoglycerate (G3P) and pyruvate as the starting substrates. In the non-natural isopentenol utilization pathway (IUP), IPP and DMAPP can be synthesized by sequential phosphorylation steps using isopentenol isomers isoprenol and prenol as precursors. (**b**) Carbon-chain elongation pathway. IPP and DMAPP are condensed to form C_{10}, C_{15}, C_{20}, and other longer terpenoid precursors for monoterpenes, sesquiterpenes, diterpenes, sesterterpenes, triterpenes, and tetraterpenes. (**c**) The biosynthesis of stevioside is presented as an example to demonstrate terpenoid synthesis. C_5 building blocks condensed to form GGPP as the precursor of diterpenes, then terpene synthases or cyclases helped to form the terpene skeleton, followed by the modification step, such as oxidation and glycosylation.

Designing and constructing cell factories for high-value terpenoids by utilizing metabolic engineering and synthetic biology strategy, combined with fermentation process control, is gradually demonstrating its unique advantages and potential. Microorganisms serve as key executors in biosynthetic reactions, and their growth physiological characteristics directly influence production efficiency and commercial potential. In this review, we highlight several prokaryotic and eukaryotic microorganisms suitable as chassis for generating edible terpenoids, along with their applications. Subsequently, we present a comprehensive review of the most recent advancements in cell factories for terpenoids utilized in food flavor, preservation, nutritional enhancement, and healthcare. Finally, we provide insights into the prospective directions and challenges associated with the commercial production of edible terpenoids based on microbial cell factories.

2. Microbial Chassis as Cell Factories for Edible Terpenoids

Traditionally, terpenoids are extracted from various sources, including plants, animals, microorganisms, and marine organisms [14,15]. However, due to the complex mixture of compounds present in cells, these extracts often contain a diverse array of components. When these extracts or essential oils are used directly as food additives, they may introduce non-target components, potentially leading to unexpected flavors or functions. For instance, steviol glycosides (SGs) extracted from *Stevia rebaudiana* are rich in stevioside, which gives SGs a bitter taste. Furthermore, the yield of active substances obtained from natural materials is often low, failing to meet the high demand. Therefore, chemists continually strive to develop chemical synthesis routes based on terpenoid structures. However, chemical synthesis often fails to fully replicate the intricate configuration of the active ingredients. For example, β-carotene synthesized by chemical methods exhibits an all-trans configuration, which differs from its natural counterpart, and high-dose intake indicates carcinogenic activity [16].

Microorganisms, noted for their rapid reproduction rate, short growth cycle, easily controlled cultivation conditions, mature fermentation processes, and independence from geographical climate influences, have been employed in the synthesis of a variety of compounds. In fact, some microorganisms in nature are capable of producing nutritious and healthy microbial foods via fermentation. However, the variety of terpenoids in wild strains is restricted and insufficient to satisfy diverse human needs. Advancement of omics research has led to the identification of terpenoid pathways and key synthesis enzymes, including cyclases, from various sources and functions. This has provided convenient conditions for the heterologous synthesis of terpenoids [17], thus bringing the microbial synthesis of natural terpenoids into public awareness. Over recent decades, scientists have successfully integrated terpenoid synthesis pathways into both typical and non-typical microbial chassis, achieving the biosynthesis of an array of terpenoids through synthetic biology strategies [18,19].

In 2003, Keasling and colleagues introduced the yeast MVA pathway and amorpha-4,11-diene synthase into *E. coli*, leading to the synthesis of sesquiterpene olefin precursor to artemisinin, namely amorphadiene [20]. A decade later, the cytochrome b_5 cDNA sequence and dehydrogenase gene were introduced into yeast, resulting in a high-yield engineered strain for producing artemisinic acid, a precursor of artemisinin. This patented technology was subsequently employed by Sanofi to produce tens of tons of artemisinic acid annually. When combined with chemical methods, this achievement becomes a significant milestone in the successful application of synthetic biology [21]. In terms of isoprene, Martos assessed various factors, including materials, energy efficiency, economic feasibility, and land use, and revealed that the *E. coli* cell factory is economically viable compared to conventional petrochemical alternatives [22].

2.1. Prokaryotic Microorganisms

Prokaryotic microbial cells, characterized by their simple cellular structures, rapid reproduction, robustness, ease of genetic manipulation, and metabolic regulation, are frequently used in the construction of cell factories. Most prokaryotic microorganisms, including gram-negative bacteria and cyanobacteria, possess an endogenous MEP pathway capable of producing IPP and DMAPP [13]. A stoichiometric and redox balance analysis reveals that the MEP pathway exhibits high carbon efficiency. For each molecule of IPP produced, the pathway consumed 1.196 molecules of glucose, representing a carbon consumption 14% higher than the MVA pathway [23]. Consequently, prokaryotic microorganisms are extensively employed for the synthesis of terpenoid compounds (Table 1).

Table 1. Production of terpenoids in prokaryotic cell factories.

Organism.	Chemicals	Strategy	Yield	Reference
Escherichia coli	Lycopene	Multiplex automated genome engineering (MAGE) of twenty endogenous genes to increase the flux of DXP pathway.	9 mg/g DCW	[24]
	Taxadiene	Multivariate-modular approach to metabolic-pathway engineering.	1.25 g/L	[23]
	Lycopene	Reconstitution of MVA Pathway.	40 mg/L	
	Geraniol	(1) Screening of different geranyl diphosphate synthetase (GGPS) and geraniol synthetase (GES); (2) Screening of *E. coli* hosts; (3) Using truncated leader peptides as fusion tags to improve geraniol production.	2.124 g/L	[25]
	Geraniol	(1) Introducing an alcohol acyltransferase; (2) Using another *E. coli* expressing acetyl esterase, which hydrolyzes geranyl acetate to geraniol.	13.19 g/L	[26]
	Myrcene	(1) Using a linalool dehydratase isomerase (LDI); (2) Aerobic-anaerobic two-stage fermentation.	1.25 g/L	[27]
	Germacrene A, β-Elemene	(1) Screening for germacrene A synthase; (2) RBS engineering of cyanobacteria-derived terpene synthase (NS); (3) Fusion expression of IspA and NS; (4) Rewriting the native central carbon metabolism of acetyl-coA and pyruvate; (5) High-throughput screening of NS using lycopene color as an indicator; (6) Overexpressing two membrane transporters, TolC and MsbA, to pump compounds extracellularly.	β-elemene, 3.52 g/L germacrene A, 2.13 g/L	[28]
	α-Ionone	(1) Fusing the carotenoid cleavage dioxygenase 1 (CCD1) with thioredoxin (TrxA); (2) Applying directed evolution of CCD1; (3) Over-expressing the alkyl hydroperoxide reductase (*ahp*C/F).	700 mg/L	[29]
	Dihydro-β-ionone	(1) Identifying an enoate reductase (DBR1) gene that could transform β-ionone to dihydro-β-ionone; (2) Constructed a cell-free synthesis system containing CCD1, DBR1, and glucose dehydrogenase (GDH).	13.34 mg/L	[30]
	Dihydro-β-ionone	(1) Integrating the de novo β-ionone biosynthesis pathway into *E. coli*; (2) Constructing the DBR1 module in *S. cerevisiae*; (3) Coculturing *E. coli* and *S. cerevisiae*.	27 mg/L	[31]
Corynebacterium glutamicum	Sesquarterpene	(1) Cloning of the novel carotenoid gene cluster to *C. glutamicum*; (2) HPLC analysis of the extracted pigment.	-	[32]
	α-Carotene	(1) Constructing a basal lycopene-producing strain; (2) Identifying a bi-functional OluLCY; (3) Integrating a heterologous MVA pathway; (4) Utilizing a multi-copy integration method to integrate OluLCY.	1.054 g/L	[33]
	Astaxanthin Astaxanthin-β-d-diglucoside	(1) Precursor pathway engineering; (2) Optimization of the fusion enzyme CrtZ~W; (3) Balancing of terminal CrtZ and CrtW expression; (4) Additional expression of a glycosyltransferase, CrtX.	103 mg/L 27 mg/L	[34]
	Coenzyme Q_{10}	(1) Enhancing the supply of precursor FPP and its conversion to DPP; (2) Enhancing the supply of aromatic precursor pHBA by knocking out the side-product synthesis pathway and replacing the key synthetic enzyme with a mutant that eliminates feedback inhibition; (3) Assembling the precursor-enhancing pathway, the 10-HB condensation pathway, and subsequent modification pathways in one host.	-	[35]
Bacillus subtilis	Menaquinone-7 (Vitamin K2-7), β-Carotene	(1) Designing pyruvate and malonyl-CoA-responsive genetic circuits; (2) Remodeling the MEP and MVA metabolism by using the two-input-multi-output (TIMO) circuit-assisted framework; (3) Introducing AND gate to regulate heptamyl diphosphate synthase to biosynthesize MK-7 and β-carotene.	467.2 mg/L 1.9-fold increase	[36]
	Amorphadiene	(1) Fusing GFP at the N-terminus of amorphadiene synthetase (ADS) to increase the efficiency of translation; (2) Overexpressing all MEP pathway genes together with *ispA* gene; (3) Optimizing the medium.	416 mg/mL	[37]
Synechococcus elongatus PCC 7942	1,8-Cineole	Expressing codon-optimized cineole synthase (CnsA).	105.6 μg/g WCW	[38]
Synechococcus sp. PCC 7002	Pinene	Overexpression of *Abies grandis* PS (*Ag*PS).	1.525 mg/L	[39]
Synechocystis sp. PCC 6803	Valencene	(1) Knocking out squalene synthase and squalene hopane cyclase coding genes for the FPP branch pathway; (2) Knocking down the key enzyme CrtE in the GGPP derivative pathway; (3) Fusing IspA and valencene synthase (*Cn*VS) for co-expression.	19 mg/g DCW	[40]

2.1.1. *E. coli*

E. coli, possessing a well-defined genetic background and comprehensive genetic manipulation techniques, serves as a model prokaryotic microorganism. Presently, diverse strategies have been employed to enhance the titer of terpenoids in *E. coli* [41].

In 2009, Church et al. optimized and combined the RBS of 20 endogenous genes in *E. coli* to enhance the flux of the DXP pathway and achieved a 5-fold increase in lycopene production [24]. The multivariate-modular approach to metabolic-pathway engineering strategy was used to optimize the endogenous MEP pathway for IPP synthesis and the heterologous pathway for taxadiene synthesis, achieving a 15,000-fold increase in taxadiene production [23]. Nevertheless, dependence solely on the endogenous MEP pathway remains suboptimal. A study reported that by reconstituting five genes in the MVA pathway and the *idi* gene in *E. coli*, a three-fold increase in lycopene titer was realized [42]. In recent years, several naturally occurring plant-derived terpenoids have also been biosynthesized in *E. coli*.

In Li's lab, they screened for an active geranyl diphosphate synthetase (GGPS) and fused a truncated leader peptide to promote the soluble expression of geraniol synthetase (GES). The geraniol titer of engineered *E. coli* increased to 2.12 g/L [25]. In order to alleviate the toxicity of the end product, they introduced an alcohol acyltransferase to esterify geraniol to geranyl acetate, and then they used another *E. coli* expressing acetyl esterase, which hydrolyzes geranyl acetate to geraniol. By fed-batch fermentation in a 10 L bioreactor, the titer of geraniol was reached to 13.19 g/L [26]. Further, they recruited a truncated linalool dehydratase isomerase (LDI) to isomerize geraniol to linalool and achieve de novo biosynthesis of myrcene at 1.25 g/L by using aerobic-anaerobic two-stage fermentation [27]. In order to efficiently produce plant-derived sesquiterpenes β-elemene in *E. coli*, Bai and his colleagues systematically designed the metabolic pathway. They screened for a germacrene A synthase (NS), which could transform farnesyl pyrophosphate (FPP) to β-elemene via germacrene A, engineered the RBS of NS, and linked NS with farnesyl diphosphate synthase (IspA). After rewriting the native central carbon metabolism of acetyl-CoA and pyruvate and overexpressing two membrane transporters, TolC and MsbA, to pump compounds extracellularly, they achieved the highest reported titer of 3.52 g/L of β-elemene and 2.13 g/L of germacrene A [28].

E. coli cell factories can not only elongate carbon chains but also cleave carbon chains. As we know, carotenoid cleavage dioxygenases (CCDs) could catalyze the cleavage and oxidation of α-carotene or ε-carotene to form ionone. Huang et al. used an evolved thioredoxin-CCD1 fusion protein in a lycopene-producing strain to synthesize α-ionone. However, they found that the by-product H_2O_2 of CCD1 will break down the precursor lycopene. So, they overexpressed the alkyl hydroperoxide reductase (*ahp*C/F) to reduce oxidative stress. As a result, they achieved 700 mg/L α-ionone in a 5 L bioreactor [29]. For dihydro-β-ionone, Zhao and his colleagues identified an enoate reductase (DBR1) gene from *Artemisia annua*, which can transform β-ionone to dihydro-β-ionone. Then, they constructed a cell-free system containing CCD1, DBR1, and glucose dehydrogenase (GDH) for dihydro-β-ionone production. The molar conversion was 85.8% [30]. Furthermore, they constructed a de nove β-ionone biosynthesis platform in *E. coli*. When co-cultured with *S. cerevisiae* harboring DBR1, the final yield of dihydro-β-ionone was achieved at 27 mg/L [31].

2.1.2. *Corynebacterium glutamicum*

C. glutamicum, well-known for rapid growth, broad substrate spectrum, and resistance to bacteriophage infections, and can be cultivated on a large scale, is generally regarded as a safe (GRAS) organism [43,44]. Therefore, it is often used in the food industry, particularly in the large-scale production of amino acids [45]. With the improvement of genetic manipulation techniques, its production spectrum has expanded, and it is often developed into a cellular factory for terpenoids.

C. glutamicum contains glycosylated C_{50} carotenoids decapreoxanthin, which is a yellow pigment. It indicated that *C. glutamicum* has an abundance of FPP and GGPP precursors. In order to extend the ability to synthesize other carotenoids, Choi et al. integrated their newly identified gene cluster *crtNaNcM* from *Planococcus martimus* into the *C. glutamicum* genome. Various novel C_{35} carotenoids were detected in the cells. It indicates that *C. glutamicum* can serve as a promising carotenoid cell factory [32]. Li et al. knocked out the key genes involved in the native C_{50} carotenoid synthesis, ranging from *crtE* to *crtYe*, and strengthened the expression of CrtE/B/I to increase the lycopene level. Additionally, they discovered a bifunctional enzyme, OluLCY, exhibiting the highest ε-cyclase activity and the lowest β-cyclase activity, and performed multi-copy integration of *oluLCY* to the host. By further strengthening the MVA pathway to increase the supply of IPP, they obtained a strain with an α-carotene titer of 1.054 g/L. It was the first report of α-carotene synthesis in *C. glutamicum* [33]. Henke and co-workers assembled lycopene β-cyclase, β-carotene hydroxylase, and β-carotene ketolase in a lycopene-producing *C. glutamicum* platform to produce astaxanthin via β-carotene. By optimizing and balancing the expression of CrtZ and CrtW, an astaxanthin titer of 103 mg/L was achieved in a 2 L fed-batch bioprocess. Furthermore, the pathway was expanded towards astaxanthin-β-d-diglucoside biosynthesis by additional expression of a glycosyltransferase CrtX [34].

C. glutamicum is a non-ubiquinone-containing bacterium. To produce Coenzyme Q_{10}, Burgardt et al. increased the supply of precursor FPP and decaprenyl diphosphate (DPP) by expressing FPP synthase (IspA) and DPP synthase (DdsA). Then, they knocked out the side-product synthesis pathway and replaced the key synthetic enzyme with a mutant that eliminates feedback inhibition, thereby increasing the flux into the shikimic acid pathway to synthesize pHBA. After they enhanced the precursor supply pathway, the 10 HB condensation pathway, and subsequent modification pathways into *C. glutamicum*, the resultant strain UBI413 was shown to synthesize CoQ_{10} [35].

2.1.3. *Bacillus subtilis*

B. subtilis, a model strain of gram-positive bacteria, possesses clear genetic information. It can be easily manipulated genetically and shows a strong ability to utilize inexpensive substrates. Since it has a fast growth rate and robust expression and secretion capabilities, it is extensively used in the production of chemicals and protein products. In addition, it has an endogenous MEP synthesis pathway and is considered GRAS [46]. Thus, it is a potential chassis for edible terpenoids.

The introduction of heterologous MVA pathways in bacteria presents an effective strategy for enhancing IPP supply. However, the static regulatory strategies may potentially impact the titer and yield of the cell factories because of the accumulation of toxic intermediates and metabolic imbalances. To address this, researchers have devised a dynamic regulation strategy aimed at balancing the metabolic flux between the MEP and MVA pathways in *B. subtilis*. In this research, a series of genetic circuits activated by pyruvate and inhibited by malonyl-CoA were designed to respond to intracellular levels of pyruvate and malonyl-CoA, and an AND gate was used to balance the flux of the MEP and MVA pathways to reshape the IPP synthesis pathway. In the strain, BS17-P20 synthesizing menaquinone-7 (MK-7, Vitamin K2-7) with dynamically regulated central metabolic pathways, IPP synthesis pathways, and terpenoid synthesis pathways, the level of IPP was increased by 4-fold, and the MK-7 titer reached 467.2 mg/L. This dynamic cascade regulation system is also applicable to the biosynthesis of β-carotene; the titer of carotenoids and lycopene in the engineered bacteria BS17-P13 increased by 1.9 and 10 times, respectively [36]. In another case, all the genes in the MEP pathway were integrated on a plasmid to increase the flux of the MEP pathway. The precursor FPP supply was increased by enhancing the expression of IspA. Then, GFP was fused to the N-terminus of the amorphadiene synthase (ADS) to improve its translation efficiency. These metabolic strategies combined with medium optimization enabled *Bacillus subtilis* to synthesize amorphadiene with a yield of 416 mg/L [37].

2.1.4. *Cyanobacteria*

Cyanobacteria are a kind of photosynthetic autotrophic microorganism that can grow using sunlight and trace minerals without an additional carbon source. Due to their endogenous MEP pathway, cyanobacteria have been recognized as novel promising hosts as a large-scale platform for light-driven terpenoid biosynthesis [47,48].

There are three model cyanobacteria, *Synechococcus elongatus* PCC 7942, *Synechococcus* sp. PCC 7002, and *Synechocystis* sp. PCC 6803 [38]. Scientists compared the potential of two models, *S.* PCC 7002 and *S.* PCC 7942, in producing terpenes. They expressed the corresponding terpene synthase by the same strong promoter in the plasmid into the two hosts and measured their terpene yield. It was found that different chassis have different product spectrums. It is suggested that the choice of host is a crucial step when designing and constructing synthesizing terpene cell factories [49]. Using *S.* PCC 7942 as the host, it only requires the expression of codon-optimized cineole synthase (CnsA) to directly convert intracellular GPP into 1,8-cineole [38]. Similarly, when the gene of pinene synthase originated from *Abies grandis* (*Ag*PS) was integrated into the chromosome, the pinene titer of *S.* PCC 7002 was achieved at 1.525 mg/L under the condition of CO_2 aeration and continuous illumination [39]. In *S.* PCC 6803, Dietsch et al. knocked out squalene synthase and squalene hopene cyclase coding sequences for the FPP branch pathway to avoid squalene by-products. Furthermore, they knocked down the key gene *crtE* in the GGPP derivative pathway and introduced a heterologous FPP synthase (IspA) to increase the supply of precursor FPP. In order to further improve the conversion rate, they fused IspA and valencene synthase (*Cn*VS) for co-expression. The resulting strain produced 19 mg/g DCW of valencene [40]. Unlike other work producing terpenoids using traditional enzyme engineering and pathway engineering strategies, Rodrigues and co-workers focused on optimizing cultivation conditions in the isoprene production in cyanobacteria. They found that the yield of isoprene increased under violet light and elevated temperature. This provides a new strategy for other cyanobacteria-based cell factories to produce terpenoids [50].

2.2. *Eukaryotic Microorganisms*

In eukaryotic microorganisms, the MVA pathway is used to synthesize terpenoid precursors IPP and DMAPP. Compared with prokaryotic microorganisms, the structure of eukaryotic chassis is more complex, providing a necessary place for compartmentalized metabolic regulation and facilitating the expression of transmembrane proteins. The post-translational modification capabilities in eukaryotic cells are beneficial for the correct expression and function of enzymes derived from plants and other eukaryotic cells. In addition, eukaryotic microorganisms have strong adaptability and tolerance to environmental conditions, making them more stable and efficient in specific industrial environments (Table 2).

Table 2. Production of terpenoids in eukaryotic cell factories.

Organism	Chemicals	Strategy	Yield	Reference
Saccharomyces cerevisiae	β-Myrcene	(1) Screening for heterologous β-myrcene synthase (MS) and truncated expression of MS; (2) Rationally designing and obtaining a high-activity mutant tQiMS^{D534N}; (3) Reverse fusing of t*Qi*MS^{D534N} and ERG20^{WW}; (4) Deletion of the YPL062W gene and replacing yeast endogenous ERG20 with ERG20^{W}; (5) A fed-batch fermentation in a 5 L bioreactor.	142.64 mg/L	[51]
	(-)-Borneol	(1) Identifying a high-efficient (-)-bornyl diphosphate synthase (*Bb*TPS3); (2) Codon-optimizing and N-terminus truncated expressing of *Bb*TPS3; (3) A fed-batch fermentation in a 5 L Bioreactor.	148.59 mg/L	[52]
	Retinol	(1) Integration of the β-carotene dioxygenase (BLH) gene into a β-carotene-producing strain; (2) Increasing the supply of precursor β-carotene and cofactor NADPH; (3) Introducing two reductases, Env9 and YbbO, in order to reduce retinal to retinol; (4) Adding butylated hydroxytoluene (BHT) to inhibit retinol oxidation.	2.479 g/L	[53]
	Carnosic acid	(1) Fuse-expressing *Sm*CPS-*Sm*KSL and BTS1-ERG20^{F96C}; (2) Expressing the related modifying enzymes CYP76AK6, CYP76H24, and CYP76AH1; (3) Fused-expressing cytochrome P450 reductase (CPR) and NADPH-cytochrome B5 (Cytb5); (4) Enhancing the supply of NADPH and heme; (5) Activating the transcription of proteins folding-related genes in the ER.	75.18 mg/L	[54]
Yarrowia lipolytica	Lycopene	(1) Introducing IUP pathway by expressing CHK and IPK; (2) Over-expressing ACC1 and DGA1 to promote lipid accumulation; (3) Feeding exogenous palmitic acid to accumulate acetyl-CoA from β-oxidation.	4.2 g/L	[55]
	Geraniol	(1) Expressing the truncated plant-derived geraniol synthase (t*Cr*GES); (2) Integrating ERG10, HMGS, tHMG1, and IDI into the chromosome; (3) Processing parameters optimization.	1 g/L	[56]
	Sclareol	(1) Overexpressing tHMG1 and ERG20; (2) Co-expressing the key enzymes t*Ss*LPPS and *Ss*SCS of sclareol synthesis pathway; (3) Introducing *Pa*Ggpps and *Ss*Ggpps to simultaneously utilize IPP, DMAPP, and FPP to increase GGPP supply; (4) Assembling t*Ss*LPPS and *Ss*SCS into multi-enzyme complexes using protein scaffold peptide tags RIAD and RIDD to facilitate substrate trafficking.	12.9 g/L	[57]
	Squalene	(1) Knocking out Ku70 and co-expressing RAD52 and RAD59 to facilitate homologous recombination; (2) Enhancing the MVA pathway flux; (3) Increasing the supply of acetyl-CoA and NADPH; (4) Weakening the pathway for squalene degradation to ergosterol.	35 g/L	[58]
	β-Ionone	(1) Promoter engineering and rDNA-mediated multi-round iterative transformations of carotenoid cleavage dioxygenase (*Of*CCD1); (2) Co-expressing the key enzymes for xylose metabolism, xylose reductase (*Ss*XR), xylitol dehydrogenase (*Ss*XDH), and endogenous xylulose kinase (XK).	-	[59]
Ogataea polymorpha	β-Farnesene	(1) Screening for β-farnesene synthase (*Aa*FS) and fusion-expressing ERG20-*Aa*FS; (2) Restoring the fatty acyl-CoA synthetase (FAA1) encoding gene in M16 and generating M16F, which blocks the accumulation of FFA; (3) Enhancing the flux of the MVA pathway and the supply of precursor acetyl-CoA; (4) Fed-batch fermentation using methanol as a carbon source.	14.7 g/L	[60]
	β-Elemene	(1) Evaluating different germacrene A synthase in *O. polymorpha;* (2) Modulating MVA pathway; (3) Enhancing the pentose phosphate pathway (PPP) to increase the supply of cofactor NADPH; (4) Fusion expressing phosphoketolase (PK) and phosphotransacetylase (PTA) to synthesize more precursor acetyl-CoA; (5) Downregulating the competitive pathway by fusing a degradation peptide CLN2^{PEST} to ERG9; (6) Fed-batch fermentation.	4.7 g/L	[61]
Pichia pastoris	Germacrene A	(1) Fusion-expressing protein ERG20-(PT)4P-GAS; (2) Overexpressing MVA biosynthesis pathway genes (i.e., IDI1, tHMG1, and ACS); (3) Introducing acetyl-CoA synthetase gene *ACS* to enhance the supply of acetyl-CoA; (4) Optimizing the nitrogen source in YPD.	1.9 g/L	[62]
	Santalene	(1) Integrating α-santalene synthase expression cassette TEF1p-SAS-AOX1t into the genome of STE-0; (2) Overexpressing MVA pathway key enzymes and acetyl-CoA synthase; (3) Increasing the copy number of *SAS;* (4) Optimizing the medium and fed-batch fermentation.	21.5 g/L	[63]
	Lycopene	(1) Overexpressing CrtE, CrtB, and CrtI to construct a lycopene synthetic pathway; (2) Overexpressing the endogenous HMGR and HMGS; (3) Overexpressing IDI1 from *E. coli* and increasing the copy of endogenous GGPPS gene; (4) Fed-batch fermentation.	0.714 g/L	[64]
	α-Farnesene	(1) Selecting the optimal α-farnesene synthase (*Md*AFS); (2) Strengthening the MVA biosynthetic pathway and acetyl-CoA supply pathway; (3) Introducing the IUP pathway into the peroxisome of *P. pastoris;* (4) Constructing dual-regulation strain; (5) Oleic acid and sorbitol co-feeding.	2.56 g/L	[65]
	Dammarenediol-II	Self-assembly of squalene epoxidase (ERG1) and dammarenediol synthase (*Pg*DDS) by using the PDZ domain and its corresponding ligand (PDZlig).	0.10 mg/g DCW	[66]

Table 2. *Cont.*

Organism	Chemicals	Strategy	Yield	Reference
Rhodosporidium toruloides	Astaxanthin	(1) Using gamma irradiation mutagenesis to obtain a strain with a high yield of astaxanthin; (2) Using response surface methodology to optimize the large-scale culture conditions.	1.26 mg/L	[67]
	Limonene	(1) Using carotenoid-deficient *R. toruloides* as the host to block competing pathways; (2) Increasing the flux of MVA pathway; (3) Introducing an orthogonal limonene synthesis pathway based on neryl pyrophosphate (NPP) by fusion expressing neryl pyrophosphate synthase and limonene synthase *Sl*tNPPS1-*Cl*tLS1.	393.5 mg/L	[68]
	Limonene	(1) Verifying the peroxisomal targeting signals (PTSs) via fluorescence microscopy analysis; (2) Targeting MVA and limonene biosynthetic pathways to peroxisomes; (3) Dual metabolic regulation of the synthetic pathway in the cytoplasm and peroxisome.	1.05 g/L	[69]
	ent-Kaurene	(1) Selected ent-kaurene synthase from *Gibberella fujikuroi* (*Gf*KS); (2) Balancing the expression level of GGPP synthase (GGPPS) and KS; (3) Scaling up the cultivation to a 2 L bioreactor in a medium containing 75% corn stover hydrolysate.	1.4 g/L	[70]
	Amorphadiene Bisabolene	(1) Expressing the codon-optimized bisabolene synthase (BIS) and amorphadiene synthase (ADS); (2) Testing different carbon sources.	36 mg/L, 680 mg/L	[71]
	α-Bisabolene 1,8-Cineole	(1) Integrating a second BIS expression cassette; (2) Transcriptomics and proteomics analysis in the α-bisabolene producing strains; (3) Balancing the expression levels of the two terminal enzymes: GPPS and 1,8-cineole synthase to synthesize 1,8-cineole; (4) Regulating MVA pathway; (5) Optimizing the medium and fermentation process.	2.6 g/L 1.4 g/L	[72]

2.2.1. *S. cerevisiae*

S. cerevisiae is a GRAS species with a clear genetic background and metabolic pathway, efficient tools for gene manipulation, and metabolic regulation, making it the most popular chassis in microbial cell factories. It has been used for the production of monoterpenes, diterpenes, triterpenes, and even more complex terpenoids, such as menthol, sclareol, squalene, and so on [73–75].

Shu et al. identified a β-myrcene synthase (MS) t*Qi*MS from six plant-derived sources that can induce *S. cerevisiae* to produce myrcene. Combined with rational design, reverse fusion with ERG20^{WW}, and peroxisome localization, a β-myrcene titer of 142.64 mg/L was obtained in a 5 L bioreactor [51]. Ma et al. discovered a (-)-bornyl diphosphate synthase *Bb*TPS3 based on transcriptome. With further codon optimization and fusion expression with ERG20$^{F96W\text{-}N127W}$, a (-)-borneol titer of 148.59 mg/L in a 5 L bioreactor was achieved [52]. In the case of retinol, Hu et al. integrated the β-carotene dioxygenase (BLH) gene into the previously constructed β-carotene engineering strain to constitute a complete vitamin A synthesis pathway. They further increased the supply of precursor β-carotene and cofactor NADPH through metabolic engineering. Additionally, two reductases, Env9 and YbbO, were introduced into the strain to further enhance retinol production. To reduce the oxidation of the produced retinol, butylated hydroxytoluene (BHT) antioxidant was added during high-density fermentation, ultimately yielding 2479.34 mg/L of retinol [53]. For the more complex phenolic tricyclic diterpene carnosic acid (CA), scientists not only overexpressed *Sm*CPS and *Sm*KSL to provide the terpene skeleton miltiradiene but also expressed the related modifying enzymes CYP76AK6, CYP76H24, and CYP76AH1 to synthesize carnosic acid. In order to promote electron transportation and release oxidative stress, they expressed cytochrome P450 reductase and NADPH-cytochrome B5 fusion protein (*Sm*CPR~t28*Sp*Cytb5) and *Sc*CTT1 catalase. Finally, by enhancing the supply of NADPH and heme and activating the transcription of proteins folding-related genes in the endoplasmic reticulum, they obtained the best strain, WCA11, with a CA titer of 75.18 mg/L [54].

The natural isoprenoid biosynthesis pathway is often limited by low carbon and energy efficiency. Researchers have designed a novel pathway, namely the isopentenol utilization pathway (IUP), in *E. coli*, which can generate DMAPP and IPP through sequential phosphorylation of isoprenol or prenol (Figure 2a) [76]. It was reported that the advantage of the IUP is its minimal cofactor requirement, where only two molecules of ATP are needed to synthesize one molecule of IPP. However, it was found that isopentenol in the IUP pathway inhibits the respiration of *S. cerevisiae*. In order to overcome this problem, Li et al. designed an IU pathway-dependent (IUPD) strain with growth-coupled production. By utilizing the auxotroph phenotype towards isoprenol, they developed a growth-coupled high-throughput screening method and identified two kinase mutants with higher activity, *Sm*DAGK$^{S47A/L124A}$ and *At*IPK$^{S270P/A272R}$. The resultant strains with strengthened IUP pathways had higher fluxes of IPP/DMAPP, producing more squalene, limonene, and carotenoids [77].

2.2.2. *Yarrowia lipolytica*

In order to avoid the overflow of carbon metabolism and generation of by-product ethanol, scientists have focused their attention on Crabtree-negative species. *Y. lipolytica*, as a model strain of oleaginous yeasts, is Crabtree-negative and GRAS [78]. It possesses high levels of the tricarboxylic acid (TCA) cycle and the pentose phosphate (PP) pathway. Through these two pathways, a large amount of acetyl-CoA, ATP, and NADPH can be accumulated, making it a promising and superior chassis for terpenoid compounds [79].

The level of precursors IPP and DMAPP determines the yield of subsequent metabolites. To increase the precursor level, key enzymes CHK and IPK in the IUP pathway were introduced into *Y. lipolytica*, resulting in a 15.7-fold increase in IPP and DMAPP. By overexpressing ACC1 and DGA1 to promote lipid accumulation, the intracellular hydrophobicity was modulated, thereby increasing the storage capacity for lycopene. With exogenous palmitic acid supplementation, the degradation blocks acetyl-CoA from β-oxidation further enhanced the natural MVA pathway. The final lycopene titer was 4.2 g/L in 3 L batch bioreactors [55]. This indicates that the hydrophobic environment of oleaginous yeast is beneficial for the accumulation of terpenoids. In geraniol production, Agrawal et al. expressed the truncated plant-derived geraniol synthase (t*Cr*GES) in *Y. lipolytica* and then integrated ERG10, HMGS, tHMG1, and IDI into the chromosome to enhance precursor GPP supply. The resulting strain S5 produced geraniol with a titer of 1 g/L shake-flask fermentation [56]. Sun et al. co-expressed the key enzymes t*Ss*LPPs and *Ss*SCS of sclareol synthesis pathway in *Y. lipolytica* with tHMG1 and ERG20 overexpression, enabling the host to synthesize sclareol. They further designed a two-layer GGPP accumulation pathway, introducing *Pa*GGPPS and *Ss*GGPPS to simultaneously utilize IPP, DMAPP, and FPP to increase GGPP supply. Then, by using protein scaffold peptide tags RIAD and RIDD, they assembled t*Ss*LPPS and *Ss*SCS into multi-enzyme complexes to facilitate substrate transport and reduce the production of by-products geranylgeraniol (GGOH) and labdendiol (LOH). The engineered strain produced sclareol as much as 12.9 g/L [57]. To facilitate genetic modification in *Y. lipolytica*, Xu and co-workers knocked out the non-homologous end joining protein Ku70 and co-expressed the key homologous recombination proteins RAD52 and RAD59, achieving a homologous recombination efficiency of 68% without affecting the growth. They also systematically reshaped metabolism by enhancing MVA pathway flux, increasing the supply of acetyl-CoA and NADPH, and weakening the pathway for squalene degradation to ergosterol. In a 5 L fed-batch fermentation, the engineered strain achieved a squalene titer of 35 g/L [58]. In another case, Shi et al. transformed a carotenoid cleavage dioxygenase (*Of*CCD1) with a strong multi-round promoter iteratively into *Y. lipolytica* to construct a β-ionone-producing strain. Then, they further broaden the substrate range by overexpressing the key enzymes for xylose metabolism, xylose reductase (*Ss*XR), xylitol dehydrogenase (*Ss*XDH), and endogenous xylulose kinase (XK). The final engineered strain could synthesize β-ionone using lignocellulosic hydrolysate as the carbon source [59]. It indicates that *Y. lipolytica* has the potential to utilize low-cost raw materials for the green production of high-value terpenoids.

2.2.3. Methylotrophic Yeasts

Methylotrophic microorganisms can utilize one-carbon compounds as a carbon source for growth. *Ogataea polymorpha* and *Pichia pastoris*, which are Crabtree-negative, have been receiving wide attention as hosts due to their ability to use methanol as a carbon and energy source. They possess high protein expression and secretion capabilities and allow for high-density fermentation [80]. The properties make them beneficial for biomanufacturing processes and widely used for efficient protein production. Since they are GRAS species, enzymes produced by these two yeasts, such as lipases, have been used as enzyme preparations in the food industry [81] already. Their robust five-carbon assimilation capacity makes them potential hosts to use lignocellulosic hydrolysates as raw materials. In addition, their endogenous MVA pathway can provide abundant precursors IPP and DMAPP for terpenoids, indicating that *O. polymorpha* and *P. pastoris* hold significant potential as cell factories for terpenoids [80].

Li et al. screened a farnesene synthase (*Aa*FS) with high activity and introduced it into the fatty acid-producing *O. polymorpha* strain M16 in their previous research. By further increasing the flux of the MVA pathway and the supply of precursor acetyl-CoA, the optimized engineering strain produced β-farnesene with a yield of 46 mg/g methanol, which is the highest reported production from one carbon [69]. Ye and co-authors identified an efficient germacrene A synthase, *Ls*LTC2, in *O. polymorpha*. Furthermore, they enhance the acetyl-CoA, MVA, and NADPH flux and downregulate the sterol competition flux. Through spontaneous Cope rearrangement from germacrene A. They obtained β-elemene at a titer of 4.7 g/L in the fed-batch fermentation [61].

In another methylotrophic yeast, *P. pastoris*, researchers have successfully constructed cell factories capable of producing terpenoid compounds such as germacrene A, santalene, and lycopene by introducing corresponding terpene synthases and performing conventional metabolic engineering [62–64]. In the case of producing α-farnesene, Liu et al. adopted the heterologous IUP pathway and a peroxisome-cytoplasm dual regulation strategy to enhance the α-farnesene synthetic flux. Firstly, they screened and utilized a high-activity farnesene synthase (*Md*AFS) and strengthened the MVA and acetyl-CoA pathway. Secondly, they targeted the more efficient IUP pathway to peroxisomes to promote the accumulation of IPP and DMAPP. Finally, by simultaneously regulating the IPP and α-farnesene synthetic pathway in the cytoplasm and peroxisomes, they obtained 1.4 g/L of α-farnesene in the medium co-feeding with oleic acid and sorbitol as carbon sources [65]. Peroxisome compartmentalization not only alleviates the burden on growth caused by the accumulation of cytotoxic intermediates such as IPP but also separates IPP from competitive pathways in the cytoplasm and has now been used for the construction of various cell factories. In another case, *P. pastoris* has been used to synthesize the precursor of a more complex ginsenoside precursor, dammarenediol-II. To facilitate the transfer of substrates and alleviate the interference from competing ergosterol pathway, researchers expressed squalene epoxidase (ERG1) and dammarenediol synthase (*Pg*PDD) separately using PDZ domain and its corresponding ligand (PDZlig), respectively. Ultimately, the yield of dammarenediol-II in the self-assembling multi-enzyme complex strain was 2.1-fold higher than that of the free expression strain [66].

2.2.4. *Rhodosporidium toruloides*

R. toruloides is a famous lipid yeast that can accumulate a large amount of lipids under nutrient-limited conditions. Its lipid level exceeds 70% of the dry cell weight [82], and its red phenotype is related to the natural ability to synthesize carotenoids. The high levels of lipids and carotenoids indicate that *R. toruloides* has sufficient acetyl-CoA precursors, making it an ideal candidate for terpenoid cell factories. In addition, it also has advantages such as high-density growth, excellent stress resistance, and a wide substrate spectrum, allowing for large-scale fermentation using inexpensive raw materials [83]. Although it does not possess an efficient endogenous homologous recombination system like model microorganisms such as *S. cerevisiae*, lots of gene manipulation tools have been developed recently [84], enhancing the convenience and efficiency of homologous recombination significantly. Currently, different terpenoid compounds have been synthesized by *R. toruloides*. Apart from *R. toruloides*, other oleaginous red yeasts have also been modified to synthesize terpenoids successfully [85,86].

In order to directly utilize the excellent performance of *R. toruloides* in accumulating carotenoids, Tran et al. obtained a mutant strain, G17, with increased astaxanthin content by using gamma irradiation mutagenesis. They used response surface methodology to optimize the large-scale culture conditions, enabling *R. toruloides* to produce astaxanthin with a content reaching 1.26 mg/L using molasses as a carbon source [67]. In the case of limonene, Yang and co-workers introduced an orthogonal limonene synthesis pathway based on neryl pyrophosphate (NPP) into *R. toruloides*. They used carotenoid-deficient *R. toruloides* as the host to block competing pathways, enhanced the MVA pathway flux, and simultaneously expressed the key enzyme neryl pyrophosphate synthase and limonene synthase for the downstream limonene synthesis pathway. The engineered strain achieved a final limonene yield of 393.5 mg/L [68]. Furthermore, they identified several peroxisomal targeting signals (PTSs) and used them to target MVA and limonene biosynthetic pathways to peroxisomes. Then, they carried out dual-metabolic regulation in the cytoplasm and peroxisome to enhance the supply of acetyl-CoA. A higher yield of limonene was finally reached at 1.05 g/L [69]. Similarly, Geiselman et al. utilized metabolic engineering and Design-Build-Test-Learn (DBTL) strategies to introduce the ent-kaurene biosynthesis pathway and precursor GGPP supplying pathway into *R. toruloides*. The resultant strain obtained ent-kaurene at a titer of 1.4 g/L in a 2 L bioreactor feeding corn stover hydrolysates [70]. In order to further expand the application of *R. toruloides* in the production of terpenoid compounds using corn stover hydrolysate, Gladden and colleagues introduced bisabolene synthase (BIS) and amorphadiene synthase (ADS) into *R. toruloides* to produce bisabolol and amorphadiene, and also introduced GPPS and 1,8-cineole synthase to produce 1,8-cineole. Through metabolomic and proteomic analysis, they identified the rate-limiting steps and balanced the MVA pathway and downstream pathway. Finally, they obtained strains for bisabolene, amorphadiene, and 1,8-cineole with titers of 36 mg/L, 2.6 g/L, and 1.4 g/L, respectively [71,72].

3. Application of Terpenoid Microbial Cell Factories in the Food Industry

Terpenoids have complex and diverse structures, with various functional groups and active sites, thus exhibiting different biological activities, such as special fragrances, antioxidant, antibacterial, and anticancer effects. Moreover, due to the complexity of terpenoid structures, a single terpenoid compound can play more than one role in a food product. For instance, squalene acts as both an antioxidant and an antimicrobial agent in muffin preservation [87]. In the food industry, terpenoids are highly favored due to their natural source. However, extraction and purification from natural plants or animals is cumbersome and inefficient. Nowadays, an increasing number of microbial hosts have been engineered for the biosynthesis of terpenoids, and some terpenoids have achieved gram-scale production in microorganisms. Here, we provide an overview of several representative terpenoid-producing microbial cell factories that have been used in the food industry (Table 3).

Table 3. Representative terpenoid cell factories used in food industry.

Functions	Chemicals	Host	Strategy	Yield	Reference
Flavor and fragrance	Linalool	*Pantoea ananatis*	(1) Identifying (S)-specific linalool synthase (LINS) and (R)-specific LINS; (2) Co-expressing them with mutated farnesyl diphosphate synthase (IspA*).	(S)-linalool, 5.60 g/L (R)-linalool, 3.71 g/L	[88]
	Linalool	*Saccharomyces cerevisiae*	(1) Semi-rationally designing the substrate binding pocket entrance of linalool synthase; (2) Screening for a mutant t67OMcLIS$_M$^{F447E} with higher affinity and catalytic activity; (3) Targeting the linalool biosynthetic pathway to *S. cerevisiae* peroxisomes; (4) Constructing a more stable diploid engineering strain through mating.	2.6 g/L	[89]
	Myrcene	*Escherichia coli*	(1) N-terminal truncating of linalool dehydratase isomerase (LDI); (2) Optimizing the induction temperature; (3) Site-directed mutation of LDI; (4) Aerobic-anaerobic two-stage fermentation.	1.25 g/L	[27]
	Geraniol	*E. coli*	(1) Utilizing nucleoside diphosphate-x (NUDIX) instead of geraniol synthase to produce geraniol; (2) Enhancing GPP supply by optimizing MVA pathway and GPPS; (3) Fusion-expressing the efficient GPPS and NUDIX for geraniol synthesis; (4) Knocking out alcohol dehydrogenases and adopting microaerobic cultivation to avoid the conversion of geraniol to geranial; (5) Introducing the alcohol acyltransferase *Rh*AAT1 to transform geraniol into the low-toxicity geranyl acetate.	19 g/L	[90]
Biopreservatives	Limonene	*S. cerevisiae*	(1) Integrating six copy limonene synthase (tLimS) genes into genome; (2) Enhancing MVA and GPP pathways by expressing tHMG1, IDI1, and ERG20$^{F96W\text{-}N127W}$; (3) Dynamically regulating the expression of ERG20 to restrain the competitive squalene pathway by using dCas system; (4) Increasing the supply of acetyl-CoA and NADPH by regulating the glyoxylate cycle and pentose phosphate pathway; (5) Compartmentalization engineering of MVA and limonene synthetic pathways; (6) Two-phase scale-up fermentation.	2.63 g/L	[91]
	Limonene	*Yarrowia lipolytica*	(1) Comparative transcriptomic analysis and identifying the changes in transcriptional levels of the genes in response to limonene; (2) Identifying an unknown functional protein YALI0F19492p conferring limonene tolerance; (3) Short-term adaptive laboratory evolution to improve the tolerance to and yield of limonene.	Near 8 mg/L	[92]
	Pinene	*E. coli*	(1) Identifying an effective pinene synthase *Pt*PS1^{Q457L}; (2) N-terminal truncation to improve its hydrophilicity and solubility; (3) Introducing a GPPS mutant *Ag*GPPS$^{D90G/L175P}$ to enhance the supply of precursor GPP; (4) Identifying a suitable host DH10B; (5) Adjusting fermentation conditions.	1035 mg/L	[93]
	Pinene	Cell-free system	(1) Constructing *E. coli* MEVI with an upstream MEV module and *E. coli* PINI with a downstream pinene module; (2) Constructing a cell-free system to produce pinene with MEVI and PINI; (3) Optimizing the physicochemical parameters using Plackett–Burman experimental design and the steepest ascent strategy; (4) Optimizing a single enzyme addition.	1256 mg/L	[94]
Food fortifiers and dietary supplements	Lycopene	*E. coli*	(1) Developing a high-throughput screening method dependent on the color of lycopene; (2) Obtaining a triple mutant IDI$^{L141H/Y195F/W256C}$ by random mutations and targeted mutagenesis; (3) Replacing the wild-type IDI with IDI$^{L141H/Y195F/W256C}$.	1.2 g/L	[95]
	Lycopene	*S. cerevisiae*	(1) Introducing phytoene dehydrogenase mutant (*Bt*CrtI$^{Y160F/N576S}$) to increase lycopene yield; (2) Atmospheric and room-temperature plasma (ARTP) mutagenesis combining with H_2O_2-induced adaptive laboratory evolution (ALE); (3) Identifying the rate-limiting steps through transcriptome analysis of high-yield mutant strain; (4) Adjusting the copy number of *TmCrtE* to drive GGPP synthesis; (5) Enhancing the supply of FAD and NADPH; (6) Recovering *Ura3* marker.	8.15 g/L	[96]

Table 3. *Cont.*

Functions	Chemicals	Host	Strategy	Yield	Reference
	Stevioside	*S. cerevisiae*	(1) Expressing tHMG1 and down-regulating ERG9 to strengthen the supply of precursor GGPP; (2) Selecting active diterpene synthases by using the steviol yield as an indicator; (3) Fusion-expressing two diterpene synthases; (4) Combining directed evolution and fusion expression of two glycosyltransferases to synthesize stevioside; (5) Enhancing the supply of UDP-glucose by expressing key enzymes.	1149 mg/L	[97]
	Rebaudioside M	*S. cerevisiae*	(1) Knocking out endogenous glycoside hydrolase SCW2 to inhibit the hydrolysis of stevioside; (2) Identifying an efficient glycosyltransferase, UGT91D2, and co-expressing it with UG76G1 to convert stevioside to Reb M via Red E; (3) Overexpressing UGP1 to strengthen the supply of sugar donor UDP-glucose; (4) Introducing SIR2 to prolong the growth cycle; (5) Whole-cell catalyzing the production of Reb M using stevioside as the substrate.	12.5 g/L	[98]
Medicinal health food	Germacrene A	*Y. lipolytica*	(1) Identifying the most efficient germacrene A synthase (*Dl*GAS); (2) Fused expression of *Dl*GAS–ERG20; (3) Strengthening the expression of the MVA pathway through strong promoters and multi-copy integration; (4) Restraining the downstream squalene synthesis pathway by truncating the native promoter of ERG9 (P_{ERG9}); (5) Increasing the acetyl-CoA supply by reducing the activity of ACC1 and promoting lipid degradation.	39 g/L	[99]
	β-Elemene	*Y. lipolytica*	(1) Identifying an efficient germacrene A synthase (LTC2) and integrating the sequence into genome; (2) Increasing the flux of MVA pathway by over-expressing the key enzymes; (3) Regulating the competition squalene synthetic pathway by substituting an inducible promoter that is feedback inhibited by ergosterol; (4) Adjusting the copy number of heterologous LTC2; (5) Fed-batch fermentation; (6) Centrifuging the broth and then heating the dodecane layer at 250 °C to obtain product β-elemene.	5.08 g/L	[100]
	Valencene	*S. cerevisiae*	(1) Performing adaptive laboratory evolution and obtaining a platform strain BN-91A, which can grow using mannitol as sole carbon source; (2) Introducing valencene pathway into BN-91A; (3) Decreasing the branch pathway by downregulating ERG9 (squalene synthase) and disrupting BTS1 (GGPPS); (4) Increasing the MVA pathway by over-expressing tHMG1 and ERG12 (mevalonate kinase); (5) Increasing copy number of valencene synthase (*Cn*VS); (6) Facilitating mannitol assimilation by over-expressing mannitol dehydrogenase (DSF1) and mannitol transporter (HXT13); (7) Enhancing NADPH supply by integrating NADH kinase (POS5) gene; (8) High cell-density fermentation.	5.6 g/L	[101]
	Valencene and nootkatone	*S. cerevisiae*	(1) Identifying another valencene synthase (*Eg*VS) with higher catalytic efficiency than *Cn*VS; (2) Semi-directed mutagenesis and obtaining an *Eg*VS mutant with the highest catalytic efficiency; (3) Optimizing the biosynthetic pathway by increasing the copy number of key enzymes and changing promoters; (4) Coupling cell growth and biochemical pathways by controlling URA3 expression with the GAL promoter; (5) High-cell-density fermentation and purifying valencene; (6) Chemically transforming valencene to nootkatone.	16.6 g/L a yield of 80%	[102]
	Patchoulol	*E. coli*	(1) Semi-rationally designing and obtaining a PTS mutant $PTS2^{C415F/H454A}$ with higher substrate-binding affinity; (2) Expressing FPPS-$(PT)_4$P-PTS2 fusion protein; (3) Deleting the competitive acetate synthetic pathway; (4) Enhancing efflux transporters MacAB and MsbA.	970.1 mg/L	[103]
	Patchoulol	*E. coli*	(1) Modifying and obtaining a PTS mutant PTS3mut4; (2) Deleting the cyclic AMP synthesis pathway to alleviate glucose repression; (3) Constructing a pyruvate biosensor and using it to regulate glycolysis pathway.	1675 mg/L	[104]
	Patchoulol	*S. cerevisiae*	(1) Expressing FPPS-PTS fusing protein and tHMG1; (2) Modifying off-pathway by deleting ROX1, YJL064W, and YPL062W; (3) Enhancing ergosterol pathway by over-expressing ERG11 and ERG24; (4) Regulating NADPH supply by over-expressing POS5; (5) Decreasing intracellular ROS by over-expressing cytoplasmic catalase (CTT1) and peroxisome catalase (CTA1); (6) Rational engineered PTS and obtained PTS^{T404S}, which could reduce production of the concomitant byproducts of patchoulol; (7) High-density fermentation with Mg^{2+} addition.	1632 mg/L	[105]

3.1. Acting as Flavor and Fragrance

Terpenoids, which are in forms of esters, alcohols, and oxides, are commonly found in fruits, vegetables, and nuts. Different terpenes have different fragrances. They are the main components of essential oils and are widely used as natural fragrances and flavors. Linalool, geraniol, nerolidol, and other monoterpenes and sesquiterpenes have been recognized as GRAS by the United States Food and Drug Administration (FDA) [106]. In the food industry, they have been used as food additives to provide food with a unique fragrance and taste [11]. For example, monoterpenes such as menthol and limonene are often used as flavor additives in foods like chewing gum, beverages, and candies. The boosting strategies of metabolic engineering and synthetic biology in cell factories have greatly promoted the sustainable production of terpenoid compounds [107,108].

3.1.1. Linalool

Linalool is an acyclic monoterpene and is widely used in beverages and processed foods due to its flavoring and fragrant properties.

It has two enantiomers, (S)-linalool and (R)-linalool. The two forms of linalool have different fragrance profiles and biological properties, while separation and purification are difficult when extracting from plants. Hoshino et al. identified (S)-specific linalool synthase (LINS) and (R)-specific LINS and co-expressed them with mutated farnesyl diphosphate synthase (IspA*) in *Pantoea ananatis* host. The titer of (S)-linalool and (R)-linalool reached 5.6 g/L and 3.71 g/L, respectively [91]. In order to improve the catalytic activity for linalool synthase, Zhou et al. semi-rationally designed its substrate binding pocket entrance and obtained a mutant t67OMcLIS$_M$^{F447E} with higher affinity for GPP and higher catalytic activity to synthesize linalool. Furthermore, they targeted the linalool biosynthetic pathway to *S. cerevisiae* peroxisomes and obtained a more stable diploid engineering strain through mating. The resultant strain achieved a titer of 2.6 g/L linalool in a 5 L fed-batch fermentation [89]. In another research, Wang and co-workers introduced linalool dehydratase, which can isomerize geraniol to linalool and subsequently dehydrate linalool to myrcene, into *E. coli*. The engineered strain produced myrcene at a titer of 1.25 g/L in aerobic and anaerobic two-stage fermentation [27].

3.1.2. Geraniol

Geraniol is an acyclic monoterpene alcohol with a rose-like odor and a sweet, floral, rose-like taste. It is an important raw material in the commercial fragrance industry. It is reported that various microorganisms can convert geraniol into high-value terpenoid products such as (R)-citronellol, (S)-citronellol, nerol, and neral [109]. Besides its unique scent, it also possesses biological activities like anti-microbial, anti-cancer, anti-oxidant, and anti-inflammatory. Therefore, geraniol is widely used in fields such as food, cosmetics, and pharmaceuticals, exhibiting great commercial value.

Nowadays, researchers have identified that geraniol synthase (GES) originated from various plants and assembled a geraniol de novo synthetic pathway in microorganisms, such as *E. coli* [26], *S. cerevisiae* [110], *Y. lipolytica* [56], and *Candida glycerinogenes* [111]. However, the toxicity of geraniol to cells limits the production of microbial cell factories. Conversion of geraniol to its acetate form is an effective method to reduce its toxicity. Shukal and co-workers combined metabolic engineering, enzyme engineering, ribosomal binding site, and abiotic engineering in *E. coli* to produce geranyl acetate by using glycerol as feeding stock. To circumvent potential issues of insufficient solubility or low activity of GES, they utilized nucleoside diphosphate-x (NUDIX) instead of GES to produce geraniol. After regulating the MVA pathway flux, they fusion-expressed the efficient GPPS and NUDIX for geraniol synthesis. Then, they knocked out alcohol dehydrogenases and adopted microaerobic cultivation to avoid the conversion of geraniol to geranial. Finally, by introducing the alcohol acyltransferase *Rh*AAT1, the strain transformed geraniol into the low-toxicity geranyl acetate. The final yield of geranyl acetate reached the highest reported level of 19 g/L [90].

3.2. *Acting as Biopreservatives*

With the increasing awareness of the potential health risks posed by food additives, the concept of clean-label foods gained more attention. Natural preservatives not only extend the shelf life of food but also offer additional health benefits [112]. It was reported that the global sales of natural food preservatives are estimated to be 612 million by 2024, and it might increase up to 2-fold ten years later (https://www.futuremarketinsights.com/reports/natural-food-preservatives-market (accessed on 13 February 2025)).

Due to the hydrophobic skeleton, terpenoids are often lipophilic so they can disrupt cell membranes and inhibit enzymes related to energy metabolism and DNA synthesis, leading to impaired cell viability [5]. For instance, α-pinene and limonene can inhibit the activity of *E. coli* and *Staphylococcus aureus*. On the other hand, some terpenoids act as antioxidants by scavenging free radicals and protecting food from oxidative damage [14]. For example, carotenoids are widely used as antioxidants due to their conjugated double-bond system. In this scene, terpenoids with antibacterial and antioxidant bioactivities have great potential as food preservatives [113]. In fact, terpenoids have already been commonly used as food preservatives in fruits, meat products, seafood, and dairy products, in the form of natural extracts or essential oils [112,114]. However, since extracts and essential oils contain multiple components, achieving their minimum inhibitory concentration (MIC) often requires high-level supplementation. Additional chemicals might affect the taste of the food or introduce undesirable flavors [112]. Terpenoid cell factories through metabolic engineering may meet demands for food preservation and food safety.

3.2.1. Limonene

Limonene is a natural monoterpene with antibacterial efficacy and has been widely used in the food industry as a preservative, antioxidant, and flavor. Its derivatives, such as menthol, carveol, and α-terpineol, which have pleasant fragrances, special biological activities, and physicochemical properties, are also broadly used in the fields of food, pharmaceuticals, and cosmetics [115]. Cell factories offer a new approach to the sustainable production of limonene.

In *S. cerevisiae*, Kong et al. expressed truncated limonene synthase (tLimS) to de novo synthesize limonene from GPP. They further enhanced the limonene synthesis flux through a "Push−Pull−Restrain" strategy while simultaneously increasing the supply of acetyl-CoA and NADPH by regulating the glyoxylate cycle and pentose phosphate pathway. In addition, they reconstructed the MVA and limonene synthetic pathways in mitochondria. In order to alleviate the impact of limonene accumulation on cell growth, they also performed a two-phase scale-up fermentation and obtained a limonene titer of 2.63 g/L ultimately [91]. It appears that the titer of limonene in most cell factories has reached the cells' maximum tolerance to limonene. In order to enhance the tolerance of *Y. lipolytica* to limonene, Kong et al. identified five genes using a transcriptomic analysis. They found that overexpression of an unknown protein YALI0F19492p can increase limonene tolerance and increase the titer up to 8-fold. Then, they carried out short-term laboratory adaptive evolution and obtained strains with higher tolerance. Field emission scanning electron microscopy (FE-SEM) and propidium iodide (PI) analyses revealed that the cell wall and membrane of evolved strains were intact in the presence of limonene, while both the cell wall and membrane of the starting strain were damaged [92]. It would provide new perspectives for improving the sustainable production of limonene.

3.2.2. Pinene

Pinene is a monoterpenoid derived from plants, and it is widely used as flavors, fragrances, and pharmaceuticals. Its dimerized form can serve as a high-density renewable fuel for aircraft [39]. Moreover, pinene can also be used as a raw material for the synthesis of fragrances such as terpineol and linalool.

In the monoterpene cell factories, only one heterologous monoterpene synthase is usually needed for the production of monoterpenes. In *E. coli*, Zhou et al. identified an effective pinene synthase *Pt*PS1^{Q457L} and improved its hydrophilicity and solubility by N-terminal truncation, facilitating folding and preventing degradation. Next, they introduced a GPPS mutant *Ag*GPPS$^{D90G/L175P}$ to enhance the supply of precursor GPP and optimize fermentation conditions. Ultimately, in the optimal host DH10B, a titer of 1035 mg/L pinene was produced in 1.3 L fed-batch fermentation [93]. In order to break through the limitations of mass transfer and reduce the impact of pinene on cell viability, Niu et al. achieved gram-scale production of pinene using a cell-free system. They constructed a module co-catalysis cell-free system containing an MEV upstream module and a pinene downstream module and optimized the physicochemical parameters using Plackett-Burman experimental design and the steepest ascent strategy. The final pinene titer was 1256 mg/L [94]. To some extent, cell-free systems are an effective alternative to whole-cell catalysis for synthesizing terpenoids with toxicity.

3.3. *Acting as Food Fortifiers and Dietary Supplements*

Terpenoids also have other beneficial biological effects besides flavoring and antibacterial properties. They can be used as dietary supplements to enhance the nutritional value of food and improve the health level of people. Retinol converts into retinal and retinoic acid, which are involved in the synthesis of rhodopsin on the retina [116]. Lutein plays a crucial role in protecting the retina and preventing age-related macular degeneration [117]. As functional ingredients, they are often added to infant formula foods, edible oils, and beverages. Anthocyanin is also a famous dietary supplement due to its antioxidative and hypoglycemic effects [118]. Therefore, designing and constructing microbial cell factories for functional terpenoid compounds has great application potential.

3.3.1. Lycopene

Lycopene is a red C_{40} tetraterpene, consisting of eight isoprene units (C_5) linked head-to-tail. It is capable of quenching singlet oxygen and scavenging free radicals to prevent some malignancies and support cardiovascular health [119]. It was reported that lycopene exhibits the highest antioxidant activity among the carotenoid group [120]. Thus, it is widely used as a colorant and functional food. However, the chemical synthesis of tetraterpenes faces difficulties in stereoselectivity and separation [120]. Therefore, many attempts have been made to construct sustainable production cell factories for lycopene.

In nature, some microorganisms, such as *R. toruloides* and *Blakeslea trispora*, possess carotenoid biosynthesis pathways and can synthesize mixtures of carotenoids, including lycopene, β-carotene, and astaxanthin. Traditionally, these microorganisms can be directly used for the fermentative production of lycopene. However, it is very difficult to improve the yield and purity of a particular kind of carotenoid. With the advancement in enzyme function mining and metabolic mechanism analysis, researchers focused on integrating heterologous lycopene pathways into non-carotenoid-producing hosts to enhance the titer and yield. *E. coli* and *S. cerevisiae*, as classic model microorganisms, have been used for large-scale synthesis of lycopene successfully [121].

Chen et al. developed a high-throughput screening method dependent on the color of lycopene. By using color as an indicator, they performed random mutations and targeted mutagenesis on the rate-limiting enzyme isopentenyl diphosphate isomerase (IDI) and obtained a triple mutant $IDI^{L141H/Y195F/W256C}$ with 2.53-fold higher catalytic activity. Replacing the wild-type IDI with $IDI^{L141H/Y195F/W256C}$, *E. coli* containing the complete lycopene synthetic pathway produced 1.8-fold more lycopene, exceeding 1.2 g/L [95]. In order to break the yield bottleneck, Zhou et al. performed atmospheric and room-temperature plasma (ARTP) mutagenesis and H_2O_2-induced adaptive laboratory evolution (ALE) to improve the supply of FPP. Inspired by the transcriptome analysis of the high-yield mutant strain, they integrated another copy of the *TmCrtE* gene to drive GGPP synthesis. In a 7 L fed-batch fermentation, the lycopene achieved the highest level, at a titer of 8.15 g/L [96].

3.3.2. Steviol Glycosides

The increasing prevalence of diabetes and its harm to health promote people to pay more and more attention to low-sugar diets. The use of low-calorie sweeteners instead of sugar in the daily diet is becoming popular and important. In this scene, steviol glycosides (SGs), which are GRAS, have received widespread attention [122]. They are not only used as sweeteners but also act as nutritional enhancers in the diet of diabetic patients due to their anti-diabetic properties. SGs are a type of glycoside derivative of tetracyclic diterpenoids with various sweetness and taste. In *Stevia rebaudiana*, more than 20 types of SGs are glycosylated by only four glycosyltransferases using steviol as a backbone skeleton [97]. In order to meet the demands for low-sugar and low-calorie diets, scientists are dedicated to producing special steviol glycosides via de novo synthesis or whole-cell catalysis using microbial chassis.

Due to the complex glycosyltransferase pathway and numerous by-products, attention has been focused on exploring targeted modification of glycosyltransferases [98]. They adopted a strategy of combinatorial modular operation to assemble the stevioside de novo synthetic pathway into *S. cerevisiae*. In the precursor GGPP supply module, they overexpressed the key enzyme tHMG1 and downregulated ERG9 shunt pathways. In the steviol synthesis pathway, they used steviol as an indicator and fusion-expressed the class I diterpene synthase *At*KS and the class II diterpene synthase *Sr*CPS. In the transglycosidic pathway, they combined directed evolution and fusion expression of two glycosyltransferases to synthesize stevioside. By further enhancing the supply of UDP-glucose, the final engineered strain produced stevioside at a titer of 1149 mg/L [97]. Rebaudioside M (Reb M) provides more sweet taste than stevioside and is free of bitterness. However, the concentration is significantly lower (0.4–0.5%) than stevioside (22–62%) in *Stevia rebaudiana* [122]. Li et al. constructed a whole-cell catalytic pathway for producing Reb M from stevioside. They identified an efficient glycosyltransferase, UGT91D2, and co-expressed it with UG76G1 to convert stevioside to Reb M via Red E. In order to inhibit the hydrolysis of stevioside and strengthen the supply of sugar donor UDP-glucose, they knocked out endogenous glycoside hydrolase SCW2 and overexpressed UGP1. After introducing SIR2 to prolong the growth cycle, they obtained a titer of 12.5 g/L Reb M by using 10.0 g/L stevioside as a substrate.

3.4. *Acting as Medicinal Health Food (MHF)*

The concept of "Medicine and Food Homology" originated from traditional Chinese medicine, implying that some foods not only have nutritional value but also are capable of preventing and treating diseases. People eat plants containing active compounds to treat illnesses and enhance immunity. The Chinese Ministry of Health has listed 114 kinds of items as medicinal health foods [123]. Among them, *Pogostemon cablin*, *Herba menthae*, *Flos sophorae*, and *Ganoderma lucidum* contain significant levels of terpenoids as their respective functional factors [123–125]. Nevertheless, it is worth mentioning that traditional Chinese medicine may contain multiple complex components, some of which might cause adverse reactions or allergies in certain individuals. Therefore, under the condition that the efficacy of terpenoids is known or the mechanism is clear, taking the specific active factor directly is better than taking the whole herb. For instance, artemisinin has shown significant effects in treating malaria, and extracting artemisinin for treatment might be more effective than consuming the whole *Artemisia annua herb*. With the development of synthetic biology and systems biology techniques, the biosynthetic pathways of more and more terpenoids are being elucidated, making it possible to produce these valuable terpenoids using microbial cell factories. For example, researchers achieved de novo synthesis of three terpenoid compounds in *S. cerevisiae* by integrating the whole biosynthetic pathways for valerenic acid [126], oleanane-type ginsenosides [127], and (-)-borneol [52].

3.4.1. Germacrene A and β-Elemene

Germacrene A with spicy and peppery scents can be used as flavor and fragrance in the food industry [11]. Moreover, it is one of the most critical intermediates in the biosynthesis of terpenoid compounds and can serve as a precursor of β-elemene and patchoulol. β-Elemene, extracted from the traditional Chinese herb *Rhizoma zedoariae* (Wen E Zhu) and *Curcuma longa* L. (Jiang Huang), is a volatile sesquiterpene and can be synthesized through intramolecular Cope rearrangement via germacrene A in vitro. It is an important flavor in Chinese Baijiu, not only giving a unique flavor to the liquor but also stimulating the consumer's taste buds [1]. Additionally, β-elemene has also been widely used as a clinical anti-cancer drug due to its anti-tumor activity and immune protective properties [1]. Owing to their great application value, scientists attempt to heterologously synthesize germacrene A and β-elemene in microorganisms. Currently, they have been synthesized in various microorganisms, including *E. coli* [28], *S. cerevisiae* [128], *Y. lipolytica* [99], and *P. pastoris* [62].

Liu et al. identified a germacrene A synthase (*Dl*GAS) with the highest activity in *Y. lipolytica* using a bioinformatics strategy. They further enhanced the germacrene A synthetic flux by substrate-channel engineering, MVA, and acetyl-CoA supplying. The final titer of germacrene A was achieved at 39 g/L by using glucose as a carbon source [99]. Li et al. identified another germacrene synthase, LTC2, which originated from *Lactuca sativa*, that can be used in *Y. lipolytica*, and then they integrated the LTC2-expressing cassette into the genome of Po1f-Δku70. They promoted the synthesis of germacrene A through a "push-pull-restrain" strategy successively. Finally, β-elemene was produced by shifting the temperature to 250 °C, and the titer reached 5.08 g/L. This work is the first combination of engineering *Y. lipolytica* with chemical transformation [100].

3.4.2. Valencene and Nootkatone

Valencene and nootkatone are sesquiterpenes derived from plants. Valencene provides an orange-like odor, while nootkatone has a grapefruit-like flavor. They are GRAS and widely used as food additives in the fragrances and flavors industry [129,130]. In addition, both valencene and nootkatone possess interesting biological activities, such as anti-cancer, anti-oxidant, and anti-inflammatory. Among them, nootkatone has been proven to be a major active component in traditional Chinese medicine *Alpinia oxyphylla* (Yi Zhi Ren). Because of their great market value, they were primarily produced by chemical synthesis. Recently, in the exploration of enzyme functions, researchers focused on biotransformation or heterologous synthesis to produce valencene and nootkatone [131].

For de novo valencene biosynthesizing, Zhu et al. introduced a valencene synthase (*Cn*VS) into an *S. cerevisiae* platform BN-91A, which was obtained through laboratory adaptive evolution and capable of utilizing mannitol as a sole carbon source. By comprehensively optimizing the metabolic network, a titer of 5.6 g/L of valencene was achieved in high-density cultivation with mannitol feeding [101]. In another research study, Ye and co-workers identified a more efficient valencene synthase (*Eg*VS) and performed site-directed mutation to increase its activity. By optimizing the biosynthetic pathway, they increased the titer of valencene. Furthermore, they adopted a GAL promoter to control URA3 expression in order to couple cell growth and biochemical pathways, and the titer of valencene reached 16.6 g/L in a high-density fermentation. Accompanying the following chemical conversion, the yield of nootkatone reached 80% [102]. The combination of biosynthesis and chemical transformation makes the industrial production of nootkatone possible. Recently, researchers have elucidated the natural nootkatone biosynthetic pathway in *A. oxyphylla* and achieved its total biosynthesis in an engineered *S. cerevisiae* strain [129]. This provides new resources and strategies for the subsequent production of nootkatone using green cell factories.

3.4.3. Patchoulol

Patchoulol, which has a wooden, earthy scent, is a valuable sesquiterpene with anti-inflammatory, anti-cancer, anti-depressant, and other therapeutic properties [103]. It is widely used in the food, cosmetic, and pharmaceutical industries. Patchoulol is the major active ingredient in the Chinese medicine *Pogostemon cablin* (Huo Xiang) [123]. Since the natural yield of patchoulol is limited, efforts have been made to produce patchoulol using microorganisms.

It was reported that patchoulol synthase (PTS) is a promiscuous enzyme capable of producing at least 22 different sesquiterpene compounds with various structures, including patchoulol. Scientists have conducted research on its catalytic mechanism and product spectrum [132,133]. Zhou et al. performed a semi-rational design and obtained a PTS mutant $PTS2^{C415F/H454A}$ with higher substrate-binding affinity and a 2.9-fold increase in the titer of patchoulol. After deleting competing pathways and strengthening the efflux pathways, the engineered *E. coli* strain produced nearly 1 g/L of patchoulol in a 5 L bioreactor [103]. In further research, they obtained another PTS mutant, PTS3mut4, with 11-fold higher than $PTS2^{C415F/H454A}$. By deleting the cyclic AMP synthesis pathway to alleviate glucose repression and dynamically regulating glycolysis based on pyruvate biosensor, they achieved a titer of 1675 mg/L patchoulol in *E. coli* [104]. Similarly, Liu and co-workers rationally engineered PTS and obtained PTS^{T404S}, which could reduce the production of the concomitant byproducts of patchoulol. Combining modifying off-pathway, enhancing the ergosterol pathway, regulating NADPH supply, and decreasing intracellular ROS level, they obtained an *S. cerevisiae* strain with a titer of 1632 mg/L patchoulol in a two-stage fermentation [105]. To our surprise, appropriately enhancing the by-pass synthesis of ergosterol not only did not reduce the flux of patchoulol synthesis but also increased cell density and accumulation of patchoulol in their research. This provides us with a new perspective for further breaking through the bottleneck of terpenoid synthesis.

4. Conclusions and Perspectives

Microbial cell factories have the capability to utilize cost-effective raw materials, even one-carbon materials, for the production of fine chemicals. This offers a green and sustainable platform for terpenoid production. It is essential to study natural biosynthetic pathways, analyze key enzymes and regulatory mechanisms in these pathways, and heterologously integrate them into microbial chassis cells to construct cell factories. Large-scale production via fermentation is an important strategy to meet the market demand for edible terpenoids. In recent years, scientists have been dedicated to engineering typical or non-typical microbial hosts to produce various types of terpenoids and their derivatives [10,107,134,135]. To further exploit the potential of microbial cells in edible terpenoid production, the following aspects should be taken into consideration.

First and foremost, the selection of the host is paramount. An excellent chassis can potentially double outcomes with half exertion. Selecting strains that are GRAS can, to a certain extent, avoid evaluations on food safety. For instance, *Lactococcus lactis*, a common probiotic found in fermented food, can serve as a food supplement to directly deliver active molecules to the intestine. Researchers have already constructed a pathway for lycopene production in *L. lactis* and preliminarily verified its potential in treating intestinal oxidative damage [136]. On the other hand, the selection of a precursor-rich chassis host can significantly reduce trial and error costs. Strains with abundant acetyl-CoA derivatives not only supply sufficient precursors for the terpenoid pathway but also, the metabolites derived from acyl-CoA derivatives, such as lipids, provide a hydrophobic environment for the cells conducive to the accumulation of terpenoids. Terpenoids such as monoterpenes, sesquiterpenes, and diterpenes have their respective common important intermediates in the synthetic pathway. Direct metabolic switching in existing cell factories can accelerate the strain construction process.

Secondly, reprogramming the biosynthetic pathway of terpenoids is essential. Terpenoids have diverse and complex structures. During evolution, in order to optimize the metabolic efficiency of organisms and reduce the number of unnecessary enzymes, terpenoid synthases responsible for forming the terpenoid skeleton and downstream modification enzymes often exhibit promiscuity or are bifunctional enzymes. Hence, even when a terpenoid synthesis pathway is heterologously introduced into a microbial host, there remains a potential for unexpected components. Mining enzymes with more specific functions can enhance the yield of target products and facilitate subsequent separation and purification. Nowadays, numerous strategies have been developed for mining the biosynthetic gene clusters from natural products [137]. On the other hand, to establish novel terpenoid synthesis pathways, it is crucial to identify enzymes from various species or to develop multifunctional enzymes. The rise of artificial intelligence technology enables the customization of enzymes with specific catalytic functions [138]. Moreover, preliminary advancements have been achieved in discovering and reconstructing non-typical terpenoid compound biosynthetic pathways by deep learning [139–141]. Considering that the formation and modification of terpenoid skeletons require multi-enzyme cascade reactions, strategies such as compartmentalization, membrane-less organelles, and protein scaffolds can potentially elevate metabolic rates [142,143]. In addition, high-throughput screening techniques offer rapid identification of strains with high-yield and high-titer. Distinct from traditional metabolite extraction followed by mass spectrometry detection, screening approaches based on biosensors [144] and growth-coupling [77] have significantly increased the screening throughput of microbial libraries.

Finally, it is noteworthy that the entire production process is suitable for commercial production. Some kinds of terpenoids with cytotoxicity could inhibit the production in the fermentation process. Adaptive evolution [145] combined with multi-omics analysis can help us quickly obtain tolerant strains and resolve tolerance mechanisms. Biocompatible and efficient extraction systems [15] and a derivatization strategy can alleviate the toxicity. Beyond the inherent impact of the product on its production, prolonged large-scale industrial production often encounters reductions in yield, titer, and productivity. Population dynamics at a single-cell resolution can provide insights into consistently maintaining the phenotype by increasing the robustness [146]. In the future, AI-assisted control algorithms integrating cellular physiology and fluid dynamics will facilitate process scaling-up optimization [147].

Terpenoids play significant roles in processing food for food quality and nutritional behavior. Theoretically, compounds with the same structure and conformation have the same physicochemical properties and physiological functions. Therefore, terpenoid compounds, whether naturally sourced or artificially biosynthesized, can be used to improve food quality. However, prior to employing the terpenoid compounds generated in cell factories as food additives, it is imperative to conduct rigorous assessments, including toxicity testing, allergenicity tests, and relevant legal and regulatory reviews. Moreover, explicitly labeling the source and dosage of terpenoid additives contributes to better support for transparency of food labels, thereby facilitating consumers to make more informed decisions.

In conclusion, the production of edible terpenoids using microbial cell factories not only requires technical developments for the chassis and fermentation processes but also the enhancement of the corresponding evaluation system and legislative support. Collaboration among industry, university, and research can expedite the commercialization of more edible terpenoid cell factories.

Author Contributions: Literature retrieval and information collection, Z.Z., R.L., Z.L. and X.L.; Analyzed the data: M.W. and Z.Z., writing—original draft preparation, Z.Z., R.L., Z.L. and X.L.; writing—review and editing, M.W.; project administration, M.W.; funding acquisition, M.W. All authors have read and agreed to the published version of the manuscript.

Funding: This research was funded by the Fundamental Research Funds for the Central Universities, grant number 3132023528.

Institutional Review Board Statement: Not applicable.

Informed Consent Statement: Not applicable.

Data Availability Statement: No new data were created or analyzed in this study. Data sharing is not applicable to this article.

Conflicts of Interest: The authors declare that they have no conflicts of interest.

Abbreviations

The following abbreviations are used in this manuscript:

AcAc-CoA	acetoacetyl-CoA
HMG-CoA	3-hydroxy-3-methylglutaryl-CoA
MVA	mevalonate
MVA-5P	mevalonate-5-phosphate
MVA-5PP	mevalonate-5-diphosphate
DXP	1-deoxy-D-xylulose-5-phosphate
MEP	2-C-methyl-D-erythritol 4-phosphate
CDP-ME	4-diphosphocytidyl)-2-C-methyl-D-erythritol
CDP-MEP	2-phospho-4-diphosphocytidyl-2-C methyl-D-erythritol
MEC	2-C-methyl-D-erythritol 2,4-cyclodiphosphate
HMBPP	1-hydroxy-2-methyl-2-butenyl 4-diphosphate
IPP	isopentenyl pyrophosphate
DMAPP	dimethylallyl pyrophosphate
NPP	neryl diphosphate
GPP	geranyl diphosphate
FPP	farnesyl diphosphate
GGPP	geranylgeranyl diphosphate
GFPP	geranylfarnesyl diphosphate
HMGS	HMG-CoA synthase
HMGR	HMG-CoA reductase

MVK	mevalonate kinase
PMK	MVA-5P kinase
MDD	MVA-5-PP decarboxylase;
DXS	DXP synthase
DXR	DXP reductoisomerase
MCT	CD-ME cytidylyltransferase
CMK	CD-ME kinase
MDS	MEC synthase
HDS	HMBPP synthase
HDR	HMBPP reductase
NPPS	NPP synthase
GPPS	GPP synthase
FPPS	FPP synthase
GGPPS	GGPP synthase
GFPPS	GFPP synthase
MTS	monoterpene synthase
STS	sesquiterpene synthase
DTS	diterpene synthase
SS	squalene synthase

References

1. Fan, M.; Yuan, S.; Li, L.; Zheng, J.; Zhao, D.; Wang, C.; Wang, H.; Liu, X.; Liu, J. Application of terpenoid compounds in food and pharmaceutical products. *Fermentation* **2023**, *9*, 119. [CrossRef]
2. Camara, J.S.; Perestrelo, R.; Ferreira, R.; Berenguer, C.V.; Pereira, J.A.M.; Castilho, P.C. Plant-derived terpenoids: A plethora of bioactive compounds with several health functions and industrial applications-A comprehensive overview. *Molecules* **2024**, *29*, 3861. [CrossRef]
3. Avalos, M.; Garbeva, P.; Vader, L.; van Wezel, G.P.; Dickschat, J.S.; Ulanova, D. Biosynthesis, evolution and ecology of microbial terpenoids. *Nat. Prod. Rep.* **2022**, *39*, 249–272. [CrossRef] [PubMed]
4. Hu, W.; Meng, X.; Wu, Y.; Li, X.; Chen, H. Terpenoids, a rising star in bioactive constituents for alleviating food allergy: A review about the potential mechanism, preparation, and application. *J. Agric. Food Chem.* **2024**, *72*, 26599–26616. [CrossRef] [PubMed]
5. Lyu, X.; Lee, J.; Chen, W.N. Potential natural food preservatives and their sustainable production in yeast: Terpenoids and polyphenols. *J. Agric. Food Chem.* **2019**, *67*, 4397–4417. [CrossRef]
6. De Los Santos, Z.A.; Wolf, C. Optical terpene and terpenoid sensing: Chiral recognition, determination of enantiomeric composition and total concentration analysis with late transition metal complexes. *J. Am. Chem. Soc.* **2020**, *142*, 4121–4125. [CrossRef]
7. Maharjan, A.; Kim, M.-R.; Choi, W.; Kim, H.-C.; Park, J.-H. Advanced biofuels: A path to sustainable energy. *Sustain. Energy Fuels* **2024**, *8*, 3247–3265. [CrossRef]
8. Luna-Flores, C.H.; Wang, A.; von Hellens, J.; Speight, R.E. Towards commercial levels of astaxanthin production in *Phaffia rhodozyma*. *J. Biotechnol.* **2022**, *350*, 42–54. [CrossRef]
9. Meadows, A.L.; Hawkins, K.M.; Tsegaye, Y.; Antipov, E.; Kim, Y.; Raetz, L.; Dahl, R.H.; Tai, A.; Mahatdejkul-Meadows, T.; Xu, L.; et al. Rewriting yeast central carbon metabolism for industrial isoprenoid production. *Nature* **2016**, *537*, 694–697. [CrossRef] [PubMed]
10. Ma, C.; Zhang, K.; Zhang, X.; Liu, G.; Zhu, T.; Che, Q.; Li, D.; Zhang, G. Heterologous expression and metabolic engineering tools for improving terpenoids production. *Curr. Opin. Biotechnol.* **2021**, *69*, 281–289. [CrossRef] [PubMed]
11. Jiang, H.; Wang, X. Biosynthesis of monoterpenoid and sesquiterpenoid as natural flavors and fragrances. *Biotechnol. Adv.* **2023**, *65*, 108151. [CrossRef]
12. Navale, G.R.; Dharne, M.S.; Shinde, S.S. Metabolic engineering and synthetic biology for isoprenoid production in *Escherichia coli* and *Saccharomyces cerevisiae*. *Appl. Microbiol. Biotechnol.* **2021**, *105*, 457–475. [CrossRef]
13. Kant, G.; Pandey, A.; Hasan, A.; Bux, F.; Kumari, S.; Srivastava, S. Cell factories for methylerythritol phosphate pathway mediated terpenoid biosynthesis: An application of modern engineering towards sustainability. *Process Biochem.* **2024**, *139*, 146–164. [CrossRef]
14. Masyita, A.; Mustika Sari, R.; Dwi Astuti, A.; Yasir, B.; Rahma Rumata, N.; Emran, T.B.; Nainu, F.; Simal-Gandara, J. Terpenes and terpenoids as main bioactive compounds of essential oils, their roles in human health and potential application as natural food preservatives. *Food Chem. X* **2022**, *13*, 100217. [CrossRef] [PubMed]

15. Wang, Y.; Huang, J.; Lin, X.; Su, W.; Zhu, P.; Yang, N.; Adams, E. Recent progress in the extraction of terpenoids from essential oils and separation of the enantiomers by GC-MS. *J. Chromatogr. A* **2024**, *1730*, 465118. [CrossRef] [PubMed]
16. Singh, R.V.; Sambyal, K. An overview of β-carotene production: Current status and future prospects. *Food Biosci.* **2022**, *47*, 101717. [CrossRef]
17. Luo, P.; Huang, J.H.; Lv, J.M.; Wang, G.Q.; Hu, D.; Gao, H. Biosynthesis of fungal terpenoids. *Nat. Prod. Rep.* **2024**, *41*, 748–783. [CrossRef] [PubMed]
18. Jahn, L.J.; Rekdal, V.M.; Sommer, M.O.A. Microbial foods for improving human and planetary health. *Cell* **2023**, *186*, 469–478. [CrossRef] [PubMed]
19. Graham, A.E.; Ledesma-Amaro, R. The microbial food revolution. *Nat. Commun.* **2023**, *14*, 2231. [CrossRef] [PubMed]
20. Martin, V.J.; Pitera, D.J.; Withers, S.T.; Newman, J.D.; Keasling, J.D. Engineering a mevalonate pathway in *Escherichia coli* for production of terpenoids. *Nat. Biotechnol.* **2003**, *21*, 796–802. [CrossRef] [PubMed]
21. Paddon, C.J.; Westfall, P.J.; Pitera, D.J.; Benjamin, K.; Fisher, K.; McPhee, D.; Leavell, M.D.; Tai, A.; Main, A.; Eng, D.; et al. High-level semi-synthetic production of the potent antimalarial artemisinin. *Nature* **2013**, *496*, 528–532. [CrossRef] [PubMed]
22. Matos, C.T.; Gouveia, L.; Morais, A.R.C.; Reis, A.; Bogel-Łukasik, R. Green metrics evaluation of isoprene production by microalgae and bacteria. *Green. Chem.* **2013**, *15*, 2854–2864. [CrossRef]
23. Ajikumar, P.K.; Xiao, W.H.; Tyo, K.E.; Wang, Y.; Simeon, F.; Leonard, E.; Mucha, O.; Phon, T.H.; Pfeifer, B.; Stephanopoulos, G. Isoprenoid pathway optimization for Taxol precursor overproduction in *Escherichia coli*. *Science* **2010**, *330*, 70–74. [CrossRef]
24. Wang, H.H.; Isaacs, F.J.; Carr, P.A.; Sun, Z.Z.; Xu, G.; Forest, C.R.; Church, G.M. Programming cells by multiplex genome engineering and accelerated evolution. *Nature* **2009**, *460*, 894–898. [CrossRef] [PubMed]
25. Wang, X.; Chen, J.; Zhang, J.; Zhou, Y.; Zhang, Y.; Wang, F.; Li, X. Engineering *Escherichia coli* for production of geraniol by systematic synthetic biology approaches and laboratory-evolved fusion tags. *Metab. Eng.* **2021**, *66*, 60–67. [CrossRef] [PubMed]
26. Wang, X.; Xiao, L.; Zhang, X.; Zhang, J.; Zhang, Y.; Wang, F.; Li, X. Combined bioderivatization and engineering approach to improve the efficiency of geraniol production. *Green. Chem.* **2022**, *24*, 864–876. [CrossRef]
27. Wang, X.; Wang, J.; Zhang, X.; Zhang, J.; Zhou, Y.; Wang, F.; Li, X. Efficient myrcene production using linalool dehydratase isomerase and rational biochemical process in *Escherichia coli*. *J. Biotechnol.* **2023**, *371–372*, 33–40. [CrossRef] [PubMed]
28. Fordjour, E.; Liu, C.L.; Hao, Y.; Sackey, I.; Yang, Y.; Liu, X.; Li, Y.; Tan, T.; Bai, Z. Engineering *Escherichia coli* BL21 (DE3) for high-yield production of germacrene A, a precursor of β-elemene via combinatorial metabolic engineering strategies. *Biotechnol. Bioeng.* **2023**, *120*, 3039–3056. [CrossRef] [PubMed]
29. Huang, C.N.; Lim, X.; Ong, L.; Lim, C.; Chen, X.; Zhang, C. Mediating oxidative stress enhances α-ionone biosynthesis and strain robustness during process scaling up. *Microb. Cell Fact.* **2022**, *21*, 246. [CrossRef] [PubMed]
30. Qi, Z.; Tong, X.; Zhang, X.; Lin, H.; Bu, S.; Zhao, L. One-pot synthesis of dihydro-beta-ionone from carotenoids using carotenoid cleavage dioxygenase and enoate reductase. *Bioprocess. Biosyst. Eng.* **2022**, *45*, 891–900. [CrossRef]
31. Qi, Z.; Tong, X.; Ke, K.; Wang, X.; Pei, J.; Bu, S.; Zhao, L. De novo synthesis of dihydro-β-ionone through metabolic engineering and bacterium-yeast coculture. *J. Agric. Food Chem.* **2024**, *72*, 3066–3076. [CrossRef]
32. Ravikumar, S.; Woo, H.M.; Choi, J.I. Analysis of novel antioxidant sesquarterpenes (C_{35} Terpenes) produced in recombinant *Corynebacterium glutamicum*. *Appl. Biochem. Biotechnol.* **2018**, *186*, 525–534. [CrossRef] [PubMed]
33. Li, C.; Swofford, C.A.; Ruckert, C.; Chatzivasileiou, A.O.; Ou, R.W.; Opdensteinen, P.; Luttermann, T.; Zhou, K.; Stephanopoulos, G.; Jones Prather, K.L.; et al. Heterologous production of α-carotene in *Corynebacterium glutamicum* using a multi-copy chromosomal integration method. *Bioresour. Technol.* **2021**, *341*, 125782. [CrossRef]
34. Gottl, V.L.; Meyer, F.; Schmitt, I.; Persicke, M.; Peters-Wendisch, P.; Wendisch, V.F.; Henke, N.A. Enhancing astaxanthin biosynthesis and pathway expansion towards glycosylated C40 carotenoids by *Corynebacterium glutamicum*. *Sci. Rep.* **2024**, *14*, 8081. [CrossRef] [PubMed]
35. Burgardt, A.; Moustafa, A.; Persicke, M.; Spross, J.; Patschkowski, T.; Risse, J.M.; Peters-Wendisch, P.; Lee, J.H.; Wendisch, V.F. Coenzyme Q_{10} biosynthesis established in the non-ubiquinone containing *Corynebacterium glutamicum* by metabolic engineering. *Front. Bioeng. Biotechnol.* **2021**, *9*, 650961. [CrossRef]
36. Xu, X.; Lv, X.; Cui, S.; Liu, Y.; Xia, H.; Li, J.; Du, G.; Li, Z.; Ledesma-Amaro, R.; Chen, J.; et al. Remodeling isoprene pyrophosphate metabolism for promoting terpenoids bioproduction. *Engineering* **2023**, *28*, 166–178. [CrossRef]
37. Pramastya, H.; Xue, D.; Abdallah, I.I.; Setroikromo, R.; Quax, W.J. High level production of amorphadiene using *Bacillus subtilis* as an optimized terpenoid cell factory. *New Biotechnol.* **2021**, *60*, 159–167. [CrossRef] [PubMed]
38. Sakamaki, Y.; Ono, M.; Shigenari, N.; Chibazakura, T.; Shimomura, K.; Watanabe, S. Photosynthetic 1,8-cineole production using cyanobacteria. *Biosci. Biotechnol. Biochem.* **2023**, *87*, 563–568. [CrossRef]
39. Yang, R.; Zhu, L.; Li, T.; Zhu, L.Y.; Ye, Z.; Zhang, D. Photosynthetic conversion of CO_2 into pinene using engineered *Synechococcus* sp. PCC 7002. *Front. Bioeng. Biotechnol.* **2021**, *9*, 779437. [CrossRef]
40. Dietsch, M.; Behle, A.; Westhoff, P.; Axmann, I.M. Metabolic engineering of *Synechocystis* sp. PCC 6803 for the photoproduction of the sesquiterpene valencene. *Metab. Eng. Commun.* **2021**, *13*, e00178. [CrossRef] [PubMed]

41. Rinaldi, M.A.; Ferraz, C.A.; Scrutton, N.S. Alternative metabolic pathways and strategies to high-titre terpenoid production in *Escherichia coli*. *Nat. Prod. Rep.* **2022**, *39*, 90–118. [CrossRef] [PubMed]
42. Kang, M.K.; Nguyen, M.P.; Yoon, S.H.; Jayasundera, K.B.; Son, J.W.; Wang, C.; Kwon, M.; Kim, S.W. Reconstitution of the mevalonate pathway for improvement of isoprenoid production and industrial applicability in *Escherichia coli*. *J. Microbiol. Biotechnol.* **2024**, *34*, 2338–2346. [CrossRef] [PubMed]
43. Wolf, S.; Becker, J.; Tsuge, Y.; Kawaguchi, H.; Kondo, A.; Marienhagen, J.; Bott, M.; Wendisch, V.F.; Wittmann, C. Advances in metabolic engineering of *Corynebacterium glutamicum* to produce high-value active ingredients for food, feed, human health, and well-being. *Essays Biochem.* **2021**, *65*, 197–212. [CrossRef] [PubMed]
44. Zha, J.; Zhao, Z.; Xiao, Z.; Eng, T.; Mukhopadhyay, A.; Koffas, M.A.; Tang, Y.J. Biosystem design of *Corynebacterium glutamicum* for bioproduction. *Curr. Opin. Biotechnol.* **2023**, *79*, 102870. [CrossRef] [PubMed]
45. Tsuge, Y.; Matsuzawa, H. Recent progress in production of amino acid-derived chemicals using *Corynebacterium glutamicum*. *World J. Microbiol. Biotechnol.* **2021**, *37*, 49. [CrossRef] [PubMed]
46. Guan, Z.; Xue, D.; Abdallah, I.I.; Dijkshoorn, L.; Setroikromo, R.; Lv, G.; Quax, W.J. Metabolic engineering of *Bacillus subtilis* for terpenoid production. *Appl. Microbiol. Biotechnol.* **2015**, *99*, 9395–9406. [CrossRef]
47. Cheng, J.; Zhang, C.; Zhang, K.; Li, J.; Hou, Y.; Xin, J.; Sun, Y.; Xu, C.; Xu, W. Cyanobacteria-mediated light-driven biotransformation: The current status and perspectives. *ACS Omega* **2023**, *8*, 42062–42071. [CrossRef]
48. Rautela, A.; Kumar, S. Engineering plant family TPS into cyanobacterial host for terpenoids production. *Plant Cell Rep.* **2022**, *41*, 1791–1803. [CrossRef] [PubMed]
49. Chenebault, C.; Blanc-Garin, V.; Vincent, M.; Diaz-Santos, E.; Goudet, A.; Cassier-Chauvat, C.; Chauvat, F. Exploring the potential of the model cyanobacteria *Synechococcus* PCC 7002 and PCC 7942 for the photoproduction of high-value terpenes: A comparison with *Synechocystis* PCC 6803. *Biomolecules* **2023**, *13*, 504. [CrossRef]
50. Rodrigues, J.S.; Kovacs, L.; Lukes, M.; Hoper, R.; Steuer, R.; Cerveny, J.; Lindberg, P.; Zavrel, T. Characterizing isoprene production in cyanobacteria—Insights into the effects of light, temperature, and isoprene on *Synechocystis* sp. PCC 6803. *Bioresour. Technol.* **2023**, *380*, 129068. [CrossRef] [PubMed]
51. Shu, Y.; Dong, T.; Zhou, X.; Wang, H.; Liu, H.; Yao, M.; Wang, Y.; Xiao, W. Systematic engineering to enhance β-myrcene production in yeast. *J. Agric. Food Chem.* **2024**, *72*, 19395–19402. [CrossRef] [PubMed]
52. Ma, R.; Su, P.; Ma, Q.; Guo, J.; Chen, S.; Jin, B.; Zhang, H.; Tang, J.; Zhou, T.; Xiao, C.; et al. Identification of (-)-bornyl diphosphate synthase from *Blumea balsamifera* and its application for (-)-borneol biosynthesis in *Saccharomyces cerevisiae*. *Synth. Syst. Biotechnol.* **2022**, *7*, 490–497. [CrossRef]
53. Hu, Q.; Zhang, T.; Yu, H.; Ye, L. Selective biosynthesis of retinol in *S. cerevisiae*. *Bioresour. Bioprocess.* **2022**, *9*, 22. [CrossRef]
54. Wei, P.; Zhang, C.; Bian, X.; Lu, W. Metabolic engineering of *Saccharomyces cerevisiae* for heterologous carnosic acid production. *Front. Bioeng. Biotechnol.* **2022**, *10*, 916605. [CrossRef] [PubMed]
55. Luo, Z.; Liu, N.; Lazar, Z.; Chatzivasileiou, A.; Ward, V.; Chen, J.; Zhou, J.; Stephanopoulos, G. Enhancing isoprenoid synthesis in *Yarrowia lipolytica* by expressing the isopentenol utilization pathway and modulating intracellular hydrophobicity. *Metab. Eng.* **2020**, *61*, 344–351. [CrossRef] [PubMed]
56. Agrawal, A.; Yang, Z.; Blenner, M. Engineering *Yarrowia lipolytica* for the biosynthesis of geraniol. *Metab. Eng. Commun.* **2023**, *17*, e00228. [CrossRef]
57. Sun, M.-L.; Han, Y.; Yu, X.; Wang, K.; Lin, L.; Ledesma-Amaro, R.; Ji, X.-J. Constructing a green oleaginous yeast cell factory for sustainable production of the plant-derived diterpenoid sclareol. *Green. Chem.* **2024**, *26*, 5202–5210. [CrossRef]
58. Xu, M.; Yang, N.; Pan, J.; Hua, Q.; Li, C.X.; Xu, J.H. Remodeling the homologous recombination mechanism of *Yarrowia lipolytica* for high-level biosynthesis of squalene. *J. Agric. Food Chem.* **2024**, *72*, 9984–9993. [CrossRef]
59. Shi, J.T.; Wu, Y.Y.; Sun, R.Z.; Hua, Q.; Wei, L.J. Synthesis of β-ionone from xylose and lignocellulosic hydrolysate in genetically engineered oleaginous yeast *Yarrowia lipolytica*. *Biotechnol. Lett.* **2024**, *46*, 1219–1236. [CrossRef]
60. Li, J.; Gao, J.; Ye, M.; Cai, P.; Yu, W.; Zhai, X.; Zhou, Y.J. Engineering yeast for high-level production of β-farnesene from sole methanol. *Metab. Eng.* **2024**, *85*, 194–200. [CrossRef]
61. Ye, M.; Gao, J.; Zhou, Y.J. Global metabolic rewiring of the nonconventional yeast *Ogataea polymorpha* for biosynthesis of the sesquiterpenoid β-elemene. *Metab. Eng.* **2023**, *76*, 225–231. [CrossRef] [PubMed]
62. Cheng, J.; Zuo, Y.; Liu, G.; Li, D.; Gao, J.; Xiao, F.; Huang, L.; Xu, Z.; Lian, J. Development of a *Pichia pastoris* cell factory for efficient production of germacrene A: A precursor of β-elemene. *Bioresour. Bioprocess.* **2023**, *10*, 38. [CrossRef] [PubMed]
63. Zuo, Y.; Xiao, F.; Gao, J.; Ye, C.; Jiang, L.; Dong, C.; Lian, J. Establishing *Komagataella phaffii* as a cell factory for efficient production of sesquiterpenoid α-santalene. *J. Agric. F Chem.* **2022**, *70*, 8024–8031. [CrossRef] [PubMed]
64. Zhang, X.; Wang, D.; Duan, Y.; Zheng, X.; Lin, Y.; Liang, S. Production of lycopene by metabolically engineered *Pichia pastoris*. *Biosci. Biotechnol. Biochem.* **2020**, *84*, 463–470. [CrossRef]
65. Liu, H.; Chen, S.L.; Xu, J.Z.; Zhang, W.G. Dual regulation of cytoplasm and peroxisomes for improved α-farnesene production in recombinant *Pichia pastoris*. *ACS Synth. Biol.* **2021**, *10*, 1563–1573. [CrossRef]

66. Zhao, C.; Gao, X.; Liu, X.; Wang, Y.; Yang, S.; Wang, F.; Ren, Y. Enhancing biosynthesis of a ginsenoside precursor by self-assembly of two key enzymes in *Pichia pastoris*. *J. Agric. F Chem.* **2016**, *64*, 3380–3385. [CrossRef] [PubMed]
67. Tran, T.N.; Tran, N.-T.; Tran, T.-A.; Pham, D.-C.; Su, C.-H.; Nguyen, H.C.; Barrow, C.J.; Ngo, D.-N. Highly activea astaxanthin production from waste molasses by mutated *Rhodosporidium toruloides* G17. *Fermentation* **2023**, *9*, 148. [CrossRef]
68. Liu, S.; Zhang, M.; Ren, Y.; Jin, G.; Tao, Y.; Lyu, L.; Zhao, Z.K.; Yang, X. Engineering *Rhodosporidium toruloides* for limonene production. *Biotechnol. Biofuels* **2021**, *14*, 243. [CrossRef]
69. Gao, Q.; Dong, Y.; Huang, Y.; Liu, S.; Zheng, X.; Ma, Y.; Qi, Q.; Wang, X.; Zhao, Z.K.; Yang, X. Dual-regulation in peroxisome and cytoplasm toward efficient limonene biosynthesis with *Rhodotorula toruloides*. *ACS Synth. Biol.* **2024**, *13*, 2545–2554. [CrossRef]
70. Geiselman, G.M.; Zhuang, X.; Kirby, J.; Tran-Gyamfi, M.B.; Gladden, J.M. Production of ent-kaurene from lignocellulosic hydrolysate in *Rhodosporidium toruloides*. *Microb. Cell Fact.* **2020**, *19*, 24. [CrossRef] [PubMed]
71. Yaegashi, J.; Kirby, J.; Ito, M.; Sun, J.; Gladden, J.M. *Rhodosporidium toruloides*: A new platform organism for conversion of lignocellulose into terpene biofuels and bioproducts. *Biotechnol. Biofuels* **2017**, *10*, 241. [CrossRef] [PubMed]
72. Kirby, J.; Geiselman, G.M.; Yaegashi, J.; Kim, J.; Zhuang, X.; Tran-Gyamfi, M.B.; Prahl, J.P.; Sundstrom, E.R.; Gao, Y.; Munoz, N.; et al. Further engineering of *R. toruloides* for the production of terpenes from lignocellulosic biomass. *Biotechnol. Biofuels* **2021**, *14*, 101. [CrossRef]
73. Lv, X.; Zhou, X.; Ma, J.; Tao, M.; Liu, Y.; Li, J.; Du, G.; Liu, L. Engineered *Saccharomyces cerevisiae* for the de novo biosynthesis of (-)-menthol. *J. Fungi* **2022**, *8*, 982. [CrossRef] [PubMed]
74. Cao, X.; Yu, W.; Chen, Y.; Yang, S.; Zhao, Z.K.; Nielsen, J.; Luan, H.; Zhou, Y.J. Engineering yeast for high-level production of diterpenoid sclareol. *Metab. Eng.* **2023**, *75*, 19–28. [CrossRef]
75. Zhang, Y.; Wang, W.; Wei, W.; Xia, L.; Gao, S.; Zeng, W.; Liu, S.; Zhou, J. Regulation of ethanol assimilation for efficient accumulation of squalene in *Saccharomyces cerevisiae*. *J. Agric. Food Chem.* **2023**, *71*, 6389–6397. [CrossRef] [PubMed]
76. Chatzivasileiou, A.O.; Ward, V.; Edgar, S.M.; Stephanopoulos, G. Two-step pathway for isoprenoid synthesis. *Proc. Natl. Acad. Sci. USA* **2019**, *116*, 506–511. [CrossRef] [PubMed]
77. Li, G.; Liang, H.; Gao, R.; Qin, L.; Xu, P.; Huang, M.; Zong, M.H.; Cao, Y.; Lou, W.Y. Yeast metabolism adaptation for efficient terpenoids synthesis via isopentenol utilization. *Nat. Commun.* **2024**, *15*, 9844. [CrossRef]
78. Sun, M.L.; Gao, X.; Lin, L.; Yang, J.; Ledesma-Amaro, R.; Ji, X.J. Building *Yarrowia lipolytica* cell factories for advanced biomanufacturing: Challenges and solutions. *J. Agric. Food Chem.* **2024**, *72*, 94–107. [CrossRef]
79. Zhang, T.L.; Yu, H.W.; Ye, L.D. Metabolic engineering of *Yarrowia lipolytica* for terpenoid production: Tools and strategies. *ACS Synth. Biol.* **2023**, *12*, 639–656. [CrossRef]
80. Yang, L.; Liu, H.; Jin, Y.; Liu, J.; Deng, L.; Wang, F. Recent advances in multiple strategies for the synthesis of terpenes by engineered yeast. *Fermentation* **2022**, *8*, 615. [CrossRef]
81. Shrivastava, A.; Pal, M.; Sharma, R.K. *Pichia* as yeast cell factory for production of industrially important bio-products: Current trends, challenges, and future prospects. *J. Bioresour. Bioprod.* **2023**, *8*, 108–124. [CrossRef]
82. Zhao, Y.; Song, B.; Li, J.; Zhang, J. *Rhodotorula toruloides*: An ideal microbial cell factory to produce oleochemicals, carotenoids, and other products. *World J. Microbiol. Biotechnol.* **2021**, *38*, 13. [CrossRef] [PubMed]
83. Yu, Y.; Shi, S. Development and perspective of *Rhodotorula toruloides* as an efficient cell factory. *J. Agric. F Chem.* **2023**, *71*, 1802–1819. [CrossRef] [PubMed]
84. Jiao, X.; Zhang, Y.; Liu, X.; Zhang, Q.; Zhang, S.; Zhao, Z.K. Developing a CRISPR/Cas9 system for genome editing in the basidiomycetous yeast *Rhodosporidium toruloides*. *Biotechnol. J.* **2019**, *14*, 1900036. [CrossRef] [PubMed]
85. Gong, G.; Wu, B.; Liu, L.; Li, J.; He, M. Engineering oleaginous red yeasts as versatile chassis for the production of oleochemicals and valuable compounds: Current advances and perspectives. *Biotechnol. Adv.* **2024**, *76*, 108432. [CrossRef]
86. Chen, Q.; Lyu, L.; Xue, H.; Shah, A.M.; Zhao, Z.K. Engineering a non-model yeast *Rhodotorula mucilaginosa* for terpenoids synthesis. *Synth. Syst. Biotechnol.* **2024**, *9*, 569–576. [CrossRef] [PubMed]
87. Kumar, L.R.G.; Sanath Kumar, H.; Tejpal, C.S.; Anas, K.K.; Nayak, B.B.; Sarika, K.; Greeshma, S.S.; Chatterjee, N.S.; Mathew, S.; Ravishankar, C.N. Exploring the physical and quality attributes of muffins incorporated with microencapsulated squalene as a functional food additive. *J. Food Sci. Technol.* **2021**, *58*, 4674–4684. [CrossRef] [PubMed]
88. Hoshino, Y.; Moriya, M.; Matsudaira, A.; Katashkina, J.I.; Nitta, N.; Nishio, Y.; Usuda, Y. Stereospecific linalool production utilizing two-phase cultivation system in *Pantoea ananatis*. *J. Biotechnol.* **2020**, *324*, 21–27. [CrossRef]
89. Zhou, P.; Zhou, X.; Yuan, D.; Fang, X.; Pang, X.; Yuan, K.; Li, A.; Wang, X. Combining protein and organelle engineering for linalool overproduction in *Saccharomyces cerevisiae*. *J. Agric. Food Chem.* **2023**, *71*, 10133–10143. [CrossRef]
90. Shukal, S.; Ong, L.; T, R.; Chen, X.; Zhang, C. Microaerobic fermentation enables high-titer biosynthesis of the rose monoterpenes geraniol and geranyl acetate in *Escherichia coli*. *ACS Sustain. Chem. Eng.* **2024**, *12*, 3921–3932. [CrossRef]
91. Kong, X.; Wu, Y.; Yu, W.; Liu, Y.; Li, J.; Du, G.; Lv, X.; Liu, L. Efficient synthesis of limonene in *Saccharomyces cerevisiae* using combinatorial metabolic engineering strategies. *J. Agric. Food Chem.* **2023**, *71*, 7752–7764. [CrossRef] [PubMed]

92. Li, J.; Zhu, K.; Miao, L.; Rong, L.; Zhao, Y.; Li, S.; Ma, L.; Li, J.; Zhang, C.; Xiao, D.; et al. Simultaneous improvement of limonene production and tolerance in *Yarrowia lipolytica* through tolerance engineering and evolutionary engineering. *ACS Synth. Biol.* **2021**, *10*, 884–896. [CrossRef]
93. Zhou, Y.; Wang, J.; Wang, X.; Wang, F.; Li, X. Efficient production of α-pinene through identifying the rate-limiting enzymes and tailoring inactive terminal of pinene synthase in *Escherichia coli*. *Fuel* **2023**, *343*, 127872. [CrossRef]
94. Niu, F.X.; Huang, Y.B.; Shen, Y.P.; Ji, L.N.; Liu, J.Z. Enhanced production of pinene by using a cell-free system with modular cocatalysis. *J. Agric. Food Chem.* **2020**, *68*, 2139–2145. [CrossRef]
95. Chen, H.; Li, M.; Liu, C.; Zhang, H.; Xian, M.; Liu, H. Enhancement of the catalytic activity of isopentenyl diphosphate isomerase (IDI) from *Saccharomyces cerevisiae* through random and site-directed mutagenesis. *Microb. Cell Fact.* **2018**, *17*, 65. [CrossRef] [PubMed]
96. Zhou, K.; Yu, C.; Liang, N.; Xiao, W.; Wang, Y.; Yao, M.; Yuan, Y. Adaptive evolution and metabolic engineering boost lycopene production in *Saccharomyces cerevisiae* via enhanced precursors supply and utilization. *J. Agric. Food Chem.* **2023**, *71*, 3821–3831. [CrossRef] [PubMed]
97. Huang, W.; Liu, Y.; Ma, X.; Ma, C.; Jiang, Y.; Su, J. Rational design for the complete synthesis of stevioside in *Saccharomyces cerevisiae*. *Microorganisms* **2024**, *12*, 1125. [CrossRef]
98. Li, S.; Luo, S.; Zhao, X.; Gao, S.; Shan, X.; Lu, J.; Zhou, J. Efficient conversion of stevioside to rebaudioside M in *Saccharomyces cerevisiae* by a engineering hydrolase system and prolonging the growth cycle. *J. Agric. F Chem.* **2024**, *72*, 8140–8148. [CrossRef]
99. Liu, Q.; Zhang, G.; Su, L.; Liu, P.; Jia, S.; Wang, Q.; Dai, Z. Reprogramming the metabolism of oleaginous yeast for sustainably biosynthesizing the anticarcinogen precursor germacrene A. *Green. Chem.* **2023**, *25*, 7988–7997. [CrossRef]
100. Li, W.; Mai, J.; Lin, L.; Zhang, Z.G.; Ledesma-Amaro, R.; Dong, W.; Ji, X.J. Combination of microbial and chemical synthesis for the sustainable production of β-elemene, a promising plant-extracted anticancer compound. *Biotechnol. Bioeng.* **2023**, *120*, 3612–3621. [CrossRef] [PubMed]
101. Zhu, C.; You, X.; Wu, T.; Li, W.; Chen, H.; Cha, Y.; Zhuo, M.; Chen, B.; Li, S. Efficient utilization of carbon to produce aromatic valencene in *Saccharomyces cerevisiae* using mannitol as the substrate. *Green. Chem.* **2022**, *24*, 4614–4627. [CrossRef]
102. Ye, Z.; Huang, Y.; Shi, B.; Xiang, Z.; Tian, Z.; Huang, M.; Wu, L.; Deng, Z.; Shen, K.; Liu, T. Coupling cell growth and biochemical pathway induction in *Saccharomyces cerevisiae* for production of (+)-valencene and its chemical conversion to (+)-nootkatone. *Metab. Eng.* **2022**, *72*, 107–115. [CrossRef]
103. Zhou, L.; Wang, Y.; Han, L.; Wang, Q.; Liu, H.; Cheng, P.; Li, R.; Guo, X.; Zhou, Z. Enhancement of patchoulol production in *Escherichia coli* via multiple engineering strategies. *J. Agric. Food Chem.* **2021**, *69*, 7572–7580. [CrossRef]
104. Zhou, L.; Wang, Q.; Shen, J.; Li, Y.; Zhang, H.; Zhang, X.; Yang, S.; Jiang, Z.; Wang, M.; Li, J.; et al. Metabolic engineering of glycolysis in *Escherichia coli* for efficient production of patchoulol and τ-cadinol. *Bioresour. Technol.* **2024**, *391*, 130004. [CrossRef] [PubMed]
105. Liu, M.; Lin, Y.C.; Guo, J.J.; Du, M.M.; Tao, X.; Gao, B.; Zhao, M.; Ma, Y.; Wang, F.Q.; Wei, D.Z. High-level production of sesquiterpene patchoulol in *Saccharomyces cerevisiae*. *ACS Synth. Biol.* **2021**, *10*, 158–172. [CrossRef]
106. Liang, Z.; Zhi, H.; Fang, Z.; Zhang, P. Genetic engineering of yeast, filamentous fungi and bacteria for terpene production and applications in food industry. *Food Res. Int.* **2021**, *147*, 110487. [CrossRef]
107. Chen, X.; Zhang, C.; Lindley, N.D. Metabolic engineering strategies for sustainable terpenoid flavor and fragrance synthesis. *J. Agric. Food Chem.* **2020**, *68*, 10252–10264. [CrossRef] [PubMed]
108. Schempp, F.M.; Drummond, L.; Buchhaupt, M.; Schrader, J. Microbial cell factories for the production of terpenoid flavor and fragrance compounds. *J. Agric. Food Chem.* **2018**, *66*, 2247–2258. [CrossRef]
109. Chen, W.; Viljoen, A.M. Geraniol—A review of a commercially important fragrance material. *S. Afr. J. Bot.* **2010**, *76*, 643–651. [CrossRef]
110. Guo, Y.; Li, F.; Zhao, J.; Wei, X.; Wang, Z.; Liu, J. Diverting mevalonate pathway metabolic flux leakage in *Saccharomyces cerevisiae* for monoterpene geraniol production from cane molasses. *Biochem. Eng. J.* **2022**, *181*, 108398. [CrossRef]
111. Zhao, C.; Wang, X.H.; Lu, X.Y.; Zong, H.; Zhuge, B. Metabolic engineering of *Candida glycerinogenes* for sustainable production of geraniol. *ACS Synth. Biol.* **2023**, *12*, 1836–1844. [CrossRef] [PubMed]
112. Munekata, P.E.S.; Pateiro, M.; Dominguez, R.; Nieto, G.; Kumar, M.; Dhama, K.; Lorenzo, J.M. Bioactive compounds from fruits as preservatives. *Foods* **2023**, *12*, 343. [CrossRef] [PubMed]
113. Teneva, D.; Denev, P. Biologically active compounds from probiotic microorganisms and plant extracts used as biopreservatives. *Microorganisms* **2023**, *11*, 1896. [CrossRef] [PubMed]
114. Sojic, B.; Milosevic, S.; Savanovic, D.; Zekovic, Z.; Tomovic, V.; Pavlic, B. Isolation, bioactive potential, and application of essential oils and terpenoid-rich extracts as effective antioxidant and antimicrobial agents in meat and meat products. *Molecules* **2023**, *28*, 2293. [CrossRef]
115. Ren, Y.; Liu, S.; Jin, G.; Yang, X.; Zhou, Y.J. Microbial production of limonene and its derivatives: Achievements and perspectives. *Biotechnol. Adv.* **2020**, *44*, 107628. [CrossRef] [PubMed]

116. Zhang, Y.; Yu, H.; Ye, L. From β-carotene to retinoids: A review of microbial production of Vitamin A. *J. Agric. F Chem.* **2024**, *72*, 20752–20762. [CrossRef]
117. Patel, A.; Rova, U.; Christakopoulos, P.; Matsakas, L. Microalgal lutein biosynthesis: Recent trends and challenges to enhance the lutein content in microalgal cell factories. *Front. Mar. Sci.* **2022**, *9*, 1015419. [CrossRef]
118. Jin, Y.; Arroo, R.R.J. Application of dietary supplements in the prevention of type 2 diabetes-related cardiovascular complications. *Phytochem. Rev.* **2021**, *20*, 181–209. [CrossRef]
119. Roy, S.; Deshmukh, R.K.; Tripathi, S.; Gaikwad, K.K.; Das, S.S.; Sharma, D. Recent advances in the carotenoids added to food packaging films: A review. *Foods* **2023**, *12*, 4011. [CrossRef] [PubMed]
120. Fordjour, E.; Liu, C.L.; Yang, Y.; Bai, Z. Recent advances in lycopene and germacrene a biosynthesis and their role as antineoplastic drugs. *World J. Microbiol. Biotechnol.* **2024**, *40*, 254. [CrossRef]
121. Pinheiro, M.J.; Bonturi, N.; Belouah, I.; Miranda, E.A.; Lahtvee, P.J. Xylose metabolism and the effect of oxidative stress on lipid and carotenoid production in *Rhodotorula toruloides*: Insights for future biorefinery. *Front. Bioeng. Biotechnol.* **2020**, *8*, 1008. [CrossRef] [PubMed]
122. Okonkwo, C.E.; Adeyanju, A.A.; Onyeaka, H.; Nwonuma, C.O.; Olaniran, A.F.; Alejolowo, O.O.; Inyinbor, A.A.; Oluyori, A.P.; Zhou, C. A review on rebaudioside M: The next generation steviol glycoside and noncaloric sweetener. *J. Food Sci.* **2024**, *89*, 6946–6965. [CrossRef] [PubMed]
123. Hou, Y.; Jiang, J.G. Origin and concept of medicine food homology and its application in modern functional foods. *Food Funct.* **2013**, *4*, 1727–1741. [CrossRef] [PubMed]
124. Gong, X.; Ji, M.; Xu, J.; Zhang, C.; Li, M. Hypoglycemic effects of bioactive ingredients from medicine food homology and medicinal health food species used in China. *Crit. Rev. Food Sci. Nutr.* **2020**, *60*, 2303–2326. [CrossRef]
125. Galappaththi, M.C.A.; Patabendige, N.M.; Premarathne, B.M.; Hapuarachchi, K.K.; Tibpromma, S.; Dai, D.Q.; Suwannarach, N.; Rapior, S.; Karunarathna, S.C. A review of *Ganoderma* triterpenoids and their bioactivities. *Biomolecules* **2022**, *13*, 24. [CrossRef] [PubMed]
126. Wong, J.; d'Espaux, L.; Dev, I.; Cas, V.D.H.; Keasling, J. De novo synthesis of the sedative valerenic acid in *Saccharomyces cerevisiae*. *Metab. Eng.* **2018**, *47*, 94–101. [CrossRef]
127. Zhang, H.; Hua, X.; Zheng, D.; Wu, H.; Li, C.; Rao, P.; Wen, M.; Choi, Y.E.; Xue, Z.; Wang, Y. De novo biosynthesis of oleanane-type ginsenosides in *Saccharomyces cerevisiae* using two types of glycosyltransferases from *Panax ginseng*. *J. Agric. F Chem.* **2022**, *70*, 2231–2240. [CrossRef]
128. Zhang, W.; Guo, J.; Wang, Z.; Li, Y.; Meng, X.; Shen, Y.; Liu, W. Improved production of germacrene A, a direct precursor of β-elemene, in engineered *Saccharomyces cerevisiae* by expressing a cyanobacterial germacrene A synthase. *Microb. Cell Fact.* **2021**, *20*, 7. [CrossRef]
129. Deng, X.; Ye, Z.; Duan, J.; Chen, F.; Zhi, Y.; Huang, M.; Huang, M.; Cheng, W.; Dou, Y.; Kuang, Z. Complete pathway elucidation and heterologous reconstitution of (+)-nootkatone biosynthesis from *Alpinia oxyphylla*. *New Phytol.* **2024**, *241*, 779–792. [CrossRef]
130. Food Safety Commission of Japan. "Valencene" produced using *Rhodobacter Sphaeroides* 168 strain (Genetically modified foods and feeds). *Food Saf.* **2023**, *11*, 34–35. [CrossRef]
131. Song, Y.; Liu, H.; Quax, W.J.; Zhang, Z.; Chen, Y.; Yang, P.; Cui, Y.; Shi, Q.; Xie, X. Application of valencene and prospects for its production in engineered microorganisms. *Front. Microbiol.* **2024**, *15*, 1444099. [CrossRef] [PubMed]
132. Frister, T.; Hartwig, S.; Alemdar, S.; Schnatz, K.; Thons, L.; Scheper, T.; Beutel, S. Characterisation of a recombinant patchoulol synthase variant for biocatalytic production of terpenes. *Appl. Biochem. Biotechnol.* **2015**, *176*, 2185–2201. [CrossRef] [PubMed]
133. Ekramzadeh, K.; Bramer, C.; Frister, T.; Fohrer, J.; Kirschning, A.; Scheper, T.; Beutel, S. Optimization of factors influencing enzyme activity and product selectivity and the role of proton transfer in the catalytic mechanism of patchoulol synthase. *Biotechnol. Prog.* **2020**, *36*, e2935. [CrossRef]
134. Fordjour, E.; Mensah, E.O.; Hao, Y.; Yang, Y.; Liu, X.; Li, Y.; Liu, C.L.; Bai, Z. Toward improved terpenoids biosynthesis: Strategies to enhance the capabilities of cell factories. *Bioresour. Bioprocess.* **2022**, *9*, 6. [CrossRef] [PubMed]
135. Zhang, C.; Hong, K. Production of terpenoids by synthetic biology approaches. *Front. Bioeng. Biotechnol.* **2020**, *8*, 347. [CrossRef]
136. Wu, J.; Tian, X.; Xu, X.; Gu, X.; Kong, J.; Guo, T. Engineered probiotic *Lactococcus lactis* for lycopene production against ROS stress in intestinal epithelial cells. *ACS Synth. Biol.* **2022**, *11*, 1568–1576. [CrossRef] [PubMed]
137. Zhai, X.; Yao, L.; Zhou, Y.J. Construction of microbial chassis for terpenoid discovery. *Synth. Syst. Biotechnol.* **2022**, *7*, 1181–1182. [CrossRef]
138. Li, F.; Yuan, L.; Lu, H.; Li, G.; Chen, Y.; Engqvist, M.K.M.; Kerkhoven, E.J.; Nielsen, J. Deep learning-based k_{cat} prediction enables improved enzyme-constrained model reconstruction. *Nat. Catal.* **2022**, *5*, 662–672. [CrossRef]
139. Herisson, J.; Duigou, T.; du Lac, M.; Bazi-Kabbaj, K.; Sabeti Azad, M.; Buldum, G.; Telle, O.; El Moubayed, Y.; Carbonell, P.; Swainston, N.; et al. The automated Galaxy-SynBioCAD pipeline for synthetic biology design and engineering. *Nat. Commun.* **2022**, *13*, 5082. [CrossRef]

140. Zheng, S.; Zeng, T.; Li, C.; Chen, B.; Coley, C.W.; Yang, Y.; Wu, R. Deep learning driven biosynthetic pathways navigation for natural products with BioNavi-NP. *Nat. Commun.* **2022**, *13*, 3342. [CrossRef] [PubMed]
141. Ryu, G.; Kim, G.B.; Yu, T.; Lee, S.Y. Deep learning for metabolic pathway design. *Metab. Eng.* **2023**, *80*, 130–141. [CrossRef] [PubMed]
142. Cheah, L.C.; Liu, L.; Plan, M.R.; Peng, B.; Lu, Z.; Schenk, G.; Vickers, C.E.; Sainsbury, F. Product profiles of promiscuous enzymes can be altered by controlling in vivo spatial organization. *Adv. Sci.* **2023**, *10*, e2303415. [CrossRef] [PubMed]
143. Liu, N.; Dong, W.; Yang, H.; Li, J.H.; Chiu, T.Y. Application of artificial scaffold systems in microbial metabolic engineering. *Front. Bioeng. Biotechnol.* **2023**, *11*, 1328141. [CrossRef]
144. Henke, N.A.; Austermeier, S.; Grothaus, I.L.; Gotker, S.; Persicke, M.; Peters-Wendisch, P.; Wendisch, V.F. *Corynebacterium glutamicum* CrtR and its orthologs in actinobacteria: Conserved function and application as genetically encoded biosensor for detection of geranylgeranyl pyrophosphate. *Int. J. Mol. Sci.* **2020**, *21*, 5482. [CrossRef] [PubMed]
145. Jia, Y.L.; Li, J.; Nong, F.T.; Yan, C.X.; Ma, W.; Zhu, X.F.; Zhang, L.H.; Sun, X.M. Application of adaptive laboratory evolution in lipid and terpenoid production in yeast and microalgae. *ACS Synth. Biol.* **2023**, *12*, 1396–1407. [CrossRef] [PubMed]
146. Olsson, L.; Rugbjerg, P.; Torello Pianale, L.; Trivellin, C. Robustness: Linking strain design to viable bioprocesses. *Trends Biotechnol.* **2022**, *40*, 918–931. [CrossRef] [PubMed]
147. Wang, G.; Haringa, C.; Noorman, H.; Chu, J.; Zhuang, Y. Developing a computational framework to advance bioprocess scale-up. *Trends biotechnol.* **2020**, *38*, 846–856. [CrossRef]

MDPI AG
Grosspeteranlage 5
4052 Basel
Switzerland
Tel.: +41 61 683 77 34

Foods Editorial Office
E-mail: foods@mdpi.com
www.mdpi.com/journal/foods

www.ingramcontent.com/pod-product-compliance
Lightning Source LLC
Chambersburg PA
CBHW061444210926
51850CB00008B/1611
9783725877539